A Natural History of the New World

GEOLOGIC COLUMN

(Mesozoic and Cenozoic Eras)

Quaternary Period
   Holocene Epoch   11,500
   Pleistocene   2.6 MA
Tertiary
   Pliocene
     Piacenzian Stage   3.6
     Zanclean   5.3
   Miocene
     Messinian   7.2
     Tortonian   11.6
     Serravallian   13.6
     Langhian   15.9
     Burdigalian   20.4
     Aquitanian   23.0
   Oligocene
     Chattian   28.4
     Rupelian   33.9
   Eocene
     Priabonian   37.2
     Bartonian   40.4
     Lutetian   48.6
     Ypresian   55.8
   Paleocene
     Thanetian   58.7
     Seldanian   61.7
     Danian   65.5

(Cenozoic Era)

Cretaceous
   Senonian
     Maastrichtian   70.6
     Campanian   83.5
     Santonian   85.8
     Coniacian   89.3
   Gallic
     Turonian   93.5
     Cenomanian   99.6
     Albian   112.0
     Aptian   125.0
     Barremian   130.0
   Neocomian
     Hauterivian   136.4
     Valanginian   140.2
     Berriasian   145.5
Jurassic
Triassic

(Mesozoic Era)

A Natural History
of the New World

The Ecology and Evolution
of Plants in the Americas

ALAN GRAHAM

The University of Chicago Press
Chicago & London

ALAN GRAHAM is curator of paleobotany and palynology at the Missouri Botanical Garden. His many books include *Late Cretaceous and Cenozoic History of Latin American Vegetation and Terrestrial Environments, Floristics and Paleofloristics of Asia and Eastern North America*, and *Late Cretaceous and Cenozoic History of North American Vegetation*.

The University of Chicago Press, Chicago 60637
The University of Chicago Press, Ltd., London
© 2011 by The University of Chicago
All rights reserved. Published 2011
Printed in the United States of America

20 19 18 17 16 15 14 13 12 11   1 2 3 4 5
ISBN-13: 978-0-226-30679-7   (cloth)
ISBN-13: 978-0-226-30680-3   (paper)
ISBN-10: 0-226-30679-8   (cloth)
ISBN-10: 0-226-30680-1   (paper)

Library of Congress Cataloging-in-Publication Data

Graham, Alan.
    A natural history of the New World : the ecology and evolution of plants in the Americas / Alan Graham.
        p. cm.
    Includes bibliographical references and index.
    ISBN-13: 978-0-226-30679-7 (hardcover : alk. paper)
    ISBN-13: 978-0-226-30680-3 (pbk. : alk. paper)
    ISBN-10: 0-226-30679-8 (hardcover : alk. paper)
    ISBN-10: 0-226-30680-1 (pbk. : alk. paper)   1. Paleobotany—America—Tertiary.   2. Paleobotany—America—Cretaceous.   3. Paleobotany—America—Cenozoic.   4. Plant ecology—America.   5. Plants—Evolution—America.
    6. Plants, Fossil—America.   I. Title.
    QE926.G73   2011
    561.097—dc22
                                                                2010001292

♾ The paper used in this publication meets the minimum requirements of the American National Standard for Information Sciences—Permanence of Paper for Printed Library Materials, ANSI Z39.48-1992.

To the many biologists, geologists, and other scientists who have given
so much and benefited so mightily from study of the New World and
its natural history and to those who continue to work to protect its legacy

In the Amazon, researchers documenting the ways of native peoples join forces with an embattled chief to stop illegal loggers and developers from destroying the earth's most precious wilderness. After tribal chief Almir Surui encouraged the mapping of resources on his people's preserve—from medicinal plants to ancestral burial grounds—loggers put a $100,000 price on his head. Says a Brazilian legislator: "Almir made his people understand the value of their culture and their land." The Amazon loses 8,800 acres a day to deforestation.

—JOSHUA HAMMER, "Rain Forest Rebel," 2007

Scientists in Brazil are considering strike action or "civil disobedience of another form" in protest over the imprisonment of a renowned primatologist [Marc van Roosmalen]. The case, they say, is a visible part of a wider government crackdown on scientists working in tropical areas around the world.

—ERIKA CHECK and THOMAS HAYDEN,

"Strike Threat over Jailed Primatologist," 2007

Contents

Preface

The Simojovel resident lowered his gun, smiled broadly if somewhat sheepishly, and said, "Puedes beber de mi pozo." His wel-
come culminated a long morning of discussion near a bluff bordering the village cemetery. Although unsettling, there had been a
faint overlay of conviviality throughout the encounter—the villag-
ers knew we probably meant no harm, and we guessed they probably were not going to execute us, however convenient the site.
We had come to the village of Simojovel to collect rock samples.
The bluff was an exposure of the La Quinta Formation in an area
of Chiapas, Mexico, famous for the plant and animal inclusions
of its amber deposits. Mining rights to the amber had been ceded
to local residents, and although there was no amber at this site,
our presence was an opportunity to demonstrate rightful authority, and to pass an otherwise uneventful morning conversing and
good-naturedly harassing strangers from afar. It took awhile to explain that we really had come thousands of miles to collect rocks of
no apparent value from which we would extract spores and pollen
to reconstruct climates and vegetation that had not existed for millions of years. Convinced we were probably more harmless than
devious, we were eventually accepted with the local greeting, "You
may drink from my well."

Generations of resident scientists and foreign visitors have been
observing and studying the natural history of the New World for
five hundred years. This book summarizes the results of that re-

search as they apply to the formation of ecosystems over time in response
to changes in geology and climate acting on evolutionary processes. A convenient organizational framework is provided by three drops in temperature that divide the past 100 million years into four time segments: Late
Cretaceous through the early Eocene (100 million to 50 million years ago),
middle Eocene through the early Miocene (50 million to 16.3 million years
ago), middle Miocene through the Pliocene (16.3 million to 2.6 million
years ago), and the Quaternary (2.6 million years ago to the present). Each
was an eventful period during which great changes took place in the Earth's
flora, fauna, landscape, and climate. The record of these changes begins
here with a Cretaceous world dramatically different from the present, and
one from which the evolution of new life-forms, fluctuations in the environment, and gradual uniformitarian and sudden catastrophic events shaped
the Earth's surface and its living envelope. It has been a lengthy, immensely
complex, delicately balanced, and highly interactive process.

In this account I focus on vegetation, for three principal reasons. First,
plants are the conspicuous, conveniently stationary component of the
Earth's ecosystems and are most often used to name or characterize the
systems (boreal forest, mangroves, savanna). Second, plant fossils are abundant and have been used extensively to reveal the history of lineages and
communities—their origin, evolution, migration, and the time relationships between them—and to reconstruct the Earth's paleoenvironments.
Last, much of the literature and many natural history presentations intended for general audiences understandably feature charismatic, exotic,
dangerous, attacking animals harassed by film crews and wannabe celebrity
naturalists. This makes for good theater, and it admirably focuses attention on selected treasures of the biological kingdom, but on another level,
it gives only a partial picture of the Earth's ecosystems, and it is not a broad
enough basis for a meaningful consideration of their structure, function,
importance, conservation, and sustained management. Plants provide the
oxygen we breathe, the food we consume, the fossil fuel that both supports
and threatens our civilization, and the habitats within which all life coex-
ists. For those seeking a fuller understanding of the Earth's web of life and a
rational agenda for ensuring its continuation, this should be emphasized.

In *The Eternal Frontier: An Ecological History of North America and Its Peoples* (2001), Tim Flannery offers a complementary account of the biota of
North America, describing especially well the history of its fauna and people. The mammals are equally well summarized by Michael Woodburne
in *Late Cretaceous and Cenozoic Mammals of North America* (2004), and by

Christine Janis and others in the two-volume *Evolution of Tertiary Mammals of North America* (1998, 2008). My account here stresses the critical role of plants and time.

In two earlier summaries I address North America north of Mexico (Graham 1999, or I) and Latin America (Graham 2010, or II). These works contain detailed information and references and may be consulted as needed for background to the present text. To reduce repetitious citation, they are referenced here as I and II. In addition, there are edited summaries of the floristic relationships with eastern Asia (Graham 1972) and the vegetation and vegetation history of Latin America (Graham 1973, 2006). The present text is a combined version of these and other works, especially the 1999 and 2010 treatments, and it draws freely on them. This book differs in that although it serves as a summary for specialists in vegetation history, it is also intended for a broader audience of specialists in other fields, as well as students who have an interest in the New World and how it got that way. It brings the references and Web addresses from 1999 and 2010 up to date, although protocol identifiers, domains, and pathways are subject to change. The text also includes new information on climates, fossil faunas, mountain uplift, global sea level change, and paleotemperatures. There have been intriguing new discoveries, such as tapir and crocodilian lineages in the early Eocene of the High Arctic, and the gargantuan boid *Titanoboa cerrejonensis*, thirteen meters long and weighing over a ton from the Paleocene of Colombia (Head et al. 2009). The latter find suggests that the mean annual temperature (MAT) in the neotropical lowlands may have been considerably warmer than the present, about 28°C (but see Sniderman 2009). If so, this has implications for the steepness of Paleogene thermal gradients, ocean circulation, and the migration of some tropical and subtropical species out of an equatorial zone that may have been too warm. There is also recent evidence for an explosive emission of methane gas from the Norwegian Sea that goes a long way in explaining the extraordinary rise in global temperature around 55 million years ago (Ma). A current trend, still to be confirmed in its details, is to recognize uplift of the Rocky Mountains and the Sierra Nevada to greater heights, at least locally, and earlier (in the Eocene, about 45 Ma) than suggested in the older literature (primarily in the Oligocene and Miocene, 30 Ma onward). By contrast, the great heights of the Andes Mountains (altitudes above about 2000 m) are now generally acknowledged as coming comparatively late in geologic time (after about 10 Ma). In addition, Cretaceous seas may not have been as high as earlier thought (possibly at times only 100 m or more higher than at present,

rather than uniformly reaching 300 m), perpetuating some still mostly un-
substantiated models predicting extensive Cretaceous glaciations.

This text is unique in its coverage of the New World as a single geographic unit. This treatment is desirable because the continental interchange of biotic elements helped shape the ecosystems of northern North
America, Mexico, the Antilles, Central America, and South America. The
text is also a single authored treatment, rather than an edited summary.
This affords the advantage of a unified perspective while acknowledging
the many inherent challenges in that approach.

As an aid to following the development of New World environments and
biotas through time, a geologic time scale appears as the frontispiece (after
Gradstein et al. 2005). The status of the Quaternary has recently been the
subject of considerable debate, including a proposal that it be abandoned
entirely and included in the Late Cenozoic (Giles 2005; Mascarelli 2009).
The International Commission on Stratigraphy recommended in May 2009
that the boundary be set at 2.6 Ma. For the moment, the Quaternary is
defined as follows:

Quaternary 2.5/2.6–0 Ma
   Holocene (or Recent) 11.5–0 kyr (thousand years BP)
   Pleistocene 2.5/2.6 Ma–11.5 kyr
     Late glacial 21–11.5 kyr
     Last Glacial Maximum (LGM) 21–18 kyr (19/20–26.5 kyr; Clark et al.
     2009)

The arrangement of continents at 100 Ma is shown in figure 2.19, a global
sea level curve in figure 3.1, and a global paleotemperature in figure 3.4.
Conversion tables for measurements are available at several Web sites (e.g.,
www.onlineconversion.com).

Persons providing assistance and courtesies during previous studies have
been acknowledged in those earlier publications on northern North America and Latin America. In the production of this volume, many others have
generously contributed additional information, permission to use published and unpublished material, and editorial and technical expertise.
My sincere gratitude to Stephen Barker (Cardiff University); Mike Blom-
berg, Trish Distler, Iván Jiménez, and Fred Keusenkothen (Missouri Botanical Garden, St. Louis); Taciana Calvalcanti (CENARGEN/EMBRAPA,
Brasilia); Thure Cerling (University of Utah), Henry Fricke (Colorado Col-
lege), and Bruce MacFadden (University of Florida), for information on

C3 and C4 plants; Omar Colmenares (INTEVEP, Caracas) for the chance to study material from the Maracaibo region of Venezuela; Christine Janis (Brown University) for reading chapter 10 and for comments on the mammalian faunas; Andrew Knoll (Harvard University) for updates on earliest Precambrian life; the National Evolutionary Synthesis Center (NESCENT) for the opportunity to participate in the Biological Diversification on the West Indian Archipelago Conference hosted by the Punta Cana Ecological Foundation, Santo Domingo, Dominican Republic; David Roubik for the invitation to present lectures and interact with the staff of the Smithsonian Tropical Research Institute (STRI), Panama, as part of the Smithsonian's Senior Visiting Scholars Program; Eldredge Bermingham, director, and Carlos Jaramillo, Center for Tropical Paleoecology and Archaeology, for incorporating and databasing my research material and associated literature into the Smithsonian's permanent collections at STRI; and Thomas van der Hammen for discussions during my tenure as visiting research scientist at the University of Amsterdam. Reviewers, executive editor Christie Henry and editorial assistant Abby Collier at the University of Chicago Press, and Carlisle Rex-Waller provided numerous helpful suggestions during preparation of the book. The National Science Foundation has supported my studies over many years, and the National Geographic Society funded fieldwork in Bolivia. Photographs not otherwise credited are by the author.

For most of us there is something fascinating in how communities of living systems came about and the vast eons of time it took for them to reach their present status. There is usually some curiosity about how the information is obtained and an admiration for the persons who obtained it once their often heroic efforts are known. It is hoped, as the world's ecosystems become better understood, that there will also be an enhanced feeling of respect for what they represent and a growing insistence that they be sustained.

: :   **REFERENCES**   : :

Clark, P. U., et al. (eight coauthors). 2009. The Last Glacial Maximum. *Science* 325:710–14.

Flannery, T. 2001. *The eternal frontier: An ecological history of North America and its people.* Atlantic Monthly Press, New York.

Giles, J. 2005. Geologists call time on dating dispute. *Nature* 435:865.

Gradstein, F. M., J. Ogg, and A. Smith. 2005. *A geologic time scale, 2004.* Cambridge University Press, Cambridge.

Graham, A., ed. 1972. *Floristics and paleofloristics of Asia and eastern North America*. Elsevier Science Publishers, Amsterdam.

———, ed. 1973. *Vegetation and vegetational history of northern Latin America*. Elsevier Science Publishers, Amsterdam.

———. 1999. *Late Cretaceous and Cenozoic history of North American vegetation*. Oxford University Press, Oxford.

———. 2006. Introduction to *Latin American biogeography—causes and effects*. Proceedings of the 51st Annual Systematics Symposium of the Missouri Botanical Garden (Alan Graham, organizer). *Ann. Mo. Bot. Gard.* 93:173–358.

———. 2010. *Late Cretaceous and Cenozoic history of Latin American vegetation and terrestrial environments*. Missouri Botanical Garden Press, St. Louis.

Head, J. J., J. I. Bloch, A. K. Hastings, and J. R. Bourque. 2009. Giant boid snake from the Palaeocene neotropics reveals hotter past equatorial temperatures. *Nature* 457:715–17.

Janis, C. M., G. F. Gunnell, and M. D. Uhen, eds. 2008. *Evolution of Tertiary mammals of North America*. Vol. 2. Cambridge University Press, Cambridge.

Janis, C. M., K. M. Scott, and L. L. Jacobs, eds. 1998. *Evolution of Tertiary mammals of North America*. Vol. 1. Cambridge University Press, Cambridge.

Mascarelli, A. L. 2009. Quaternary geologists win timescale vote. *Nature* 459:624.

Sniderman, J. M. K. 2009. Biased reptilian palaeothermometer? *Nature* 460:E1–E5.

Woodburne, M. O., ed. 2004. *Late Cretaceous and Cenozoic mammals of North America*. Columbia University Press, New York.

Additional Readings and Updates

Barrionuevo, A. 2008. Forest plan in Brazil bears the traces of an activist's vision. *New York Times*, 22 December 2008. ["Twenty years ago, a Brazilian environmental activist and rubber tapper was shot to death at his home in Acre State by ranchers opposed to his efforts to save the Amazon rain forest. After his death at age 44, Franciso Alves Mendes, better known as Chico, became a martyr for a concept that is only now gaining mainstream support here: that the value of a standing forest could be more than the value of a forest burned and logged in the name of development."]

Blood in the jungle. 2009. *Economist*, 11 June 2009. [Oil and land rights in Peru.]

Pfanner, E. 2009. UNESCO puts world's major works online. *New York Times*, 21 April 2009. [The project described in this article is the World Digital Library and is supported by UNESCO, the U.S. Library of Congress, and Google, with participation by the Bibliothèque Nationale de France, the National Library of Brazil, and others.]

Abbreviations

BCE	before the common era
BP	before present
CARs	community assembly rules
CE	common era
CLAMP	Climate-Leaf Analysis Multivariate Program
dbh	diameter at breast height
DEM	Digital Elevation Model
DSDP	Deep Sea Drilling Project
EECL	Early Eocene Climatic Optimum
Ga	billion years BP
GIS	Geographic Information System
GPS	Global Positioning System
ha	hectare (2.47 acres)
ITCZ	Intertropical Convergence Zone
K/T	Cretaceous/Tertiary boundary (65 Ma)
kyr	thousand years BP
LGM	Last Glacial Maximum (21–18 kyr)
LMA	leaf margin analysis
LPTM	Late Paleocene Thermal Maximum
Ma	million years BP
MAP	mean annual precipitation
MAT	mean annual temperature

MMCO	Middle Miocene Climatic Optimum
NALMA	North American Land Mammal Age
Neogene	Miocene and Pliocene
NLR	nearest living relative, or modern analog method, for reconstructing paleoclimates from the ecology of similar living species
Paleogene	Paleocene, Eocene, and Oligocene
PEMEX	Petróleos de Mexicanos
PETM	Paleocene-Eocene Thermal Maximum
PGCO	Post-glacial Climatic Optimum
ppbv	parts per billion by volume
ppmv	parts per million by volume
t-c-t	Taxodiaceae-Cupressaceae-Taxaceae

Getting Started

1

The role of the historian is to explain how did our familiar world develop from one very unfamiliar.

—GORDON WOOD, *The Purpose of the Past: Reflections on the Uses of History*, 2008

The Earth is the result of 15 billion years of physical evolution, from the Big Bang through the modern-day Holocene, or Recent, epoch. James Hutton (1726–97), often regarded as the founder of modern geology, belonged to the deist, or eternal, school of philosophy—"everything has existed forever." He is most widely remembered for his famous quote, "We find no vestige of a beginning, no prospect of an end." Hutton's world is currently under the stewardship of a population estimated at 6.5 billion people, with over 1 billion living in the New World. South America has 370 million people, North America (north of Mexico) 335 million, Mexico 107 million, Central America 37 million, and the Antilles

35 million. It seems inconceivable that only sixty years ago the world population was 2.5 billion; within the lifespan of most of us living today it will reach 9 billion people, and it is expected to crest at midcentury at around 10 billion. The Earth is clearly approaching the limits of its economic, political, religious, social, civil, and environmental tolerances, and momentous events are surely in store in the forthcoming decades. The beginning of James Hutton's famous statement is no longer true, and the latter part is no longer certain.

Between its ancient beginning and its present state, a convenient starting point for tracing the natural history of the modern New World is around 100 Ma. This is when outlines of its present-day physical features, climates, lineages, plant communities, and ecosystems can just be recognized, and when they began the latter stages of modernization. The angiosperms, or flowering plants, today constituting the principal vegetation cover of the Earth, were in the early stages of their diversification and radiation. The great western cordilleras of the Alaska Range, Brooks Range, Rocky Mountains, Sierra Madre of Mexico, and the Andes of South America were just being uplifted. The Sierra Nevada, Transvolcanic Belt, the islands of the Antilles, and the land bridge connecting the New World continents through Central America had not yet appeared. An epicontinental sea extended from the Arctic Ocean south through the interior lowlands, and across Mexico, inundating the Central Plateau, the Isthmus of Tehuantepec, and the Yucatán Peninsula. In South America, shallow Cretaceous seas covered parts of the present-day Amazon Basin and other low-lying regions from the base of the Mato Grosso Plateau in Brazil, south through the pampas, Gran Chaco, Patagonia, and the Paraguay-Uruguay-Paraná inlet. The sea was beginning to retreat about 100 Ma, preserving along its margin some of the most extensive fossil plant and animal assemblages in the world. Climates were gradually cooling from mid-Cretaceous highs, putting pressure on the giant reptilian herbivores at the top of the food chain that had dominated the Earth's fauna since the Triassic period at 250 Ma. Toward the end of this Mesozoic cooling trend at 65 Ma, an asteroid landed on the Yucatán Peninsula. The combined effect of climate and impact was to end the age of reptiles and begin the age of mammals and, ultimately, the anthropogene, or the age of humans. Geological, environmental, and biological histories are a continuum, but 100 Ma is a suitable time to begin.

Several assumptions have guided the writing, selection of subjects, organization, and sequence of topics in this work. One is that the biological units under consideration must be the Earth's ecosystems. As defined here, ecosystems are the extensive, recognizable, natural units of the Earth's surface consisting of the plants, animals, and microorganisms interacting with the physical factors of climate and geological processes; in other words, each is "the whole ball of wax" for regions within which we as humans have evolved and must exist. Examples are the tundra, prairie, savanna, desert, deciduous forest, and rain forest. At the alpha level of research, the complexity of organisms, the technologies used to study them, and the massive amounts of available information require specialization into myriad of subdisciplines. Eventually, however, the data on plants and animals must be integrated into a broader view of the biota. In turn, understanding how biotas function and interact requires consideration of the climate and physical environment, and this raises the approach to the level of ecosystems. The number of fossil floras and faunas studied in association with geologic and climatic data, and interpreted in the context of global trends, now allows the reconstruction of ecosystem history for several places and times in North America and in some regions of tropical America. The number of such sites in Latin America is increasing especially rapidly with continued collaborative studies and particularly with new projects by student and resident scientists. A measure of the progress being made is that it is now possible to predict with some accuracy the type of paleocommunity likely to be present given the locality and the age of the deposits. This suggests that the database is adequate to provide a context for assessing environmental interpretations made from individual assemblages. A broad level of predictability also certifies the overall reliability of the various methods being used to reconstruct ecosystems.

Another assumption, emerging from my own fifty years of fascination with the natural history of the New World, is that understanding present-day ecosystems must include the concept of time. It is possible to gain some understanding of the biological world by describing sequences of fossil floras and faunas, or by studying the taxonomy, distribution, ecology, and processes operating within and between extant organisms and communities. However, the fullest understanding of this world, its origin, current status, trends, and future will come from the seamless integration of neobiology with paleobiology, and a knowledge of the environmental dynamics that drive the processes of change. Like an Annie Leibovitz photograph that captures more than the moment, modern ecosystems provide a snapshot of a scene that is a composite of all events bearing upon its subjects over the years. The importance of time in assembling broad unifying summaries in biology was early demonstrated by Charles Lyell's contribution to Charles Darwin's thinking on evolution. A geologist, Lyell brought to the conscious forefront of Darwin's thoughts an image of species projecting back into the past, thus providing the opportunity for evolutionary mechanisms such as competition, mutation, hybridization, extinction, and geographic and reproductive isolation to produce changes in organisms over time—the Darwinian concept of evolution. In modern phraseology, "diversity within clades . . . depends on age and thus the time available for accumulating species" (Ricklefs 2007, 601). Communities of organisms, loosely aggregated through similarities in ecological requirements, and with the opportunity to assemble, also project into the past, and their history offers clues to their present status (Fine and Ree 2006) and analogs for anticipating their future.

A third assumption is that achieving an understanding of ecosystems requires, as the name implies, a systems approach. To a nonspecialist reading some of the biological literature, it may occasionally seem that elevating single factors to the status of causal events, sanctifying one approach over another, bypassing previous studies to create an illusion of originality, or presenting multiple factors in a "versus" context, is the modus operandi. This is evident in such debates as climate versus the asteroid impact in the demise of the dinosaurs, vicariance versus dispersal in biogeography, fossils versus molecules in dating the divergence of lineages (II, chap. 8; Gillman et al. 2009; Gill 2009), and other topics. In the initial euphoria of newfound methods, their shortcomings and the priority and enduring strengths of other approaches are sometimes minimized, and this paves the way for their own marginalization with the next inevitable innovation. For whatever reason—whether it be a need to simplify to convey a sense of understanding or just ego-emissions—it is limiting to canonize the importance of a single factor, approach, or person in explaining immensely complicated processes when multiple factors are clearly involved. To paraphrase the journalist H. L. Mencken, "For every complex problem there is a simple solution. And it is always wrong." That is, if the intricacy of the approach or explanation does not match the intricacy of the problem, it is probably an oversimplification. As noted by Gröcke and Wortmann with reference to the use of stable isotopes in investigating past environments, "The Earth is an extremely dynamic system, with intimate links and feedbacks between the hydrosphere, atmosphere, biosphere, lithosphere and convecting mantle. Integrative science is the future of paleoclimatic reconstructions, and through such integration a better understanding of the dynamics of the Earth System will emerge" (2008, 1). The same is true for understanding the biogeography and evolutionary relationships of organisms. The need for a rigorously balanced assessment of all relevant factors is generally understood, but it is not always unambiguously stated and consistently followed.

If it is granted that some mental capaciousness is required for envisioning the origin and development of New World ecosystems over a hundred million years of time, this leads to a fourth and related organizational premise. Namely, that conclusions about ecosystem history must be tested within a context of information derived from multiple, independent lines of inquiry. For vegetation history, these contexts include faunal history, trends in climate, geologic events such as orogeny (mountain building) and tectonics (plate movement), the coming and going of land bridges, and geomorphological structures (e.g., sand dunes, glacial moraines, and ancient arches that formerly subdivided the landscape and its biota). Sedimentology is another important context because sediments such as coal, lignite, and bauxite form under warm temperature and plentiful moisture, while deeply oxidized red beds, cross-bedded sandstones, mud cracks, and evaporites like halite (salt) and gypsum all tell of aridity. The amount of titanium, a terrestrial element, in ocean basins reflects the extent of water flow from the land into the basin (i.e., precipitation); ophiolites mark the time and place where plates collided; and a meteoritic rise in iridium marks the time when Earth and unearthly fragments collided. These sedimentary clues to the paleoenvironment are preserved in the same rocks that contain the fossils, and they provide a valuable context for reading ecosystem history.

Another context is phylogeny—the study of evolutionary relationships between organisms based on combined morphological, cytological, and multigene molecular approaches. This flags for special attention identifications of fossils unusually old for the lineage, placing them in an unexpected part of the world or in unlikely association. For this context to be of greatest value, we must reduce uncertainties in the fossil record (e.g., misidentifications, incompleteness of the record) and in molecular-based approaches (flawed molecular clocks and the dating of nodes, proposed phylogenetic relationships in conflict with morphological and other evidence), along with occasional claims for the overriding sanctity of cladistics (as cited by Liede-Schumann and Hartmann 2009; Stuessy and König 2009). Progress is being made, primarily by way of examples provided through the studies of experienced investigators objectively incorporating a multiplicity of approaches (e.g., Manos et al. 2007). The interactions between fossils, morphology, and molecules are currently in the inevitable "debate before the resolution" phase.

The basis for many of these contexts is generally understood, and they are incorporated into discussions throughout this work. Others require some explanation, presented in separate sections or chapters. Among the latter are paleotemperature curves derived from oxygen isotope analysis pioneered by Harold Urey and Cesare Emiliani; sea level curves constructed by Peter Vail and Jan Hardenbol using Exxon's worldwide database on coastal sediments; phylogenies incorporating chemical profiling advocated and early demonstrated by Ralph Alston and B. L. Turner; and a temporal context provided by relative age and the absolute dating techniques developed by J. A. Arnold, W. F. Libby, Harold Urey, and others (see chap. 3).

Modern plant and animal communities are among the most complicated assemblages ever to occupy the planet, but fortunately, it is also a time when an impressive arsenal of innovative approaches is available for tracing their history. Consider, for example, what it is now possible to do:

- Reconstruct changes in the atmospheric concentration of CO_2, methane, and other greenhouse gases over the past 100 million years from the extent of carbonate and other sediments (e.g., limestone, $CaCO_3$) and from air bubbles trapped in glacial ice.
- Trace marine paleotemperatures and terrestrial ice volumes over a similar period from oxygen isotope ratios and, by extrapolation, the ancient temperatures of coastal and terrestrial habitats in conjunction with the fossil record.
- Detect fluctuations in the amount of heat reaching the Earth's surface over time owing to changes in the planet's position relative to the sun (Milankovitch variations).
- Estimate the paleoaltitude of mountains at different stages in their history from the MAT of fossil floras deposited at sea level, and others at high elevations, then use the worldwide average lapse rate of about 6°C/km to calculate the original altitude of the higher flora.
- Trace rainfall patterns from the width of growth rings in stalactites, and from the ecology of mites and midges often preserved in them.
- Reconstruct paleotemperatures from amino acid racemization of fossil *Emu* eggshells. Racemization is the change from levorotatory—rotating polarized light to the left—to a mixture of levorotatory and dextrorotatory forms of a molecule, and the rate of change is a function of temperature.
- Detect the change from cold glacial to warmer postglacial climates (e.g., in central Alaska between 12,000 and 10,500 BCE) by the ecology of beetle species preserved in lake sediments.
- Track the temperature record of marine waters over time from long-chain organic molecules called alkenones preserved in ocean sediments as biochemical fossils. The relative abundance and degree of saturation is temperature dependent: increasing unsaturation indicates cooling water.

- Determine past temperatures in tropical lowlands from the solubility of noble gases (argon, helium, krypton, neon, radon, xenon) in radiometrically dated groundwater. The solubility of these stable and inert gases varies with temperature, and amounts of He$+$ in groundwater from Brazil indicate temperatures were cooler by an unexpected 5°C at the Last Glacial Maximum (LGM) at 21–18 kyr.

- Trace precipitation patterns from the titanium content in marine basins. As noted, titanium is a terrestrial mineral, so its abundance in an ocean core is a function of precipitation. Sequences from offshore Venezuela reveal wet and dry periods in the Amazon Basin during the Quaternary, and some of these coincide with the rise and fall of the great ceremonial centers in ancient Latin America.

- Estimate changes in atmospheric CO_2 from stomatal density on the lower surface of leaves. Stomata are openings that allow the intake of CO_2 for photosynthesis and the outflow of water that constitutes the transpiration stream emitted from vegetation. Their number varies with the amount of CO_2 in the atmosphere (increasing stomata over time means decreasing CO_2). Since CO_2 is a greenhouse gas, relative changes in stomatal density in sequences of fossil floras indicate trends in temperature through time. The changes can be calibrated to actual temperatures by measuring stomatal density from old to recent leaf collections at sites with long meteorological records.

- Estimate paleotemperatures from the percentage of entire-margined leaves prominent in tropical environments compared to the lobate or toothed-margined leaves of temperate environments. The percentage is primarily a reflection of MAT, while other features of leaf physiognomy such as large size, thin texture, and drip tips correlate generally with high MAP.

- Track the migration patterns of seed-distributing birds from the occurrence of isotopes in their feathers. The abundance of ^2H (deuterium) in rainfall has a characteristic geographic signal that is deposited in the feathers. Most migratory birds grow new feathers before each migration, so the approximate latitudinal starting point can be determined from the ^2H/normal H ratio. When this technique is applied to potential seed-dispersing black-throated blue warblers (*Dendroica coerulescens*), for example, it documents that "Warblers wintering in the western Caribbean islands migrate from the northern United States, whereas those wintering in the eastern Caribbean islands arrive from the southern United States" (Hobson 2002; Rubenstein et al. 2002).

These are among the remarkably useful methods that augment the two tra-ditional approaches to the study of fossil plants—paleobotany dealing with plant macrofossils such as leaves, fruits, seeds, flowers, wood (dendrochro-nology), and rodent middens, and palynology dealing with spores and pol-len, phytoliths (plant crystals), starch grains adhering to tools and utensils, and other plant microfossils preserved in the rocks of the Earth.

It is exciting to contemplate that after some five hundred years of ob-servation and study, we are at a time when information from biology, ge-ology, and climatology, past and present, can be integrated into a better understanding of the Earth's ecosystems. The early founders of geology, tax-onomy, evolutionary biology, and biogeography—Lyell, Linnaeus, Darwin, Wallace, and Humboldt—would be amazed and undoubtedly immensely pleased.

THE GOAL: THE MODERN PLANT COMMUNITIES OF THE NEW WORLD

Having set 100 Ma as a convenient time for the start of our survey, it is worthwhile to define the endpoint. The goal is to trace the origin and devel-opment of the existing plant communities of the New World, and to iden-tify the various events that have guided the process. To do this, a system of vegetation classification must be developed that is suitable for the purpose. It is a work in progress because the biota is diverse; in places no natural vegetation or only remnants remain (fig. 1.1), and in the vast extent of the tropics many groups and regions are poorly known. Only two countries in South America, Peru and Ecuador, have relatively complete lists of native plants and, as inventories are completed, unexpected results are emerg-ing. About 27 percent of the plant species of the world are threatened, and among these, endemics are especially vulnerable because of their limited distribution. In an analysis of catalogs for Ecuador, an astonishing 83 per-cent of the endemic plants qualified as threatened (Pitman and Jørgensen 2002).

The hierarchy of categories used to classify vegetation are the plant forma-tions (e.g., the deciduous forest), associations (beech-maple or oak-hickory woods within the deciduous forest), and stages (a fern glade or a tempo-rary open weed community within a beech-maple wood). Several criteria were used to develop a system suitable for an introduction to New World vegetation and applicable specifically to a discussion of ecosystem history. First, the categories should be consistent across geographic, political, and linguistic boundaries. A vegetation type in Mexico called the bosque tropi-

Figure 1.1 Vestige of the tropical forest, Guatemala. Photograph by Bruce Graham.

cal subcaducifolio (tropical semideciduous woodland) is known by sixteen other names (Rzedowski 1978). The lowland neotropical rain forest is known as the bosque tropical perennifolia, selva alta perennifolia, tropical wet forest, floresta tropical húmida, and more informally the hylaea (the name used by Humboldt). To the extent possible, a single name is applied to each vegetation type throughout its range.

Second, the scheme must accommodate the ancient analogs as well as the modern derivatives of the vegetation. Names using prominent genera to define units (e.g., pine-oak association) are convenient because they often include modern representatives that are present and recognizable in the fossil record (e.g., *Pinus*, pine; and *Quercus*, oak).

Third, the system need only be as complex as required for paleovegetation analysis. For example, in Mexico the height and deciduousness of tropical forest trees varies with moisture, so it is possible to recognize a gradational series ranging from the selva alta subperennifolia (tall semievergreen forest), through several drier types, to the selva baja caducifolia (low deciduous selva). Forest height and humidity are not directly evident from fossil floras, so these selvas are grouped together here into the lower to upper montane broad-leaved forest formation.

Using these criteria, twelve modern plant formations are recognized for the purpose of tracing vegetation and terrestrial environmental history of the New World. They are listed below and defined the first time they are used in the text (see also table 1.1).

The New World at 100 Ma consisted of eight plant formations that formed the vegetation component of the terrestrial ecosystems: (1) polar broad-leaved deciduous forest, (2) notophyllous broad-leaved evergreen forest, (3) paratropical forest, (4) tropical forest, (5) mangrove, (6) aquatic, (7) herbaceous bog/marsh/swamp, and (8) beach/strand/dune (chap. 5). These formations were unique or only shadowy versions of their modern counterparts in structure, composition, ecology, and distribution. The task is to understand how, when, and why these eight ecosystems transitioned into the twelve modern ones listed in table 1.1 that now characterize the New World and support its human and other animal life. The expectation is that this information will be useful in planning for the future as fluctuations in CO_2, temperature, and rainfall approximate conditions already known from former geologic times.

I characterize each plant formation by citing representative species. These names, along with the terminology for vegetation types, can be formidable obstacles to understanding ecosystems. Even the greatest minds—Humboldt, for one—can be daunted: "He maintains a horror for the single fact—hence his distain for botanical nomenclature" (De Terra 1955). Even so, a familiarity with the types of communities, and some of their representative species, is necessary to conceptually project them back in time and to envision their change and movement over the landscape in response to evolutionary and environmental processes. There are several ways to do this. Each plant formation I mention is illustrated either in this work or in Graham 1999 (I, chap. 1) and 2010 (II, chap. 3). Table 1.1 provides a guide to all these illustrations. The Web offers additional examples as electronic illustrations retrievable through the various search engines. For example, a search for *Larrea tridentata* (creosote bush, a prominent plant of the New World deserts) will yield several images of the plant in its natural habitat. Other sources of illustrations, along with distribution maps, bibliographic, and ecological information, are the virtual herbaria of the Missouri Botanical Garden (www.mobot.org), the New York Botanical Garden (www.nybg .org), and other botanical institutions. As of 2008 the TROPICOS database at the Missouri Botanical Garden contained 1,008,141 names; 3,406,030 records; 111,521 bibliographic entries; and 70,111 digital images for the approximately 6 million plants in the collection. There are three thousand herbaria in the world, and efforts are underway to link the major ones into a single, searchable database. Consideration is also being given to the feasibility of iPhone digital field guides (Eisenberg 2009) and using barcoding

Table 1.1. Guide to text figures illustrating plant formations of the New World

| | Figure numbers | | |
Plant formation	Present text	I, chap. 1	II, chap. 3
Desert	2.6, 2.22, 2.38	1.14	3.4–10, 3.128, 3.141
Shrubland/chaparral-woodland-savanna	2.5, 2.23, 2.50	1.12	3.11, 3.12, 3.88, 3.129, 3.148
Grassland	2.9	1.11	3.16, 3.17
Mangrove	2.24	1.16	3.18, 3.74, 3.92
Beach/strand/dune			3.23
Freshwater herbaceous bog/marsh/swamp	2.51		3.24, 3.25, 3.150, 3.162
Aquatic	2.25		3.89, 3.153
Lowland neotropical rain forest	2.26, 2.46		3.26, 3.28, 3.93, 3.155–57
Amazon várzea (gallery) association			3.161
Atlantic rain forest			3.174
Lower to upper montane broad-leaved forest			3.77, 3.78, 3.144
Deciduous forest association	2.4	1.9	3.36
Floodplain association		1.10	
Transition to rain forest			3.58, 3.59
Mossy forest association			3.65
Coniferous forest			
Boreal association		1.3, 1.4	
Midaltitude (pine, pine-oak) association			3.37–39, 3.86
Mid- to low-altitude association			
(*Araucaria*) phase	2.48		3.175
Montane phase	2.16	1.5–8	3.41–43, 3.60
Alpine tundra (páramo)	2.43	1.5	3.135–37
Tundra	2.3	1.2	

Note: I = Graham 1999; II = Graham 2010.

technology to identify species (Stoeckle and Hebert 2008). In theory, a sequence of DNA from a piece of the organism could be read using a handheld analyzing device, the information relayed via satellite to a database of sequences of known organisms, and the name along with other information (e.g., distribution, ecology, uses) retrieved instantaneously. The technique is easier applied to animals than to plants, where extensive hybridization makes it difficult to find a "standard" piece of DNA for a species. Both are new technologies just on the horizon, and if successful, they would revolutionize identifying organisms and retrieving information about them. Even with all these valuable new tools, however, there is still no substitute for the wisdom gained through a career-long study of the organisms, or for field experience to avoid becoming a modern-day incarnation of what Rudwick (2005) called Buffon—an indoor savant snugly ensconced in his study (read, e.g., laboratory). As a preliminary exercise, the electronic illustrations, virtual herbaria, and general readings included in the references below can bring greater meaning to the plants, animals, communities, terms, descriptions, people, and subjects encountered in this work. As I have noted previously:

> A perception that has undoubtedly influenced my writing is that an increasing number of students seem most comfortable as passive recipients of information. This is in contrast to learning endeavors that involve active participation. One of my most challenging and ultimately satisfying educational experiences as a beginning student was reading several texts that were important in biology and geology, but that were outside my immediate specialty and for which I had only a minimum background. I struggled to a general understanding and a lasting appreciation of this material through persistence and a great deal of help from others. Although the present survey is intended to provide adequate background for the subject matter discussed, dictionaries, glossaries, additional texts, libraries, peers, and teachers in related fields do exist and are sources of additional explanatory information. A more full, lasting, and satisfying understanding of the topic can be gained by assuming the role of active participant and using the text as a cooperative venture. (I, xii–xiii)

One way to become an active participant is to have a modern atlas—for example, the 1997 *Oxford Atlas of the World*—open to the physiographic features of a particular continent or region (North America, 124–25; South America, 150–51), and a computer open to access images of geography available on Web sites like Falling Rain Genomics (Fallingrain.com/world) or NASA and NOAA. In this way information can be obtained on topics that are unfamiliar or of particular interest—the early peopling of America, the Little Ice Age, the Medieval Warm Period, El Niños, the tropical rain forest, the Atacama Desert, the Aymara people. This is a slow, methodical, and thoughtful way to approach any introductory material. However, the quest for knowledge, stimulated by curiosity and imagination, as a basis for understanding, does require effort. It is also a lot of fun and ultimately quite satisfying.

: :   REFERENCES   : :

De Terra, H. 1955. *Humboldt*. Alfred A. Knopf, New York.

Eisenberg, A. 2009. Digital field guides eliminate the guesswork. *New York Times Novelties*, 9 May 2009.

Falling Rain Genomics, Inc. Global Gazetteer, version 2.1. http://www.fallingrain.com/world/. [A directory of the world's cities and towns, with topographic maps, weather data, and selected satellite images.]

Fine, P. V. A., and R. H. Ree. 2006. Evidence for a time-integrated species-area effect on the latitudinal gradient in tree diversity. *American Naturalist* 168:796–804.

Gill, V. 2009. Evolution faster when it's warmer. BBC News, Science and Environment, 24 June 2009, http://news.bbc.co.uk/2/hi/8115464.stm. ["Climate could have a direct effect on the speed of 'molecular evolution' in mammals." See also Gillman et al. 2009.]

Gillman, L. N., D. J. Keeling, H. A. Ross, and S. D. Wright. 2009. Latitude, elevation and the tempo of molecular evolution in mammals. *Proc. R. Soc. Lond. B*, 25 June 2009, http://rspb.royalsocietypublishing.org/content/early/2009/06/24/rspb.2009.0674.full.

Graham, A. 1999. *Late Cretaceous and Cenozoic history of North American vegetation*. Oxford University Press, Oxford.

———. 2010. *Late Cretaceous and Cenozoic history of Latin American vegetation and terrestrial environments*. Missouri Botanical Garden Press, St. Louis.

Gröcke, D. R. and U. G. Wortmann, eds. 2008. *Investigating climates, environments and biology using stable isotopes*. Special issue, *Palaeogeogr. Palaeocl. Palaeoecol.* 266:1–128.

Hobson, K. A. 2002. Incredible journeys. *Science* 295:981–82.

Liede-Schumann, S., and H. E. K. Hartmann. 2009. Mesembryanthemum—back to the roots? *Taxon* 58:345–46.

Manos, P. S., et al. (seven coauthors). 2007. Phylogeny of extant and fossil Juglandaceae inferred from the integration of molecular and morphological data sets. *Syst. Bot.* 56:412–30.

Oxford Atlas of the World. 1997. 5th ed. Oxford University Press, Oxford.

Pitman, N. C. A., and P. M. Jørgensen. 2002. Estimating the size of the world's threatened flora. *Science* 298:989.

Ricklefs, R. E. 2007. Estimating diversification rates from phylogenetic information. *Trends Ecol. Evol.* 22:601–10.

Rubenstein, D. R., et al. (six coauthors). 2002. Linking breeding and wintering ranges of a migratory songbird using stable isotopes. *Science* 295:1062–65.

Rudwick, M. J. S. 2005. *Bursting the limits of time: The reconstruction of geohistory in the age of revolution.* University of Chicago Press, Chicago. [See review by Naomi Oreskes, *Science* 314 (2006): 596–97.]

Rzedowski, J. 1978. *Vegetación de México.* Editorial Limusa, Mexico City.

Stoeckle, M. Y., and P. N. D. Hebert. 2008. Barcode of life: DNA tags help classify animals. *Scientific American*, October 2008, 82–88.

Stuessy, T. F., and C. König. 2009. Classification should *not* be constrained *solely* by branching topology in a cladistic context. *Taxon* 58:347–48.

Wood, G. 2008. *The purpose of the past: Reflections on the uses of history.* Penguin Press, New York. [See also Wood's interview on C-SPAN's Q&A Web site, 13 April 2008, http://www.q-and-a.org/.]

Additional Readings and Updates

The Abominable Mystery. 2009. Special issue on Darwin, foreword by J. Jernstedt, introduction by R. A. Stockey, S. W. Graham, and P. R. Crane. *Am. J. Bot.* 96:1–381.

Adams, D. C. and M. L. Collyer. 2007. Analysis of character divergence along environmental gradients and other covariates. *Evolution* 61:510–15.

Bigelow, N. H., and M. E. Edwards. 2001. A 14,000-year paleoenvironmental record from Windmill Lake, central Alaska: Late glacial and Holocene in the Alaska Range. *Quat. Sci. Rev.* 20:203–15.

Biodiversity Heritage Library (BHL). http://www.biodiversitylibrary.org/. [The BHL site allows one to search for scientific names throughout the natural history literature. Follow the following links from the home page: Tools, NameServices, NameGetDetail, nameBankID, enter 6663950, and click Invoke.]

Blunt, W. 2002. *Linnaeus: The Complete Naturalist.* 2nd ed. Introduction by W. T. Stearn. Francis Lincoln, Ltd., London.

Burnett, D. G. 2007. *Trying Leviathan: The nineteenth-century New York court case that put a whale on trial and challenged the order of Nature.* Princeton University Press, Princeton. [See the review by H. Nicholls in *Nature* 450 (2007): 1161.]

Christian, M., M. Finer, and C. Ross. 2008. Last chance to save one of world's most species-rich regions. *Nature* 455:861, doi: 10.1038/455861c. Published online 15 October 2008.

Conniff, R. 2007. Forgotten, yes. But happy birthday anyway. *New York Times*, Travel section, 30 December 2007. [Note on the commemoration of the three hundredth birthday of Buffon at the Muséum National d'Histoire Naturelle, Paris.]

Ebach, M. C., and R. S. Tangney, eds. 2007. *Biogeography in a changing world.* CRC Press, Taylor and Francis Group, Boca Raton.

Frohlich, M. W., and M. W. Chase. 2007. After a dozen years of progress the origin of angiosperms is still a great mystery. *Nature* 450:1184–89.

Kintisch, E. 2008. Impacts research seen as next climate frontier. *Science* 322:182–83.

Kuhn, H. C., and E. Kesser. The polemics between Carl Linnaeus and Johann Georg Siegesbeck. Scricciolo: Italian Ornithological Web Site, maintained since 1996 by Alberto Masi. http://www.scricciolo.com/linnaeus_polemic.htm.

Linnaeus at 300. 2007. News Feature, *Nature* 446:247–62.

Malhi, Y., et al. (five coauthors). 2008. Climate change, deforestation, and the fate of the Amazon. *Science* 319:169–72.

Manchester, S. R., and B. H. Tiffney. 2001. Integration of paleobotanical and neobotanical data in the assessment of phytogeographic history of Holarctic angiosperm clades. *Int. J. Plant Sci.* 162:519–27.

Morehead, A. 1969. *Darwin and the Beagle.* Harper and Row, New York.

Müsch, I., R. Wellmann, and J. Rust, eds. 2001. *Cabinet of Natural Curiosities: The complete plates in color, 1734–1765,* by Albertus Seba. Taschen, Cologne.

Pennisi, E. 2009. On the origin of flowering plants. *Science* 324:28–31.

Primack, R. B., and A. J. Miller-Rushing. 2009. The role of botanical gardens in climate change research. *New Phytologist* 182:303–13.

Ramírez, S. R., B. Gravendeel, R. B. Singer, C. R. Marshall, and N. E. Pierce. 2007. Dating the origin of the Orchidaceae from a fossil orchid with its pollinator. *Nature* 448:1042–45. [A report on orchid pollen and stingless bee from the Dominican amber 12–20 Ma concludes from this relatively recent fossil record, based on molecular evidence, that "the most recent common ancestor of extant orchids lived in the Late Cretaceous (76–84 Ma)."]

Stevens, R. D. 2006. Historical processes enhance patterns of diversity along latitudinal gradients. *Proc. R. Soc. Lond. B* 273:2283–89.

Weschler, L. 1995. *Mr. Wilson's Cabinet of Wonder: Pronged ants, horned humans, mice on toast, and other marvels of Jurassic technology.* Vintage Books, New York.

Wing, S. L. 1998. Tertiary vegetation of North America as a context for mammalian evolution. In *Evolution of Tertiary mammals of North America,* vol. 1 of *Terrestrial carnivores, ungulates, and ungulatelike Mammals,* ed. C. M. Janis, K. M. Scott, and L. L. Jacobs, 37–60. Cambridge University Press, Cambridge.

Zero population growth 40 years old. 2008. *The Reporter,* June 2008, 14–15.

2

Location, Location, Location

Nomads pursuing herds of migrating animals across the frozen landscape connecting present-day Siberian Russia with Alaska were the first to glimpse the New World. The Bering Land Bridge was accessible to varying extents during the latter part of the Tertiary period; and with the onset of eighteen to twenty glacial cycles during the 2.6 million years of the Quaternary period, sea levels periodically dropped by about 120 m, providing a continuous connection over 1500 miles wide. The most recent connection was around 17 kyr near the end of the LGM. Although conditions must have been horrendous, ice-free pathways did exist, and for a nomadic people depending on hunting, there was little choice but to follow the migrating herds as they had done for generations. The footprints are blurred, and the itinerary unsettled, but people of Asian origin did enter the New World through Beringia by 15 kyr. Their descendents had established the Clovis Culture at Murray Springs, Arizona, by 13.5 kyr, as evident by the bones of bison and mammoth associated with Clovis Point arrowheads (fig. 2.1).

Figure 2.1 Bison tooth and tip of Clovis Point arrowhead from Murray Springs, Arizona. The distance between the specimens has been reduced slightly. The actual distance is about four inches. Photograph courtesy of Vance Haynes.

There is less agreement about older sites, such as Meadowcroft Rockshelter near Pittsburgh, 11,900–12,550 BCE to possibly as old as 16 kyr. Humans were present at Monte Verde in Chile by 12.5 kyr, although in all these calculations there is the possibility that early people arrived in Latin America from southeastern Asia by boat as well as by overland crossings.

The Vikings were the second group to see the New World, some of them forced to leave Scandinavia because of obnoxious and downright dangerous behavior even by Viking standards. Thorvald Asvaldsson went to Iceland after being banished from Norway for killing a man. His son, Eric the Red, killed two men in Iceland and settled in Greenland at Midjokull (Middle Glacier) to start a colony with four hundred people in 986 CE. In turn, his son Leif Ericson first landed on Baffin Island, then Labrador, and finally Vinland (Newfoundland) at a site now called L'Anse aux Meadows in about 1000 CE. They were the first Europeans to see the New World.

The temptation to try and the opportunity to succeed in settling Greenland and Newfoundland was in large measure conditioned by climatic fluctuations in the postglacial Holocene epoch. It was during the Medieval Warm Period (800–1200 CE) that the Vikings made their extended sea voyages. During the Little Ice Age that followed (1300–1850 CE), they would abandon the sites, and toward the end of that same cold period Washington's troops would suffer a near-fatal winter at Valley Forge in 1777. These climatic events, future changes in climate, and their potential consequences are recounted in Richard Alley's *The Two-Mile Time Machine* (2000), Brian Fagan's *The Little Ice Age* (2000), and Tim Flannery's *The Weather Makers* (2005). The changes were on a regional to near-global scale:

> On the last day of 1607, Edmund Shakespeare [brother to William] was buried. It was a time of almost unbearable cold. By the middle of December the Thames had frozen solid so that "many persons did walk halfway over the Thames upon the ice, and by the thirtieth of December the multitude . . . passed over the Thames in divers places." A small tent city sprang up on the ice, with wrestling bouts and football matches, barbers shops and eating-houses, trading upon the novelty of the silent and immobile river. (Ackroyd 2005, 457)

On 5 October 1492, some five hundred years after the Viking's discovery of the New World, a lookout on board the Pinta cried out "Tierra" and claimed the 5000-maravedi prize for being the first to sight land—the Caribbean island of San Salvador. In 1502 Columbus made his last voyage to the New World, sailing along the coast of Central America, which he still believed was China. He was in search of the elusive strait described by Marco Polo as a passage from China into the Indian Ocean, by which he intended to return to Spain and thereby confirm the belief already widely suspected by geographers that the Earth was a sphere. It was left to Amerigo Vespucci, on a voyage beginning on 13 May 1501, to recognize that South America was a new continent: "We arrived at a new land which, for many reasons that are enumerated in what follows, we observed to be a continent" (quoted in Boorstin 1983, 233). Gerardus Mercator published a map of the world in 1538 and it showed for the first time lands called North America and South America.

These early wanderers were concerned primarily with food and survival; suitable lands for establishing colonies; and wealth, converts to Christianity, and slaves. The crew often included prisoners racked with contagious diseases and convicted of the most heinous crimes. They were unleashed fully armored, mounted on never-before-seen horses, and armed with guns capable of frightening and seemingly magical devastation. They were dangerous, and some were fanatically religious. As Tina Rosenberg notes in *Children of Cain* (1991), destruction of the early social structure and the subsequent history of the New World accounts in part for the attitudes of many native peoples toward distant and centralized authority. They remained inhabitants of a land, but were never allowed to become citizens of a country.

Many of the crewmen had been given the choice between a pardon in the guise of the voyage or death. The odds were not good either way. The width of the Pacific Ocean was underestimated by 80 percent, and provisions were stocked accordingly. Magellan left Spain to find a passage through the southern tip of South America to the Spice Islands. Of the 250 men who departed on 20 September 1519, only 18 returned on 8 September 1522.

Logs and diaries reveal there was often little interest and less time to contemplate nature, but even so some sights were impressive. During his third voyage in 1498, Christopher Columbus recorded in the ship's log the otherworldly beauty of the Orinoco:

> I am convinced that it is the spot of the earthly paradise [Eden] whither no man can go but by God's permission. . . . I think also that the water I have described may proceed from it . . . for I have never either read or heard of fresh water coming in so large a quantity . . . and if the water of which I speak does not proceed from the earthly paradise, it still seems to be a still greater wonder, for I do not believe that there is any river in the world so large or so deep. (Quoted in Boorstin 1983, 242; see also Cohen 1992)

With further exploration, and the passage of time, the magnificence of what had been discovered became more deeply appreciated—a sense of its marvels beautifully captured in the closing paragraphs of Fitzgerald's *The Great Gatsby*: "[F]or a transitory enchanted moment man must have held his breath in the presence of this continent, compelled into an aesthetic contemplation he neither understood nor desired, face to face for the last time in history with something commensurate to his capacity for wonder."

NORTH AMERICA (NORTH OF MEXICO)

The Arctic Region

The early nomadic hunters entered a landscape already familiar from their wanderings in northern Asia (fig. 2.2). Southward from Beringia, land was exposed through and around the margins of the Laurentian-Cordilleran ice sheet over a mile in height. The unglaciated lowland was mostly a flat, treeless terrane of gravel and bogland, frozen into permafrost, with a few iso-

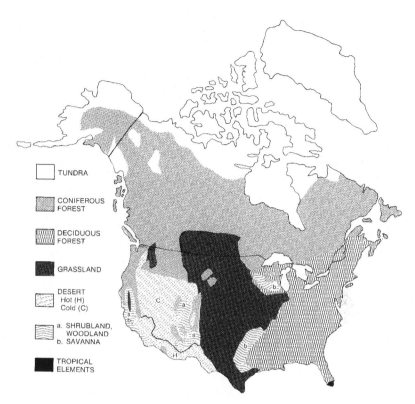

Figure 2.2 Vegetation diagram of North America north of Mexico. Some vegetation types, such as beach/strand/dune, freshwater herbaceous bog/marsh/swamp, aquatic, and alpine tundra, are too limited in extent to show on the map. Used with permission from Oxford University Press, Oxford.

lated clusters of *Picea* (spruce), *Alnus*, *Betula*, and *Salix* (dwarf alder, birch, and willow), and a herbaceous cover of lichens, mosses, grasses, sedges, and forbs (fig. 2.3; see table 1.1 above as a guide to the illustrations of plant formations). The tundra presently covers ice-free areas of the Arctic region and extends southward only to about 60°N along the west coast because of the warm Japan Current, while along the east coast it extends as far south as 55°N because of the cold Labrador Current. Inland, the southern boundary is defined by the limits of permafrost. The growing season is only some six to twenty-four weeks, frequent winds of 65 km/hr (40 mi/hr) or more may last for days, and there are twenty-four hours of near darkness for six months of the year. The lowest temperature ever recorded in North America

Figure 2.3 Tundra formation of mosses, grasses, and sedges, Devon Island, Northwest Territories, Canada. Photograph courtesy of Larry Bliss.

is −62°C (−81°F) on 3 February 1947 at Snag Airport in the Yukon. At the same time of year sixty-one years later, on 18 February 2008, the temperature at the airport was 6°C, that is, 68°C warmer.

In the coldest northern part of the Arctic, precipitation is less than in the deserts of North America and moisture in the form of meltwater is available only for about three months of the year. The northern lights shimmering eerily across the sky must have been cause for reflection among the early people. The tundra covers about 4 million km², or 19 percent of North America north of Mexico, and it extends southward at progressively higher elevations in the western mountains as alpine tundra that in Latin America is called páramo.

It is understandable that the extreme conditions, even more marginal for life during the Last Glacial Maximum, and the everyday dangers of hunting the huge megafauna, would have caused people even 15,000 years ago to seek the protection of spirits. This scenario is given credence by new suggestions that in the human brain there may be a "natural cognitive foundation for religion," potentially present in all civilizations, that eases fear by assuming an agent behind frightening and unexplained events, and endowing that agent with supernatural powers (Tremlin 2006). This posited cognitive foundation might be one factor explaining the resistance to scientific discoveries that affect understanding of such an agent, giving rise to the present dichotomy between science-based thinking and the spiritually grounded thought of members of some religious groups—a dichotomy that must have existed almost from the beginning.

The Boreal Region

Cooling molten slag began to form on the Earth's surface about 4.5 Ga (billion years ago) to constitute the first masses of granitic rock, called cratons or shields, around which the continents gradually formed. The central and eastern boreal region of North America is underlain by the Canadian Shield. The first life reported from the Precambrian was cyanophycean bacteria (blue-green algae), possibly fungi, and later bacillus bacteria in the Gunflint Chert of southern Ontario (Tyler and Barghoorn 1954). The deposits date from 1.9–1.83 Ga. Even earlier life is now known from structurally preserved fossils in rocks dated at 3.2–2.5 Ga, and from biogenic stromatolites in Australia at 3.5 Ga (Knoll 2004, pers. comm., 2009).

The vast Laurentian ice sheet was centered over Hudson Bay, and in the coldest phases of the Pleistocene, it fused with the Cordilleran ice sheet of the northern Rocky Mountains. The boreal region is mostly flat from multiple glaciations, and from the glacial outwash that has filled in much of the lowland. It is covered by numerous lakes, *Sphagnum* (peat) bogs, and a boreal coniferous forest of *Abies* (fir), *Larix* (larch), *Picea*, *Thuja* (cedar), *Tsuga* (hemlock), and a few species of *Pinus* (pine), along with *Alnus* (alder) and *Salix* (willow). In the east along its southern border there is a Lake States forest, transitional to the deciduous forest to the south, that adds *Betula alba* (white birch) and *Populus tremuloides* (quaking aspen). The plants and animals that have adapted to this environment are comparatively few, and these ten genera constitute the prominent trees of the boreal forest.

Growing conditions in the boreal region are harsh but less so than in the Arctic. Rainfall is about 500 mm a year; instead of permafrost, soil moisture is frozen for about eight months of the year; the minimum winter temperature is −35°C; and the growing season is between seven and twenty-five weeks. The boreal coniferous forest contains one-third of the world's trees, and it covers 28 percent of North America north of Mexico. It extends south as a disjunct montane-subalpine coniferous forest just below the alpine tundra in the mountains to the west, along the western coast, and at the highest elevations in the Appalachian Mountains to the east.

The Appalachian Mountains

North America is bordered on the east by the Appalachian Mountains, including the adjacent slopes or piedmont, and by the Atlantic and Gulf coastal plains. The mountains extend 3000 km from Alabama to Newfoundland and are part of an even larger system that continues into England, Scotland, Scandinavia, and in scattered regions north of the Alps as the Caledonian and Hercynian mountains. They are Paleozoic in age and have undergone about 300 million years of erosion that has resulted in relatively low elevations (Mt. Mitchell in North Carolina is the highest at 2037 m), and characteristic rounded domelike structures such as Clingman's Dome (2025 m) in the Blue Ridge Mountains of Tennessee. These ranges are mostly covered by the eastern deciduous forest, which constitutes about 11 percent of the plant cover of North American north of Mexico (fig. 2.4). With at least sixty-seven genera of trees and shrubs that form its canopy and subcanopy, the forest represents the richest of the plant formations. The most widespread association within the deciduous forest is *Quercus-Carya* (oak-hickory) on drier and sandy sites, and the richest association is *Fagus-Acer* (beech-maple) in more mesic localities. These associations reappear along the east-facing slopes of the Sierra Madre Oriental of Mexico, with the distribution interrupted by the Chihuahuan Desert of south Texas and northern Mexico, and as we shall see in chapter 9, they also occur in the forests of central China. Study of such disjunct occurrences is part of the field of biogeography, and their explanation depends on information from geology and climatic history, fossil records, a knowledge of the ecological requirements of the extant organisms, the distribution potential of their propagules, and a phylogenetically sound taxonomy (I, chap. 9).

Much of the coastal plain is covered by pine forest of *Pinus taeda* (loblolly pine), *P. echinata* (short-leaf pine), and *P. palustris* (long-leaf pine). There are grasslands in central Florida with scattered *Sabal* (palmetto) forming palm savannas, and mangroves (*Rhizophora mangle*) fringe the coast (see fig. 2.24 below). Swamp and floodplain forests of black gum (*Nyssa*) and bald cypress (*Taxodium distichum*) grow inland along the southern Gulf Coast.

One of the most fascinating environments of eastern North America is the quiet, serene, almost mysterious cypress swamp, with its festooning Spanish moss (*Tillandsia usneoides*), silently gliding alligators (*Alligator mississippiensis*), and cottonmouth water moccasins (*Agkistrodon piscivorus*). This has made it a frequent locale for films such as *Southern Comfort*, which recounts the fictitious, abusive, and ill-advised confrontation of nine National Guardsmen with the local inhabitants. My Cajun relatives in

Figure 2.4 Lower to upper montane broad-leaved forest (deciduous forest) formation: beech-maple association of *Fagus grandifolia* and *Acer saccharum*. Nicolet National Forest, Wisconsin. Photograph courtesy of the National Agricultural Library, Forest Service Photo Collection, National Archives, Washington, DC.

Louisiana are both amused and appalled at how those living in isolated parts of the swamp are often depicted—particularly with regard to their distinctive music (zydeco), culture (admittedly somewhat taciturn with strangers but joyously effusive among friends), humor (a bit earthy), language (a combination of French and Spanish), transportation (in flat-bottomed boats called pirogues), and religion (Catholic but with broad options including the still-practiced voodoo). Nonetheless, the beauty of the cypress swamps, approximating the *Glyptostrobus/Taxodium* association of the mid-Tertiary in the New World, is frequently captured in these films with dramatic effectiveness.

The Western Cordillera

The complex of mountains known as the western cordillera formed as the continent was forced westward by spreading from the Mid-Atlantic Ridge. The leading edge crumpled as it was pushed against the subducting Pacific Plate and accreted a complex mosaic of land fragments located on the plate. Volcanic material from extensive volcanism associated with these events was added later as an overlay of flood basalts, such as the Colorado Plateau (circa 24 Ma) and the Columbia Plateau (largely formed 16–14 Ma). The oldest of the highlands are the Rocky Mountains, west of which the ranges become progressively younger. In the far north along the coast of Alaska is the Brooks Range, brought to public attention because of the large oil deposits around adjacent Prudhoe Bay, construction of the 800-mile-long Alaskan pipeline, and most recently by corrosion of that pipeline after decades of inadequate maintenance. Along the south coast is the Alaska Range, with the magnificent scenery of Denali National Park, including Mt. McKinley, the highest point in North America at 6194 m, and Prince William Sound, famous site of the 1989 Exxon Valdez oil spill.

The Rocky Mountains extend from northern British Colombia to the Río Grande, but they are part of a system that continues into Mexico as the Sierra Madre Occidental and Sierra Madre Oriental. They began as a depression that was folded and uplifted to approximately 1 km above sea level by about 65 Ma, followed by a period of intense orogeny 55–40 Ma. By that time the landscape resembled the present-day High Plains along the eastern front of the mountains, which locally had reached at least half of their present altitude. The remaining height was attained afterward with continued movement of the continent away from the Mid-Atlantic Ridge and subduction of the Pacific Plate. The western cordilleras have higher

Figure 2.5 Shrubland/chaparral-woodland-savanna formation: juniper (*Juniperus occidentalis*) woodland, central Oregon, USA. Photograph courtesy of the National Archives, Washington, DC.

elevations, sharper peaks, and more dramatic scenery compared to the gentler, older, more eroded, and lower Appalachian Mountains.

There is alpine tundra on the higher peaks, and a western montane coniferous forest below that includes *Abies*, *Picea*, *Pseudotsuga mensiesii* (Douglas fir), and *Pinus ponderosa*. Toward the south at lower elevations, the coniferous forest grades into drier lowlands and woodlands of *Pinus edulis* (piñon pine) and *Juniperus* (juniper; fig. 2.5), and eventually into the western deserts (fig. 2.6). In the southern Rocky Mountains, *Pinus aristata* (bristlecone pine; I, fig. 4.8) and related species grow from central Colorado to northern New Mexico and northeastern Arizona. Some reach an age of nearly 5000 years and are the Earth's oldest organisms. When these ancient trees began their life, the Greek city-states and the disparate political units that would become Egypt were just assembling. The trees were already 3200 years old when Hannibal and his unlikely entourage of elephants crossed the Alps in 218 BCE, and they were 3000 years old at the time of Christ. Throughout

Figure 2.6 Desert formation: Sonoran Desert association of saguaro (*Carnegiea gigantea*) with creosote bush (*Larrea tridentata*), Arizona.

these millennia the width of the less-dense and lighter-colored spring wood (fig. 2.7), forming part of the annual growth ring, reflects precipitation and water availability (stream flow, water table fluctuations) and, to a lesser extent, temperature, length of the growing season, and nutrient levels. The ring patterns from living trees have been fitted into sequences from older dead trees, construction timbers at archeological sites, and fossil woods, to give a continuous year-by-year record back nearly 10,000 years. Dendrochronology, the study of tree rings (I, 128–31) provides information about trends in climate (Esper et al. 2002), fire and hurricane history, volcanic events (D'Arrigo and Jacoby 1999), waves of insect infestation, and epi-

Figure 2.7 Cross-section of bristlecone pine (*Pinus aristata*) showing light spring wood and dark summer wood of the annual rings. Photograph courtesy of Charles W. Stockton.

demic disease, the scars of which mark the year of the event. Information on wood anatomy and dendrochronology can be found on Henri Grissino-Mayer's Ultimate Tree-Ring Web Pages (http://web.utk.edu/~grissino/) and on North Carolina State University's Inside Wood site (http://inside wood.lib.ncsu.edu/).

To the west of the Rocky Mountains the land has been stretched, thinned, and weakened by uplift of the mountains. Some parts have collapsed to form down-faulted grabens (basins) and others crumpled upward to form horsts (ranges) constituting the Basin and Range Province (McPhee 1981). It is a winter-cold desert covered by *Artemisia tridentata* (sagebrush). The current trend is to recognize higher elevations that were reached earlier in time for the western cordilleras and that, if verified (and when quantified), will be important in refining our concepts of Paleocene and Eocene paleoclimates and biotas under the eastern rain shadow of the mountains. If the proto–Rocky Mountains had reached substantial heights by that time, winds off the Gulf of Mexico would have risen along these steep Paleogene slopes to create locally moist areas and a more complex humid/dry altitudinal mosaic of habitats than earlier envisioned.

Farther west is the Sierra Nevada. Radiometric dating of early volcanics suggested an origin about 33 Ma, but recent estimates are that at 50–40 Ma locally they may have been about 2200 m in elevation compared to the present 2780 m (Mulch et al. 2006). Sequoia National Park with the magnificent *Sequoiadendron giganteum* (big tree, sequoia) is located in these mountains. A related species, *Sequoia sempervirens* occurs along the coast in northern California and southern Oregon. The Coast Ranges and the Cascade Ranges, extending into southern Canada, are the westernmost and youngest of the cordilleras.

Deserts

There are four arid regions in North America, including the comparatively high-altitude, winter-cold Basin and Range Desert. The others are the warm Mojave, Sonoran (fig. 2.6), and Chihuahuan deserts. By definition a desert receives about 250 mm (10 in) per year or less of rainfall, has high evapotranspiration, often strong winds, and high radiation levels. Adaptive features of the vegetation include low stature, heavily cutinized microphyllous leaves, or leaves replaced by spines, sunken leaf stomata with overarching epidermal cells to reduce water loss, pleated stems that expand and contract like an accordion to accommodate the sudden changes in water availability (e.g., many cacti), deep taproot or extensive shallow fibrous root systems, succulence (abundant water-holding parenchyma cells), and alleopathy. The latter is an adaptation reducing competition for the limited water whereby roots or debris from the parent plant contain compounds toxic to the growth of other plants, and this contributes to the spacing often evident between desert plants (fig. 2.6; also fig. 2.22 below). The North American deserts receive slightly more than 250 mm of rainfall (300–355 mm, 10–14 in) annually and are moist deserts transitional to dry shrubland or woodland. The lowest point of land in North America is Death Valley in the Mohave Desert at 86 m below sea level, and the highest temperature for the continent was recorded there on 10 July 1913 (66°C, 134°F). The vegetation includes *Larrea tridentata* (creosote bush, fig. 2.6), *Franseria dumosa* (bur sage), *Yucca brevifolia* (Joshua tree; *Y. carnerosana*, fig. 2.22), with creosote bush occupying the driest parts of the desert. A common distribution pattern is for *Acacia*, *Prosopis juliflora* (mesquite), and other shrubs to grow near the base of hills where there is some moisture from runoff, and creosote bush to grow farther out on the driest flatlands. *Larrea* extends into the deserts of northern Mexico, then reappears as a disjunct, along with numerous other plants of the northern deserts (e.g., *Ephedra*, Mormon tea; *Juniperus*) in the dry monte vegetation of northwestern Argentina 8000 km away.

The Sonoran Desert of Arizona and adjacent northwestern Mexico is the most varied and the lowest in elevation of the American deserts (sea level to about 900 m; avg. 600 m). Rainfall is from near 0 mm at places in the west to 35 mm in the east, and it may reach 700 mm on some windward peaks. Temperatures range from winter lows of 20°C to a summer maximum of 40°C, and on bare soils the daily temperature may fluctuate by 60°C. One of the signature plants is *Carnegiea gigantea* (saguaro; fig. 2.6) showing many of the adaptations to arid environments noted earlier. Large columnar cacti are a feature of the Sonoran Desert.

The Chihuahuan Desert of southern New Mexico and west Texas is a plain between 400 m elevation along the Río Grande to 1500 m toward the south on the central Mexican Plateau. Precipitation is higher than in the Sonoran Desert (to 400 mm), temperatures are 5°C–10°C lower, and there is occasional winter frost. In addition to the widespread *Acacia*, *Larrea*, and *Prosopis*, there is *Agave* (Spanish bayonet, century plant), the source of henequen and tequila, and *Yucca* (Joshua tree).

Between the lower limits of the western coniferous forest and the desert, there is a transition zone of shrubland and woodland. Shrubland is a plant formation of low, densely spaced trees, and if there is a conspicuous component with spines, it is chaparral. The name for the protective leather leggings called chaps is derived from chaparral. Among the trees of the western shrubland are *Quercus turbinella* (scrub oak) and *Cercocarpus ledifolius* (mountain mahogany). If the crowns of the trees do not touch, the vegetation is called a woodland, the piñon pine–juniper woodland being an example (fig. 2.5). Savanna is a grassland with widely scattered trees such as *Acacia*. Savanna is not extensively developed as a natural plant formation in North America north of Mexico, and it is most familiar from African scenes showing herds of large carnivores and grazing animals where the scattered trees are often *Acacia*.

The North American dry vegetation is the result of several interacting factors. The equatorial regions receive the direct rays of the sun for the longest period of the year and are warmed more than the temperate regions to the north and south. As the warm air rises it cools, the water condenses and falls as rain. Thus, many parts of the equatorial regions are characterized by warm temperatures, low pressure systems (rising air), high rainfall, low evapotranspiration, and tropical rain forest. The cooled air descends at about 30°N and 30°S as a mountainous column of water-depleted air that warms as it approaches the Earth's surface. Thus, midlatitudes are often characterized by high pressure systems (descending air), low rainfall, fewer clouds, high temperatures, strong winds, high evapotranspiration, and desert vegetation. This circulation cell is known as the Hadley regime.

Another factor in the development of deserts is continentality; that is, the tendency for the interior parts of continents to be drier than the coasts because cool, moisture-laden winds off the oceans lose water as they move across the warmer land. A third factor is orographic deflection, that is, mountains barring moisture to the lee side. Edaphic conditions represent a fourth factor, because if the soil is coarse, any rain that falls percolates through before it can be absorbed by the roots. Last are the interrelated factors of slope, allowing surface moisture to rapidly drain away, and exposure, because in the Northern Hemisphere south-facing slopes are warmed more than north-facing ones. These elements combine at different localities and in various ways to produce dry vegetation ranging from shrubland/chaparral, to woodland, savanna, and desert.

Study of the history of arid-land vegetation is difficult because the environment is not conducive to the accumulation, transport, and preservation of organic material. Exceptions are petrified wood (I, 128–32), and the leaves, wood, fruits, seeds, and adhering pollen brought into protective rock shelters and used as nesting material by species of the packrat *Neotoma* (fig. 2.8). Packrats forage within a distance of about 50 m, so over the years the sequence of accumulating plant debris in the middens reflects from

Figure 2.8 Midden of the bushy-tailed pack rat (*Neotoma cinerea*). Photograph courtesy of Terry Vaughan.

bottom to top changes in local vegetation and climate from old to young. The packrat uses part of the midden as a urinal perch, which eventually solidifies into a shiny mass further protecting the fossil plant fragments. Comparisons of overlapping sequences from different shelters have allowed reconstruction of the North American dry-habitat communities and environments back 50,000 years (Betancourt et al. 1990).

GRASSLANDS AND THE INTERIOR LOWLANDS

In the eighteenth century, as French explorers and fur traders moved south from Canada, they encountered grassland unfamiliar to western Europeans in its vast extent. They applied the closest name they had for the vegetation, calling it *prairie*, French for "meadow" (fig. 2.9). The central part of North America is mostly a relatively low, flat to rolling region between the Appalachian Mountains and the eastern foothills of the Rocky Mountains. The physical features reflect its geologic development, beginning about 100 Ma, which contributed to conditions suitable for grasslands. At that time sea levels were higher than at present by up to 300 m, owing to the absence of glaciers, the displacement of water from tectonic activity (e.g.,

Figure 2.9 Grassland formation of *Bouteloua gracilis*, Colorado. Photograph courtesy of the National Agricultural Library, Forest Service Photo Collection, National Archives, Washington, DC.

new crust being generated from the ocean ridges), and the outpouring of lava that formed the massive Ontong Java Plateau over 30 km thick on the ocean floor off the Solomon Islands. Landscapes were low because older mountains like the Appalachians had been eroding since about 180 Ma, the Rocky Mountains and the Mexican sierras were just beginning to form, and the coastal cordilleras had not yet appeared. The result was an epicontinental sea extending from the Arctic Ocean through the central lowlands to the Gulf of Mexico (fig. 2.10). The separation of the continent into western and eastern portions is shown by fossil pollen types such as *Aquilapollenites* (fig. 2.11), found in Cretaceous and Paleocene sediments in western North America but not in the east, and the Normapolles group (fig. 2.12) found in eastern North America and extending into Europe.

The presence of an extensive midcontinent seaway created maritime climates and buffered the land from extreme seasonal changes in temperature that characterize continental climates. The present annual temperature range in North Dakota around 48°N is from 49°C (121°F) to −60°C (−51°F), while in coastal Maine at about the same latitude the range is from 12°C to 2°C. Large lakes have a similar buffering effect, and the coming and going of seas and lakes over geologic time has been a significant forcing mechanism for climatic change and biotic response.

The midcontinent sea began to drain in the Middle Cretaceous as a result of uplift of the continent, waning of plate tectonic activity, and less flood basalt pouring onto the ocean floor. By the mid-Eocene at 45 Ma, the margin of the sea ran from south of Laredo / Nuevo Laredo northward to just east of Austin, Texas, up the Mississippi Embayment to about Cairo, Illinois, around the southern end of the Appalachian Mountains, and northward along the eastern edge of the mountains. By the beginning of the Pleistocene epoch about 2.6 Ma, the coastline had retreated to its approximate

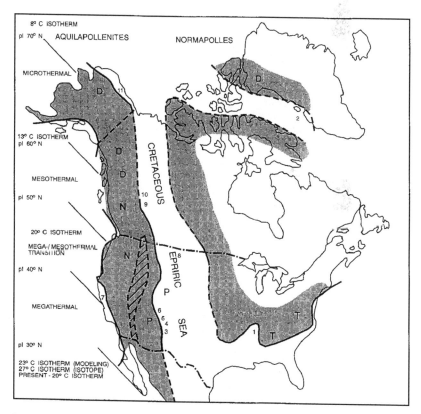

Figure 2.10 Diagram of the northern North American continent during the Middle to Late Cretaceous showing the epicontinental sea and the *Aquilapollenites* and Normapolles provinces. Numbers represent prominent fossil floras outlining the margin of the sea: (1) McNairy, (2) Lower Atanekerdluk, (3) Vermejo, (4) Raton, (5) Denver, (6) Laramie, (7) Patterson, (8) Hell Creek–Lance (Colgate), (9) Scollard, Horseshoe Canyon, (10) Coalspur, (11) Prince Creek (Coville River site). Used with permission from Oxford University Press, Oxford.

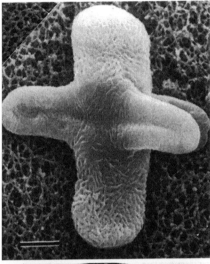

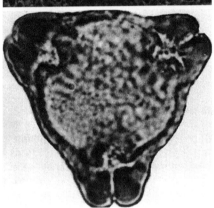

Figure 2.11 Integricorpus rigidis, a triprojectate (*Aquilapollenites*) pollen from the Campanian (Cretaceous) of Montana. From Farabee 1990. Used with permission from Elsevier Science Publishers, Amsterdam.

Figure 2.12 Trudopollis variabilis, a Normapolles-type pollen from the Campanian (Cretaceous) of Tennessee. From Tschudy 1975.

present position around Houston and New Orleans and along the Atlantic coast, and it fluctuated with the waxing and waning of glaciers. On a finer time scale, the coastlines have shifted with El Niños (reflected in present-day flooding from California to the southwest, drier in the southeast), and with the increase in hurricane numbers and intensity during periods of global warmth. The current trend toward rapidly melting glaciers, together with a slight additive effect from the thermal expansion of ocean waters, gives every indication that temperatures will rise by about 4°C and sea level will rise at least 0.18–0.59 m, and possibly as much as 1.4 m, by 2100 (IPCC 2007; "From Words to Action," 2007). In addition to the devastating and highly publicized effects of hurricane Katrina on the Gulf Coast of

North America, the Inuit people on the Arctic coast are now moving entire villages inland from the melting ice, rising seas, and loss of traditional food resources.

During the early geohistory of the interior lowlands, marine limestone was widely deposited and weathered into calcareous soils favorable to the growth of grasses. Another contributing event for grassland development was uplift of the western mountains, which blocked moisture coming in from the Pacific Ocean. Although some spring and summer moisture is brought up from low pressure systems along the Gulf Coast, overall the interior was too dry, and periodically too cold and windy, for trees and too moist for deserts. As grasses became more abundant, fires and expanding herds of browsing and grazing animals contributed to their maintenance and further expansion. Grassland has become the most extensive North American plant formation, covering 20–30 percent of the landscape. Tropical elements presently extend into southern peninsula Florida and include *Rhizophora mangle* (mangrove; see fig. 2.24 below) and some palms. Altogether, the vegetation of North America consists of about 20,000 species of vascular plants, constituting 7 percent of the Earth's flora, and they are arranged into the twelve extant plant formations described above.

The geologic history of northern North America not only accounts for the present landscape and, to an important degree, its biotic communities, but it also explains the distribution of the continent's extensive fossil plant and animal assemblages. Conditions most suitable for the preservation of organic material include the quiet waters of lakes and bays along gradually retreating shorelines, which serve as depositional basins where well-preserved and unfragmented material may accumulate. Over time these sediments harden into lacustrine or deltaic shale. There must also be rapid deposition of fine-grained sediments, such as silt that forms siltstone or ash that consolidates into volcanic shale, as opposed to porous and coarse-grained sandstone. The former allow for preservation of fine features that sometimes includes cellular and even subcellular detail, prevents distortion, and retards destruction by oxidation. Rapid accumulation of sediments also excludes microbes and fungi that break down organic material as a source of energy. Another process that reduces microbial decay is accumulation of organic matter in such abundance that the pH becomes highly acidic. Examples include stagnant swamps, bogs, lagoons, and lake bottoms where the sediments ultimately form peat, lignite, coal, and lacustrine shale. These deposits, the environments they represent, and the fossils they contain are found most abundantly in three regions of North America.

As the epicontinental sea gradually retreated, the remains of plants and animals living along the shore were preserved in lowland, swamp, and near-coastal sediments. Many of the fossil floras and faunas of central and southeastern North America actually mark the position of the sea in Cretaceous through Cenozoic times (fig. 2.10). The Cretaceous Lance Formation of Wyoming contains the first record of Mesozoic mammals in North America. The Scollard Formation of Alberta, Canada, includes the Bug Creek fauna, which represents the greatest assemblage of Mesozoic animals in the world; and in the southeast, the extensive Claiborne flora of Tennessee and Kentucky has provided much information on the biota and climates of this ocean-bordering region during middle Eocene time.

In western North America, the volcanism associated with formation of the cordilleras and plateaus also created conditions ideal for the preservation of fossils. As lava poured onto the landscape, it blocked streams, creating numerous lakes and marshes that provided habitats for the biota and basins for accumulating their remains. Ash and other sediments filled these lakes and charged the waters with silicates; there were algal blooms of silica-requiring diatoms; and the result is siltstone, diatomite, and volcanic shale (fig. 2.13) at numerous sites with exquisitely preserved fossils (fig. 2.14). The early Eocene Wind River Formation of Wyoming has more than a hundred species of vertebrates; the middle Eocene Green River Formation of Colorado and Utah has an extensive fish fauna and caddisfly mounds up to 9 m tall along the margin of the paleo-lake Gosiute (Leggitt and Cushman 2001); and both contain other kinds of insects and abundant plant fossils. Other similar assemblages are found in the Eocene/Oligocene Florissant Beds of Colorado, the middle Eocene Clarno of Oregon, the middle Eocene Republic of Washington, the middle Eocene Princeton of British Columbia, and the Miocene Trout Creek and Succor Creek floras of southeastern Oregon.

In the Arctic and boreal regions, adjacent New England, and in the upper midwestern United States, small depressions were dredged by glaciers, filled by meltwater and precipitation, and accumulated deep, low pH deposits of *Sphagnum* and other remains (fig. 2.3). The lakes and peat bogs provided innumerable sites for the preservation of spores, pollen, seeds, and plant and animal fragments that have yielded considerable information on the late Quaternary biota and environments.

The gradual retreat of the epicontinental sea from the interior lowlands and coastal plains, volcanism in the western cordillera, and glaciation in the Arctic, boreal, and alpine regions are among the geologic events creating conditions favorable for the preservation of more than two hundred fossil floras of Cretaceous and Tertiary age alone. There are many more smaller

Figure 2.13 Miocene Succor Creek flora, Harney County, Oregon. The rock is a shale composed of volcanic ash.

Figure 2.14 *Acer chaneyi*, a fossil species of maple seed from the Succor Creek flora illustrating the quality of preservation often possible in fine-grained sediments such as volcanic ash.

and partially studied floras, and even more Quaternary sites. These fossil floras and faunas range throughout the 100 Ma interval covered in this text, and provide one of the most extensive databases in the world for tracing biotic and environmental history.

MEXICO

The Cretaceous sea that covered the interior of northern North America extended along the coastal plains of Mexico to the base of the protoeastern and -western sierras, and as a mostly shallow sea through central Mexico, across the Tropic of Cancer, to the Isthmus of Tehuantepec and the Yucatán Peninsula (figs. 2.15). The waters were deep along the Chihuahuan Trough (Carciumaru and Ortega 2008), shallow over the Yucatán Peninsula, and there were barely emergent lands like the Tamaulipas Peninsula extending south from Big Bend National Park in Texas. There were reef-forming corals, and near Sabinas in the state of Coahuila, there are deposits of coal, indicating shallow seas, swamps, a low-lying landscape, and warm-temperate to subtropical environments. To the west were hills of the early Sierra Madre Occidental (fig. 2.16). These mountains have a history that is similar to and contemporaneous with the Rocky Mountains, but the exact relations are not well understood. The region is the most extensive geologic province in Mexico, over 1300 km long, and one of the world's largest volcanic fields. Activity continues into modern times as shown by the spectacular lava flows from Volcán Ceboruco in Nayarit through which Mexico Highway 15 is cut. Uplift was intense in the Cretaceous and waned in the latest Cretaceous and early in the Tertiary. This change in regional orogeny is shown with great effect in panoramas exposed in the canyons of northern Mexico. There are upturned, deep-water, ammonite-containing limestones over 300 m thick in Huasteca Canyon southwest of Monterrey (II, fig. 2.4). In contrast, later strata are flat-lying and have remained undisturbed (II, fig. 2.5). Thus, the landscape of Mexico at 100 Ma consisted of a line of low mountains on either side of a mostly submerged central plateau. The hills were covered with ferns, early conifers, and a few angiosperms that were still in their early stages of radiation. The Cretaceous floras of Mexico are mostly Late Cretaceous or Maestrichtian in age (71–65 Ma) and from the northern part of the country. They include the La Misión and Rosario assemblages of Baja California, the Huepac Chert in Sonora, the Olmos flora in Coahuila, and the Piedras Negras flora along the Río Grande in Coahuila (chap. 5). Like their Cretaceous counterparts elsewhere, they tell of warm, equable climates that extended along low thermal gradients and over discontinuous,

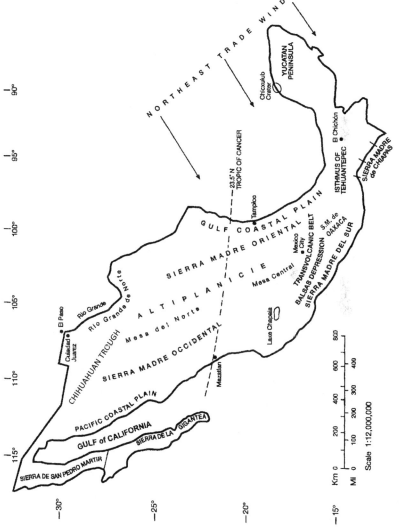

Figure 2.15 Principal physiographic features of Mexico.

mostly low-lying, and often flooded land surfaces from the equatorial regions north to the Arctic and south to the Antarctic.

In the Late Jurassic and Early Cretaceous a large fragment of land called the Maya, or Yucatán, block was located in the developing Gulf of Mexico (fig. 2.17). It was on one of the microplates that were moving to the southwest. It collided with southern Mexico in the Early Cretaceous and continued to slide to the southwest along a line marked by the Salina Cruz Fault running from Tehuantepec to the Gulf of Mexico. Yucatán was sutured on to southern Mexico by the end of the Early Cretaceous.

In the Middle and Late Cretaceous there was a volcanic island arc off the coast of northwestern Mexico. With subduction of the Pacific plate and movement of North America westward, it became accreted onto the continent as present-day Baja California in the Oligocene (35–23 Ma). A northward-spreading center developed between Baja California and the mainland about 29 Ma and was completed by 5 Ma to form the Gulf of Cali-

Figure 2.16 Coniferous forest formation: montane coniferous forest association of *Pinus ponderosa* and *P. montezumae*, Sierra Madre Occidental, El Salto, Durango, Mexico. Small, isolated stands of *Picea* grow in the area. Photograph courtesy of Jesse Perry Jr.

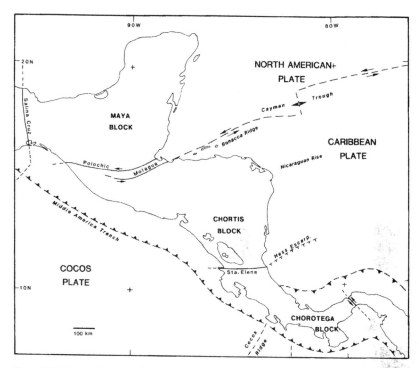

Figure 2.17 Principal tectonic features of southern Mexico and Central America. From Donnelly and others 1990.

fornia. The spreading center and its northward continuation is represented by the San Andreas fault system, which extends to about San Francisco. The sliver of land to the west of the fault belongs to the Pacific plate and it is moving northward at about 5 cm a year. At this rate Los Angeles will subduct into the Aleutian Trench in another 50 million years. One result of the movement relevant to biotic history is that the Miocene Mint Canyon flora, now located in southern coastal California, has been transported 300 km from the south; it actually represents the vegetation growing in the Sonora-Nayarit region of Mexico in the late Tertiary.

Yucatán

It was a dark and stormy night . . . about three years long

What minor evils might arise from the contact were points of elaborate question. The learned spoke of slight geological disturbances, of probable alterations in

climate, and consequently in vegetation; of possible magnetic and electric influences. Many held that no visible or perceptible effect would in any manner be produced. While such discussions were going on, their subject gradually approached,
growing larger in apparent diameter, and of more brilliant luster. Mankind grew
paler as it came. All human operations were suspended.

—EDGAR ALLAN POE, "The Conversation of Eiros and Charmion," 1850

Sixty-five million years ago, the Cretaceous period closed with one of the
most geologically spectacular and biologically altering events in Earth history, a dark and stormy night of some three years duration. About 160 Ma
the explosion of a parent asteroid called Baptistina, 170 km in diameter
and located in the inner asteroid belt, created a flux of terrestrial impacts
(Bottke et al. 2007). One of the fragments entered the Earth's atmosphere
from the southeast traveling at 90,000 km per hour. It landed in the shallow waters of the Yucatán Peninsula with a force of about 100 million
megatons of TNT, or 2 million times the largest H-bomb, leaving a crater
180 km in diameter around the present village of Chicxulub. The local effects were obviously catastrophic, but shocked (striated) quartz was splashed
out as far as Haiti, and a tsunami over 1 km in height deposited debris near
Tampico, Mexico, and at Recife in northeastern Brazil. Farther afield there
were wildfires, acid rain, near darkness, reduced rates of photosynthesis,
and diminished plant biomass for the large herbivores and carnivores at the
top of the food chain. The dinosaurs became extinct, except for their avian
descendents, and smaller animals were left to diversify, eventually changing the Mesozoic age of reptiles into the Cenozoic age of mammals (but
see Keller et al. 2009). The Cretaceous-Tertiary contact (K/T boundary)
is marked by what John McPhee (1981) calls "an unearthly concentration
of iridium"—an element rare in Earth deposits, but common in meteor-
ites and widespread in K/T boundary sediments at concentrations up to
160 times that in strata above and below. Immediately after the impact,
abundant organic material became available for saprophytic fungi. Early
colonizers and recovery vegetation followed and included numerous ferns.
These events are recorded in the Raton Basin of New Mexico and Colorado
(fig. 2.18) as a sequence of K/T beds containing dinosaurs, a clay layer rich
in iridium, Tertiary beds without dinosaurs, a coal layer of organic debris
with abundant saprophytic fungi, and a "fern spike" of recovery vegetation. Near ground zero, the destruction of terrestrial and floating marine
communities was complete, and the history of the modern Gulf-Caribbean
biota begins with the Tertiary period at 65 Ma.

The configuration of the New World and adjacent plates at the time

Figure 2.18 K/T boundary at Raven Ridge, Colorado, 6 December 2008. Photograph ISS018-E-11127, Gateway to Astronaut Photography of Earth, Lyndon B. Johnson Space Center, NASA.

of Pangaea's breakup in the Middle Jurassic around 175 Ma is shown in figure 2.19 and at present in figure 2.20. Toward the end of the Cretaceous and into the early Tertiary, the North American and South American plates continued to separate, and the Pacific plate and its subplates were subducting under Mexico and Central America along the Middle American Trench. The result of these tectonic events was compression, extension, and uplift that created the Transvolcanic Belt beginning in the early Miocene about 23 Ma. The region includes many highlands, such as the Pico de Orizaba (5650 m), Volcán Popocatépetl (5452 m), and Ixtacihuatl (5286 m). The site of Paricutín in the state of Michoacán emerged to great fanfare in the field of a Mexican farmer in 1943. Ash completely buried the villages of Paricutín and San Juan Parangaricutiro, leaving only the steeple of a church pointing symbolically out of the ash toward heaven. Another prominent volcano in Mexico is El Chichón, 650 km southeast of Mexico City. It erupted in 1982, ejecting ash into the atmosphere and creating the spectacular sunsets of that year. It also focused attention on the climatic effects of volcanic eruptions. Initially temperatures warm because of the greenhouse effect of the ash layer, then cool from deflection of incoming solar energy by ash and droplets of sulfuric acid. The effects are short-term (a few months to a few years), but with massive eruptions such as Toba in Sumatra around 74,000 BP

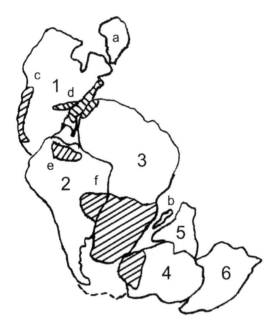

Figure 2.19 The New World and adjacent plates near the time of breakup in the Middle Jurassic circa 175 Ma: 1 = North American plate; 2 = South American plate; 3 = African plate; 4 = Antarctic plate; 5 = Indian plate; 6 = Australian plate; a = Greenland; b = Madagascar; c = proto–Rocky Mountains and the western Mexican cordillera; d = eastern Mexican cordillera; e = Guiana highlands; f = Brazilian highlands; hatched areas represent low mountains.

(Oppenheimer 2002), Tambora in Indonesia in 1815, resulting in "the year without a summer," Krakatoa in Indonesia in 1883, Mount St. Helens in the Cascade Mountains of Washington in 1980, and Mt. Pinatubo in the Philippines in 1991, climates may cool by more than 1°C for a somewhat longer period of time. During periods of active plate movement, 12–15 cm a year when volcanism is greatest, as it was in the Jurassic and Early to Middle Cretaceous, this may cause some cooling, but it is counterbalanced by the warming effect of increased CO_2.

Plate movement has decreased since the Middle Cretaceous and currently averages about 2–3 cm a year. Volcanic activity in the Transvolcanic Belt continues to the present; in 1985 Mexico City was damaged by a severe earthquake of 7.8 magnitude in which an estimated 10,000 people were killed and more than two hundred buildings were destroyed. Volcán Popocatépetl began it latest eruption on 30 June 1997, videos of which can be seen on the Web site of the Centro Nacional de Prevención de Desastres (www.cenapred.unam.mx/mvolcan.html). In the eastern Transvolcanic Belt, there is an assemblage of plant macro- and microfossils in the late Oligocene Pié de Vaca Formation being studied by Sergio Cevallos-Ferriz, Susana Magallon-Puebla, Enrique Martinez-Hernández, and colleagues at the Universidad Nacional Autónoma de México (UNAM). It is an important flora because it is one of the few in northern Latin America that reveals conditions during the transition between the hothouse interval of the early Tertiary and the icehouse interval of later Tertiary and Quaternary times.

The Transvolcanic Belt includes basins, such as the Basin of Mexico where Mexico City is located on former Lake Texcoco. Water is now being removed at a rate of 60 m^3/sec causing severe problems for the 20 million inhabitants of a city where the population increases by about 30,000 people per year. Cores are necessary to test the foundation for construction like the Torre Latinoamericana and to rectify the notable sinking of existing structures like the Palacio de Bellas Artes (images of both buildings are available at several sites on the Web). The cores from these water-saturated sediments, and from adjacent basins, contain fossil spores and pollen. They have been used by Socorro Lozano Garcia and colleagues at UNAM, and by Sara Metcalfe and coworkers at the University of Nottingham, to

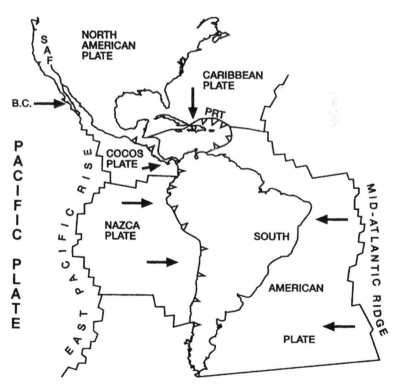

Figure 2.20 Principal plates and subduction zones from southern North America to southern South America: B.C.= Baja California; PRT = Puerto Rican Trench; SAF= San Andreas Fault.

reconstruct the Quaternary paleoenvironments of central and northern Mexico.

To the south of the Transvolcanic Belt is the Balsas Depression through which the Balsas River flows. It has been dammed and is now a major source of hydroelectric power for the region. Farther south are the hot, windy lowlands of the Isthmus of Tehuantepec. The average elevation is about 80 m, and it was one of four sites considered for construction of the transoceanic canal. The others were in Nicaragua, the Darién region of easternmost Panama and adjacent Colombia, and the eventual site across central Panama. In a twist of fate with great political and economic consequences, the United States Congress met on 9 May 1902 to choose between the final candidates of Nicaragua and Panama. The day before, on 8 May, Mt. Pelée on the Lesser Antilles island of Martinique erupted, and with the devastation so vividly evident, the more quiescent Panama was selected as the safer route.

Just north of the Isthmus of Tehuantepec is the Sierra Madre del Sur. These mountains probably originated as an offshore island arc and were added onto the mainland in the Late Cretaceous. It is a rugged region, with elevations to 3703 m north of Acapulco, and there have been numerous earthquakes, including one of 7.8 magnitude on 29 November 1978. Just south of the isthmus is the Sierra Madre de Chiapas. It originated by compression from the south, primarily during the Miocene, to form a series of plateaus, depressions, and steep-faced cuestas evident in the distinctive up-and-down topography along the road from Tonalá to the Guatemalan border. East of these mountains is the Central Depression through which the Grijalva River flows and across which Mexico's largest dam has been built. Most of the region was submerged in the Cretaceous and emerged in the middle and late Tertiary. To the east is the Coastal Plain and the Yucatán Platform. The position of the shoreline in Eocene and later times is marked by a series of lignites containing *Rhizophora* and other vegetation typical of coastal environments. These sediments extend northwest to southeast between the villages of Simojovel and Ixtapa and are a full 90 km inland from the present coast. The shoreline to the north in the middle Eocene is identified by lignites and associated sediments of the Laredo Formation of Texas and the Burgos Basin of Nuevo Laredo, Mexico, containing remains of sharks, skates, rays, *Tarpon*, Crocodylidae, and the present-day Old World mangrove palm *Nypa*. Between Nuevo Laredo and Simojovel, the retreating shoreline had reached to within 1–2 km of its present position by the middle Pliocene (4–3 Ma). This is shown by *Rhizophora*-containing lignites, alternating with near-coastal sediments containing marine dinoflagellate algae, in the Paraje Solo Formation near Coatzacoalcos, Veracruz.

The Yucatán Platform of southern Mexico, Belize, and the Petén region of northern Guatemala has an average elevation of about 200 m with a karst topography, that is, a rough porous limestone without surface drainage, with numerous underground streams and caves, and deep circular water-filled openings called cenotes. The cenotes occur in a ring, and they are now known to be collapsed structures surrounding the Chichulub Crater. The Yucatán carbonate platform is important for several reasons, in addition to being the site of the K/T asteroid impact. One is that the organic debris accumulating in the shallow waters has formed petroleum, and the platform contains one of the richest deposits of hydrocarbons in Latin America. As a result, the geology is known in great detail, including the age and location of lignite deposits. Lignites are economically important as a low-cost source of fuel, and they often contain abundant plant microfossils. The Zona del Sur Division of Petróleos de Mexicanos (PEMEX) is located at Coatzacoalcos, and when sites were being sought for our early studies on tropical vegetation and environmental history, PEMEX provided information about lignites in the middle Pliocene Paraje Solo Formation.

An important legacy of the cenotes resulting from the asteroid impact is that they provided water, served as sites for religious ceremonies, and preserved relics of the great Lowland Maya civilization that reached its classic period between 250 and 900 CE. Karst environments are marginal at best for supporting an increasing population that in the Maya region at its height may have reached 750 or more people per square mile. As noted by Jared Diamond in chapter 5 of *Collapse: How Societies Choose to Fail or Succeed*, this compares with modern population densities in Rwanda and Burundi of 750 and 540 people per square mile, respectively. For the New World the number for El Salvador is 327 and for Haiti it is 307 (United Nations World Populations Prospects Report for 2005). The Maya lowlands were a densely populated region, and by about 750 CE, the landscape had been modified to the extent that obtaining fuel, water, and food was becoming difficult. As is often the case with civilizations on the edge of sustainability, an environmental event can be the tipping point that decides their fate. For the Maya, this event was the Medieval Warm Period between about 800 and 1200 CE, as revealed by limnological, vegetation, and archeological studies by Mark Brenner, Gerald Haug, David Hodell, Barbara Leyden, and others. It can be documented also by a variety of other innovative methods now being applied to the study of Earth history. As noted previously, the amount of titanium in different layers of marine sediments is an indication of rates of river flow into the basins and, hence, precipitation. Cores from the Cariaco Basin off northern Venezuela show low percentages of titanium during the Medieval Warm Period, and exceptionally low amounts between 750 and 900 CE, revealing a series of severe droughts, each three to nine years in duration. Extensive land use in a marginal environment, erosion, deterioration of habitats, a decline in resources, intensifying competition, warfare, and climate change eventually combined to exceed the technology and the will of the people to continue. For the Maya this occurred around 750 CE and the civilization collapsed.

Modern Vegetation

The present plant communities of Mexico include about 20,000 species of vascular plants arranged into eleven communities or plant formations (fig. 2.21). The desert (fig. 2.22), and the shrubland/chaparral-woodland-savanna (fig. 2.23) cover about 40 percent of the country. The driest part consists of *Larrea*, *Agave*, *Yucca*, and numerous cacti in the Sonoran and Chihuahuan deserts. There is also *Leucophyllum frutescens*, the purple sage of cowboy literature, and species of *Cassia* covered with masses of conspicuous yellow flowers. The desert continues southward through the state of San Luis Potosí, with outliers around Tehuacán in the state of Puebla.

The Sierra Madre Occidental is covered at the highest elevations by a coniferous forest (western montane association) that includes *Abies*, *Pinus*, and the southernmost present-day distribution of *Picea* (fig. 2.16). This plant association, without *Picea*, is found again in the highlands of the Transvolcanic Belt. On the dry eastern slopes, in the Isthmus of Tehuantepec, Balsas Basin, and the Yucatán Peninsula, there is a shrubland/chaparral-woodland-savanna formation that includes *Acacia*, *Prosopis*, and the distinctive red-barked *Bursera* (Mexican mahogany). On the eastern side of the Sierra Madre Occidental and to the south, in more mesic habitats, is grassland, called pastizal at the mid to lower elevations and zacatonal at the high elevations. Grassland constitutes about 10–12 percent of the vegetation of Mexico.

Beginning at the coast and moving upslope, the vegetation zones are mangrove (*Rhizophora*, fig. 2.24; *Acrostichum aureum*, mangrove fern; *Hibiscus tiliaceus*, II, figs. 3.19–22), beach/strand/dune (*Ipomoea*, morning glory; *Uniola paniculata*, sea oats; II, fig. 3.23), freshwater herbaceous bog/swamp/marsh (*Typha*, cattail; *Thalia geniculata*, popal; II, fig. 3.24), aquatic (*Ceratopteris* and *Salvinia*, floating ferns; the invasive South American

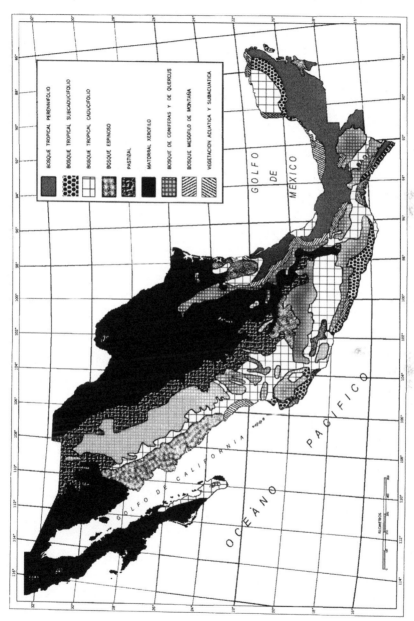

Figure 2.21 Vegetation map of Mexico. Modified from Rzedowski 1978. Used with permission from Jerzy Rzedowski.

Figure 2.22 Chihuahuan Desert, San Luis Potosí, Mexico, with the tall columnar *Yucca carnerosana*, the shrub *Larrea tridentata* (creosote bush), and the cactus *Opuntia* sp.

Figure 2.23 Shrubland/chaparral-woodland-savanna with *Juliania adstringens* (cuachalala) near Tehuantepec, Mexico. From Wagner 1964. Photograph by A. Maya, Institut de Geografía, Universidad Autónoma Nacional de México (UNAM), Mexico City.

Figure 2.24 Mangrove formation (manglar) of *Rhizophora mangle* along bay near UNAM's tropical biological station, Catemaco, Veracruz, Mexico.

Eichornia crassipes, water hyacinth, with *Lemna*, duckweed; fig. 2.25), the lowland neotropical rain forest (southern Veracruz State; fig. 2.26), and the lower to upper montane broad-leaved forest. Within the latter, on the eastern slopes of the Sierra Madre Oriental, in an altitudinal zone between 1000 and 2000 m, there is a deciduous forest association (a formation in eastern North America) of *Acer*, *Alnus* (alder), *Cornus* (dogwood), *Fagus*, *Juglans*, *Liquidambar* (sweetgum), *Magnolia*, *Platanus* (sycamore), *Populus*, *Quercus*, *Salix* (willow), and *Ulmus* (elm). Another association, included within the coniferous forest formation or within the deciduous forest formation, is the midaltitude *Pinus-Quercus* (pine-oak) forest. This is the most widespread vegetation type in Mexico, and it covers about 14 percent of the landscape. The deciduous forest grades upward into a zone of fog, and the vegetation is called the cloud forest association. Farther up, there are cold winds and shorter trees, often with twisted trunks covered with lichens, ferns, and mosses. This is the elfin forest association. Both are found in the Sierra Madre Oriental, the Transvolcanic Belt, and the southern states of Oaxaca and Chiapas. On the highest peaks of the Transvolcanic Belt there is a limited treeless vegetation of alpine tundra, or páramo. Of the twelve vegetation types recognized here for the New World, eleven (all but the lowland tundra) are found in Mexico.

Figure 2.25 (a) Aquatic formation of the floating fern *Ceratopteris pteridoides* with the water lily *Nymphaea*, Turialba, Costa Rica. From Tryon and Tryon 1982. Used with permission from Springer Science and Business Media, Dordrecht. (b) Spore from the modern *Ceratopteris cornuta*. From Tryon and Lugardon 1991. Used with permission from Springer Science and Business Media, Dordrecht. (c) Fossil spore of *Ceratopteris* about 20 Ma from the Miocene Cucaracha Formation of Panama. Comparison between modern spores and pollen and micro-fossils allows identification of the latter. Information on the ecology of the modern plants from field observations and herbarium label data reveals the kind of paleoenvironment; here, an aquatic formation from Panama circa 20 Ma similar to a modern Costa Rican formation.

Figure 2.26 Remnant of the lowland neotropical rain forest formation at UNAM's tropical biological station, Catemaco, Veracruz, Mexico.

THE ANTILLES

Cuba has the same effect on Americans that the full moon used to have on werewolves.

—ALAN RYAN, *The Reader's Companion to Cuba*, 2000

The serene surface of the Caribbean Sea belies its myriad subsurface basins, ridges, deep troughs, active spreading centers, and land fragments that reveal a complex and turbulent past (figs. 2.27, 2.28). It is defined on the north by the Puerto Rican Trench, 9220 m deep and the deepest place in the Atlantic Ocean; on the east by another subduction zone on the Atlantic side of the Lesser Antilles; and on the south by the El Pilar and Oca fault system across northern South America, which joins the Middle American Trench on the west coast of Central America. The part running from the Middle American Trench eastward across Guatemala is marked by the Polochic-Motagua fault zone (fig. 2.17). This zone continues into the Caribbean Sea as the Cayman Ridge spreading center, through the Oriente region of Cuba, and connects to the Puerto Rican Trench. The Caribbean plate is situated between the jostlings of the North American, South American, and Pacific plates, creating one of the most geologically complex regions on Earth. Only

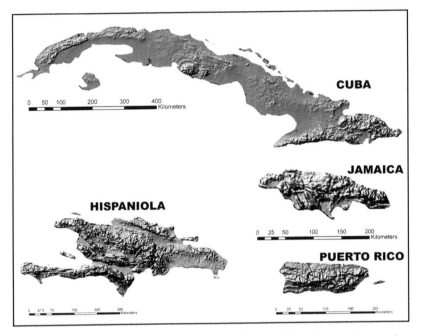

Figure 2.27 Physiography of the Antilles. Map created by Trisha Distler, Missouri Botanical Garden, St. Louis, based on the U.S. Geological Survey's Shuttle Radiography Topography Mission (SRTM) 90m DEM digital elevation database.

recently has its history been reconstructed to give a clearer picture of the origin and development of the Antilles.

In the Middle Jurassic through the Early Cretaceous (200–165 Ma) the continents were arranged as shown in figure 2.19 with proto-Venezuela situated in the developing Gulf of Mexico. North America and South America (with Africa attached) began to separate, resulting in a shallow sea with widespread evaporites and salt domes. These domes are the site of extensive salt and oil deposits in the Gulf of Mexico region, and their collapse or accidental penetration can cause dramatic happenings on the surface.

In the southeastern United States, the arched surface over these domes creates "islands," circular stands of oak surrounded by more moist vegetation, lakes, or swamps. One of these is Jefferson Island, site of the Live Oak Botanic Gardens in Lake Peigneur near New Iberia, Louisiana. New Iberia is home to an extraordinary concentration of Cajun relatives named Viator, and to the McIlhenny Tabasco Company on Avery Island. The region's botanical claim to fame is the Evangeline oak immortalized in Longfellow's

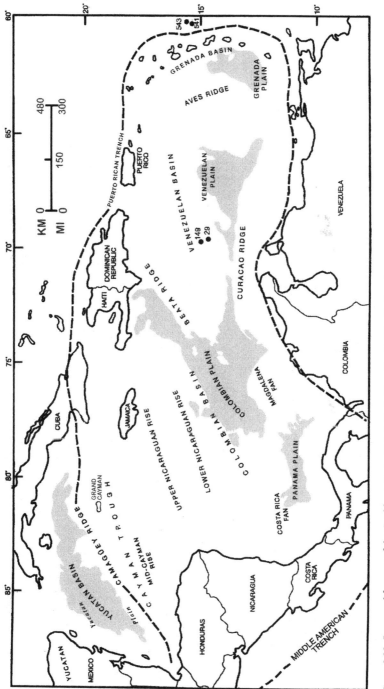

Figure 2.28 Structural features of the Caribbean Basin. Dots indicate Deep Sea Drilling Project sites.

epic poem, and the area's prehistory surfaced in a real-life saga every bit as dramatic as *Evangeline's*. On Thursday, 21 November 1980, Texaco was drilling a well at Jefferson Island, unaware that below were the caverns of the Diamond Crystal Salt Mine. At a depth of about 1228 feet, the rig began to tilt, and a whirlpool formed in the lake, eventually creating a crater 180 feet in diameter. Two Texaco oil rigs, several barges from the adjacent Delcambre Canal, a loading dock, the Live Oak Botanic Gardens, assorted greenhouses, a house trailer, trucks, and several tractors were pulled into the crater. Miraculously no one was killed; indeed, the water drained so fast that Leonce Viator Jr., who was fishing on the lake, was able to climb out of his grounded boat and run to shore.

As the continents continued to separate, the waters of the Gulf deepened and the formation of salt domes waned. The subduction zone along the coast of western Mexico was extending southward toward present-day Central America. When ocean crust descends into the Earth's interior, it becomes molten at a depth of about 200 km and may then rise to the surface through fissures to form an arc of volcanoes or volcanic islands. These conditions in the Late Cretaceous prepared the way for lands that would eventually become the Greater Antilles. An early model by Malfait and Dinkleman (1972) depicted a continuous or near-continuous stretch of land connecting or nearly connecting the continents. This land supposedly moved relatively eastward, encountered the stable Jurassic to Late Cretaceous Bahamas Platform, and broke into early versions of the present-day Greater Antilles. This geologic model has proved important to biologists because it is consistent with one of the two principal means of plant and animal distribution: vicariance and dispersal. Vicariance involves the separation of a once continuous biotic range through plate tectonics (e.g., movement of South America away from Africa) or the formation of an intervening barrier (e.g., rise of the Andes Mountains). Distribution may also occur through gradual short-distance or abrupt long-distance dispersal of seeds or other plant propagules. Thus, one means is based on movement of the land (vicariance), and the other on movement of the organism (dispersal). The early Malfait-Dinkleman model of Antillean origin was most compatible with vicariance.

The scenario for the geologic development of the Greater Antilles around which consensus is now developing is that of Pindell and Barrett (1990; Pindell 1994), augmented by the geophysical research of Paul Mann and others (e.g., Mann 1999). By this view, an arc of volcanic islands originated off present-day Pacific Central America in the Cretaceous about 130 Ma. From the beginning, the islands were separate and mostly submerged. They attained their present position by the relative motion of the three regional plates, all moving westward from the Mid-Atlantic Ridge, but at different rates. The Caribbean plate is moving at the comparatively slow rate of about 1.8–2 cm each year, North America at 3.0 cm, and South America at 3.3 cm. The result is a progressively eastward position of the Caribbean plate vis-à-vis the two other plates. After a submerged Cretaceous origin, the principal period of emergence occurred when the island arc collided with the stable Bahamas Platform in the Eocene about 45 Ma. Some islands fragmented further, others collided, and new ones emerged later, particularly in the Miocene. The prevalent view now is that the original arc was never continuous and never formed a complete land connection between the continents. This allows a greater role for dispersal in explaining the distribution and speciation of organisms in the Antilles (Heinicke et al. 2007).

Near the time of initial encounter of the volcanic island arc with the Bahamas Platform, a land fragment representing proto–northern Cuba was part of the North American Plate located along its southern edge. Another fragment on the Caribbean plate consisted of proto–southern and –western Cuba, and it collided with northern Cuba in the late Paleocene to early Eocene with the contact represented by the Pinar Fault. Parts of Cuba were above water in the middle Eocene, as shown by spores and pollen of some forty-six terrestrial plants identified from the Saramaguacán Formation. The Cauto Fault is the contact between proto-Cuba and another fragment made up of eastern Cuba, northern Hispaniola, and Puerto Rico. The Bahamas Platform is curved, so when the western edge of the volcanic arc encountered the platform, this edge slowed first, affixing eastern Cuba to the rest of the island, while the remaining part of the arc continued to move eastward. The distribution and relationships of *Adelia* (Euphorbiaceae) and other genera are consistent with the composite North American / Caribbean plate origin of Cuba (De-Nova and Sosa 2007). Northern Hispaniola (Haiti and the Dominican Republic) and Puerto Rico separated next. These events were taking place in the late Oligocene and Miocene between about 28 and 20 Ma.

An extensive flora of middle Oligocene age (30 Ma) is known from Puerto Rico and includes 91 species of macrofossils and 165 kinds of spores and pollen. There has been some question as to whether Puerto Rico had emerged by this time, but the extensive terrestrial vegetation of the San Sebastian Formation shows that at least parts were fully above water in the middle Oligocene. Southern Hispaniola, the elongated "South Island" (fig. 2.27), was added on to northern Hispaniola in the middle Miocene (15 Ma) through movement of one of the many Caribbean micro- or subplates. The contact is marked by the present Cul-de-Sac / Enriquillo Basin running northeast near Port au Prince. This region includes the lowest point of land in the Antilles, and hurricanes, so frequent in the Antilles, are often funneled through this lowland with devastating effect.

The geohistory of Jamaica is of interest to biologists because few middle Eocene to late Miocene terrestrial rocks are known from the island. This means it was mostly submerged between 42 and 10 Ma, so after the biota was annihilated at 65 Ma from the asteroid impact, it was destroyed again at 42 Ma by inundation. The modern biota is therefore of a relatively recent geologic origin. This may account, in part, for the fewer endemic plants on Jamaica (around 923 species, less than 10 million years old, and small) compared to Cuba (around 3178 endemic species, more than 45 million years old, and larger; Kier et al. 2009).

The Lesser Antilles also emerged along a subduction zone in about the middle Eocene (49–44 Ma). Earthquakes are most common in the Greater Antilles where the Caribbean plate slides past the North American plate, but with relatively limited subduction into the Puerto Rican Trench, while volcanism is more frequent in the Lesser Antilles where the North American (Atlantic) plate dips under the Caribbean plate. Mt. Pelée on Martinique erupted on 8 May 1902, killing over 30,000 people. There were two survivors, one of whom was a prisoner being held for stabbing a friend during a drunken brawl. Just before he was scheduled to be released, he showed a remarkable lack of judgment by running away, partying all night, then turning himself in the next day. He was put in solitary confinement, which happened to be in an underground cell, the day before Mt. Pelée erupted. He was badly burned from the heat and fumes but survived, and because of his ordeal he was pardoned. He later joined the Barnum and Bailey Circus and toured the world as "Sampson, the lone survivor of Mt. Pelée" (Zebrowski 2002).

Modern Vegetation

The vegetation of the Antilles consists of about 11,000 species of vascular plants (Santiago-Valentín and Francisco-Ortega 2008). Our knowledge of the vegetation comes in large part from the collections of Swedish botanist Erik Ekman in the early 1900s. He was myopic of purpose, maniacal in intensity, an admirable eccentric who traveled on foot and made exhausting trips to remote areas. He lived in poverty, was perpetually undernourished, often ill, indifferent to his physical appearance, and uniquely asocial. After one exhausting trip he fell asleep on the porch of a hut in Haiti in an especially disheveled state. He awoke to a visitor prodding him with his foot and saying to a friend, "May I present Dr. Erik Leonard Ekman, member of the Swedish Royal Academy of Sciences, Fellow of the Smithsonian Institution, and the world's greatest authority on the Caribbean flora" (Roger Lundin, pers. comm., 2002).

Many of the vegetation types of the Antilles at the level of formations are comparable to those described for North America and Mexico. Their arrangement reflects the physiography of the islands, which in most instances consists of a coastal plain rising to a central cordillera. It also reflects the distribution of rainfall brought from the east by the trade winds. The easternmost islands, and the northern side of each island, are mostly wet, while the western ends of the islands and the southern coasts are drier from the rain shadow created by the central cordilleras. There are no deserts because rainfall is too high even in the driest parts, and the island environment reduces evapotranspiration. This does not mean there is no arid vegetation, but its occurrence is usually due to edaphic conditions in the areas of lowest rainfall, and these arid-habitat communities are expanding as a result of abusive land use. The driest vegetation is on Gonâve Island of western Haiti (MAP 500 mm). Haiti is half the size of the Dominican Republic but it has about the same population (7 million) making it one of the most densely populated countries in Latin America. The environmental destruction is almost complete, and it is estimated that only about 1 percent of the natural vegetation remains (fig. 2.29). There are clear and obvious consequences when we consider the land as Columbus found it. As his son Ferdinand recorded in *The Life of the Admiral Christopher Columbus*: "The sky, air, and climate were just the same as in other places; every afternoon there was a rain squall that lasted for about an hour. The admiral writes that he attributes this to the great forests and trees of that country" (Cohen 1992, 181–82; see also Keen 1992).

A General Awakening?

Look here, in the past these mountains were densely covered with forests, and it rained a lot. You must know, May ended as it had begun, rain, rain, rain . . . ! During May it rained all the time. Not just once a day, seven and eight times a day torrential rains came down from the sky. In the past moss covered the walls of huts and sometimes good water came from the roofs. There was plenty of water. It rained day and night. It is different today. There is a lot missing here, due to the lack of water. I do not know what is wrong today.

—ROSA ENCARANCIÓN, age eighty-five, quoted in
Eberhard Bolay, *The Dominican Republic*, 1997

Figure 2.29 Political boundary between Haiti, nearly denuded of vegetation (left), and the more forested Dominican Republic (right), 25 September 2002. Scientific Visualization Studio, Landsat-5, Goddard Space Flight Center, NASA.

The floods of 24 May 2004 and the tragic loss of thousands of lives in southeastern Haiti and Jimaní, Dominican Republic, provide important lessons for scientists, conservationists, and politicians. The same storm did not have such a devastating effect in neighbouring Puerto Rico or in other regions of the Dominican Republic, mainly because the highlands are forested.

> —T. MITCHELL AIDE AND H. RICARDO GRAU, *Globalization,*
> *Migration, and Latin American Ecosystems*, 2004

I've stepped in the middle of seven sad forests
I've been out in front of a dozen dead oceans
And it's a hard rain that's gonna fall.

> —BOB DYLAN, "A Hard Rain's A-Gonna Fall," 1962
> (chosen as background music for the Jardín Botánic de la
> Universitat de Valéncia's 2008 display on the environment)

The most widespread plant formation in the Antilles is the shrubland/chaparral-woodland-savanna that grows to the lee of mountains, on south-facing slopes, on the western end of the islands, and on serpentine and coarse soils that create physiologically dry habitats. In some areas it is expanding from aridification due to abusive land use followed by abandonment of the land, while in others it is contracting because of construction, local cultivation, and the gathering of wood down to barren ground, but almost everywhere it is being modified toward near desert.

There are no natural grasslands in the Antilles, and oak and pine savannas are the result of longtime land use. The coasts are bordered by mangrove and beach/strand/dune ecosystems, and inland there are freshwater herbaceous marsh/swamp/bog, and aquatic communities. The largest swamp in the Antilles is on the south coast of Cuba at Cienaga de Zapata. One part is familiar as the Bahía de Cochinos, the Bay of Pigs. The swamp fluctuated greatly in size during each of the roughly eighteen to twenty glacial/interglacial intervals of the Pleistocene, when, for example, lower sea levels repeatedly enlarged Cuba by half, primarily from exposure of the southern coastal plains. High sea levels, as at present, temporarily bring back the modern configuration of Cuba, but the expanded version is typical of about 90 percent of the past 2.6 million years.

There is no lowland tropical rain forest in the Antilles, although the lowest wet phase of the lower to upper montane broad-leaved forest formation is similar. This formation covers an extensive area around El Verde in the Luquillo Mountains of eastern Puerto Rico. Upslope on the Dominican Republic, there is a coniferous forest of *Pinus caribaea*, used for the manufacture of paper pulp, and *P. occidentalis*, the pine on Pico Duarte, the highest point of land in the Antilles at 3098 m. On other highlands, there is an elfin or cloud forest. There is no páramo, but some physical features on Pico Durate are reminiscent of cirques and moraines that possibly indicate former limited glaciation.

All these geologic and climatic events and the resulting geomorphological features document a dynamic past for the islands of the Caribbean. They drove the processes of biotic evolution and migration that over a 45-million-year period have created the seven modern ecosystems of the Antilles: shrubland/chaparral-woodland-savanna (the last resulting mostly from human activity), mangrove, beach/strand/dune, freshwater herbaceous bog/marsh/swamp, aquatic, lower to upper montane broad-leaved forest, and coniferous forest.

CENTRAL AMERICA

During the Jurassic and Early Cretaceous, a fragment of land located in the Pacific Ocean off the coast of Mexico was moving to the southeast through plate movement. It is called the Chortis Block, and it became sutured on to the southern end of the Maya Block in the Late Cretaceous to form present-day northern Central America (figs. 2.17 and 2.30). The contact is the Polochic-Motagua fault system across northern Guatemala, which also marks the boundary between the North American and Caribbean plates. Movement along the fault causes devastating earthquakes in the region, like the one that hit Guatemala in 1976 (fig. 2.31), or the 7.3-magnitude quake off the coast of Honduras on 28 May 2009 (posted as an image of the day on NASA's Earth Observatory site, http://earthobservatory.nasa.gov/images). Ophiolites along the Polochic-Motagua fault system have been dated at 70–66 Ma, which establishes the accretion of the Chortis Block and the time of origin of northern Central America. The southern margin of the Chortis Block is the Santa Elena fault system that extends eastward into the Caribbean Sea as the Hess Escarpment just south of Largo Nicaragua.

Southern Central America (southern Costa Rica and Panama) had a different origin, and the evolution of the landscape there played an important role in the development of the biota of the New World. After the first line of volcanic islands was carried eastward to form the Greater Antilles, a second arc developed between the southern boundary of the Chortis Block and what was then northern South America (again, southern Costa Rica and Panama). Volcanic activity and uplift from subduction of two Pacific subplates called the Cocos and Nazca plates (fig. 2.20) into the Middle American Trench at about 8 cm a year, along with compression forces from the movements of North and South America, eventually established a continuous land connection. The subsequent geological, oceanographic, and climatological consequences were extraordinary. There was a reorganization of middle Miocene deep water circulation (Burton et al. 1997; Nisancioglu et al. 2003; Schneider and Schmittner 2006), a late/middle Miocene drop in $CaCO_3$ deposition in the eastern Pacific (Lyle et al. 1995; Roth et al. 2000; Newkirk and Martin 2009), and a regulatory effect on Northern Hemisphere glaciations via transport of heat and moisture from the tropics (Lunt et al. 2008). The biological consequences were equally profound. South America had begun separating from Africa in the south about 120 Ma as new crust was generated along the Mid-Atlantic Ridge. The separation continued progressively northward until northern South America separated from Africa at about 90 Ma. Around 32 Ma the continent parted from

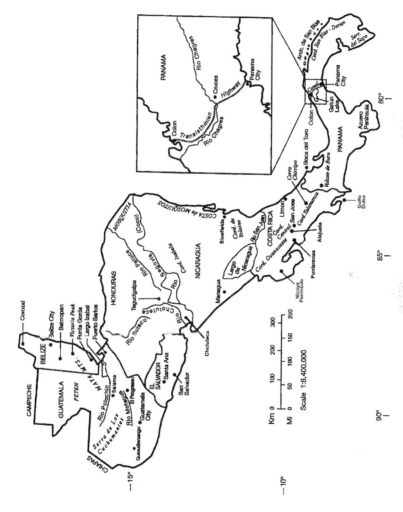

Figure 2.30 Countries and principal physiographic features of Central America.

Figure 2.31 Bent rails near Gualán, Guatemala, after earthquake on 4 February 1976 from movement along the Motagua Fault, which marks the contact between the North American and the Caribbean plates (see fig. 2.30). Earthquakes result as the Caribbean plate slides eastward relative to the North American plate. From Espinosa 1976.

Antarctica, and after that South America became an island continent and remained so for about 29 million years. Its flora and fauna developed in isolation until the Panama land bridge was established about 3.5 Ma, after which waves of plants and animals crossed in both directions between the continents (chap. 9). The last link to be forged was through the Darién region of eastern Panama and northern Colombia at about 2.5 Ma. Elevations there are still near sea level, and the connection was probably breached several times during the Quaternary. Other land fragments or exotic terranes were sutured onto the western side of Central America and now constitute the Nicoya and Osa Peninsulas of Costa Rica and the Azuero Peninsula of Panama (fig. 2.30). Also at about 2.5 Ma, the central cordillera of southern Central America had uplifted to the point that it constituted a barrier to the winds coming from the north (recall that Panama runs primarily east-west, not north-south). The plant fossil record shows that the earliest differentiation into a wetter northern coast (rainfall presently about 1500–3000 mm/yr) and a drier southern coast (1140–2290 mm/yr) began in the latest Miocene at 6–5 Ma. The few and scattered highlands also provided some habitats that were more temperate and better drained than the tropical lowlands, and where a few plants of northern origin like *Alnus* and *Quercus* grew, as they do at present.

The uplift of the Panama land bridge, connecting two continents that had been separated for about 67 million years (circa 70–3 Ma), is a watershed event in the biotic history of the New World. Details of this history are important not only because they allow a better understanding of the biogeography of the region, but because of the evolutionary implications, as many plants and animals intermingled and competed for the first time. Thus, determining the time of uplift is important. This can be established by comparing the similarity of terrestrial floras and faunas of progressively younger ages on either side of the bridge (chap. 9). It can be further estimated through study of fossil marine faunas from the Caribbean and the Pacific Ocean because, as Wendell Woodring (1966) has noted, a land bridge is also a sea barrier. Even with several kinds of paleobiological information available, however, the history is complex, and it is useful to have evidence from independent lines of inquiry. One such line is the chemical analysis of pelagic (open-ocean) sediments (Donnelly 1989).

Surface waters of the Atlantic and Pacific oceans contain relatively little silica because it is taken up by microscopic planktic (floating) marine organisms to construct their shells. When the organisms die, these shells drift down through intermediate depths, dissolve in deep waters, and the silica accumulates on the ocean floor. Upwelling brings the silica back to the surface, where it is removed again by marine organisms. This cycle means that surface waters are low in silica while intermediate and deep waters are richer. Water that flows from the Pacific to the Atlantic through the Southern Ocean is mostly surface water low in silica, and that flowing from the Atlantic to the Pacific is mostly intermediate-depth water rich in silica. As a result, bottom and intermediate waters of the Pacific have a higher silica signature than those of the Atlantic Ocean.

When deep-sea cores from the Atlantic, Pacific, and Caribbean Sea were analyzed for silica content, they showed a multipart history. In the latest Cretaceous and early in the Tertiary, all the waters have a similar silica content indicating open ocean circulation through the isthmus. In the late Eocene, the silica content between the Atlantic and Pacific becomes different, reflecting the rise of the proto–Greater Antilles separating these ocean basins. Then mid-depth and deep waters of the Atlantic and Pacific oceans become different in silica content from the Caribbean at about 15 Ma, correlated with the continuing rise of the Lesser Antilles and the development of a submerged sill in the Isthmian region, beginning the eventual formation of the Panama land bridge. Distinctions in Caribbean surface waters become evident at about 4.1 Ma, and finally the greatest differences develop between 3.7 and 3.2 Ma, indicating essentially full closure. This history is consistent with the terrestrial and with the marine paleobiological evidence. Such interaction between studies of structural geology, marine geochemistry, paleontology of terrestrial plants and marine animals, and modern biogeographic and evolutionary patterns is an example of the integrated studies necessary to understand complex ecosystems like those of Central America.

Other features of Panama are a result of its geologic history and affect the region's economic and commercial activities. One is the extensive faulting created by movements of the Cocos, Nazca, Caribbean, and South American plates that intersect in southern Central America. Along the Las Cascadas Reach of the Panama Canal, 113 faults have been identified in a distance of 3 km. Another is the inverted topography of the region. During times of extensive volcanic activity in the Tertiary, lava flowed into the stream valleys, cooled, and hardened. The softer upland sediments eventually eroded down below the lava that now rests as hard, heavy cap rock on top of soft water-logged strata (fig. 2.32). When the Panama Canal was cut through this setting, the softer sediments were squeezed into the canal, requiring perpetual dredging.

Another bane of canal operations are the El Niños that cause dry periods in central Panama, lowering navigable depths from 40 to 34 m. Ship traffic is reduced by 20 percent, and this is of considerable economic and strategic importance. The canal can accommodate ships carrying 65,000 tons of cargo, but many modern ones now carry up to 300,000 tons (Angier 2009), and by 2011 one-third of the world's container ships will be too large for the present locks. A $5 billion expansion project is underway to construct two new locks, widen the Gaillard Cut, and raise the level of Gatún Lake. These activities are providing new exposures of the Tertiary formations previously studied for their plant and animal fossils. The highest fee paid to date for use of the canal was $141,344.97 on 2 May 1997 by the passenger ship *Crown Princess*. The lowest was by Richard Halliburton, who swam the canal in 1928. He was weighed and charged $0.36. A fascinating account of the intrigues in building the canal (at the time Panama was part of Colombia) is given by David McCullough in *The Path between the Seas* (1977).

There are several fossil floras that reveal stages in the evolution of the Central American ecosystems. The oldest is the late Eocene Gatuncillo flora of the canal region of Panama. The Culebra, Cucaracha, La Boca (early to middle Miocene) and Gatun (late Miocene) floras of the former Panama

Figure 2.32 Balboa Heights, Sosa Hill, and the Interamerican Bridge from the steps of the Administration Building of the Panama Canal Company. Many hills like Sosa Hill are capped by lava that pushes the softer underlying sediments into the canal.

Canal Zone are all associated with extensive marine and terrestrial, vertebrate and invertebrate faunas. Mio-Pliocene and Pliocene floras include plants from the Padre Miguel Group and the Herrería formations of Guatemala, and the Río Banano Formation of Costa Rica. There are fossil-bearing deposits of Quaternary age throughout the region, such as those in Gatún Lake studied by Alexandra Bartlett and Elso Barghoorn of Harvard University, that reveal the latest stages in the long history of vegetation and environmental change in Central America (chaps. 6–9).

Modern Vegetation

The modern vegetation of Central America includes dry communities on the karst topography of the Petén of Guatemala. A prominent member is *Brosimum alicastrum* (ramon, or breadnut), which was a food source of the Maya for 2000 years. There are coastal mangrove and other lowland plant formations, including a lowland neotropical rain forest more extensive and diverse than its northernmost extension in Veracruz, Mexico. This community will be discussed more fully with reference to the Amazon Basin, where it reaches its greatest extent and diversity. In contrast to the rain forest, which increases in diversity to the south, the northern temperate component of the lower to upper montane broad-leaved forest decreases in number of species and extent to the south because of tropicality and few extensive highlands in southern Central America. The region around Bluefields in Nicaragua marks the southernmost extension of many northern elements including *Juniperus*, *Pinus*, *Carpinus* (hornbeam), *Liquidambar*, and *Ulmus*. A few extend farther south such as *Alnus*, *Myrica* (gale), *Juglans*, *Ostrya* (hop hornbeam), and *Quercus*. At the highest elevations, such as in the Sierra de los Cuchumantanes of western Guatemala and northern Honduras (3100–3800 m), there is a montane coniferous forest association of *Abies guatemalensis* and *Pinus hartwegii* marking the southern extent of these mostly northern genera. In Costa Rica, the Cerro Chirripó (3819 m) is glaciated at the peak, and the Cerro de la Muerte (3100 m) has stunted oaks and a grassy páramo. The completion of the Panama land bridge, the availability of the moderate highlands, and the general Neogene cooling allowed a few northern temperate plants to cross into the Andean highlands of northern South America. *Alnus* first appeared in South America at 1 Ma, and *Quercus* followed at about 330 kyr. The ecosystems of Central America include nine of the twelve described for the New World: shrubland/chaparral-woodland-savanna, mangrove, beach/strand/dune, freshwater herbaceous bog/marsh/swamp, aquatic, lowland neotropical rain forest, lower to upper montane broad-leaved forest, coniferous forest (northern Central America), and páramo. Lacking are natural desert, grassland, and tundra.

One of the historic rivers of Latin America is the Río Chagres (Minter 1948; fig. 2.33). Christopher Columbus in 1502, and later Vasco Núñez de Balboa and the infamous pirate Henry Morgan, all sailed the river. It served as the gateway to Peru, beginning the eventual conquest of the Inca by Francisco Pizarro (1531–33), and it was the principal waterway for transporting gold and other treasures to Spain from the New World. Barro Colorado Island was formed when the Río Chagres was dammed to form Gatún Lake, a waiting area for ships passing through the narrow Gaillard Cut of the Panama Canal. The research field station of the Smithsonian Tropical Research Institute (STRI) is located on Barro Colorado Island. The history of the island and some early research is recounted by Frank M. Chapman in *My Tropical Air Castle* (1929) and *Life in an Air Castle* (1938).

The administrative offices of STRI are on the grounds of the former Tivoli Hotel in Balboa (II, fig. 1.32). The hotel was rushed to near completion to house its first guest, Theodore Roosevelt, who was on an inspection tour of the Panama Canal in 1906, and it was torn down in 1971 ravaged by time

Figure 2.33 The Río Chagres, Panama.

and termites. I was there during Christmas of 1963 with my wife, Shirley, the Harvard paleobotanist Elso Barghoorn, and his secretary, Dorothy Osgood. We were going to Davíd, a small town in western Panama noted for its petrified wood. Along the way Professor Barghoorn reminded us of the many well-known benefits conferred by a Harvard education as opposed to one, say, from Michigan. The posada at Davíd was built of fossil wood, and in the courtyard there was a garden with eggshells on top of sticks (fig. 2.34). When Professor Barghoorn asked about this, the proprietor told him that when a stranger like him looks at the flowers, their glance is caught by the eggshells, and this protects them from people who might have the dreaded *ojo de malo* (the evil eye)—apparently, I reminded Barghoorn, one of the lesser-known benefits conferred by a Harvard education.

SOUTH AMERICA

The first impressions South America often makes on students of natural history are its vast extent and its diversity—in culture, landscape, climates, and biota (figs. 2.35–37). It covers about 18×10^6 km^2, or 12 percent of the Earth's surface, and extends over 8000 km through 70° of latitude. There are approximately three hundred principal ethnic groups that have survived not only the multiple waves of epidemic diseases brought through contact

Figure 2.34 Eggshells on sticks protecting a garden from the *ojo de malo* (evil eye), Posada San Sebastian, David, Panama.

with foreign explorers and early missionaries beginning in the 1500s, but more recent relocations and destruction of native habitats. The cultural diversity is likely a consequence of the environmental and landscape heterogeneity that includes elevations ranging from 13 m below sea level on the Valdés Peninsula of Argentina (fig. 2.38) to the highest peak in the New World—Cerro Aconcagua in Argentina on the border with Chile (6960 m). Rainfall varies from the hyperarid Atacama Desert in Chile where it practically never rains, to the Colombian Chacó where it practically never stops. Such heterogeneity provides an array of isolated habitats accommodating a diverse human population.

The habitats support new plants and animals evolving in situ, as well as others migrating in from elsewhere by abrupt long-distance transport and more gradually in response to environmental change. There are about 110,000 species of vascular plants in the neotropics, and a third more remain to be described. In addition, there are an unknown number of lichens, bryophytes, fungi, and microorganisms. All twelve of the plant formations recognized in this text are found in South America, although their composition and, to some extent, the nomenclature used to describe them must be distinct because of about 90 million years of separation from Africa, 67 million years from North America, and 32 million years from Australia and Southeast Asia via Antarctica, isolating South America as a continent until its reunion with North America circa 3.5 Ma.

The physical and biological diversity of South America centers on two prominent features of the landscape—the extent across nearly 8500 km of the Andes Mountains (Orme 2007a, b; Young et al. 2007) and the great basins of the continent, for example, the Amazon Basin (Furley 2007) and the

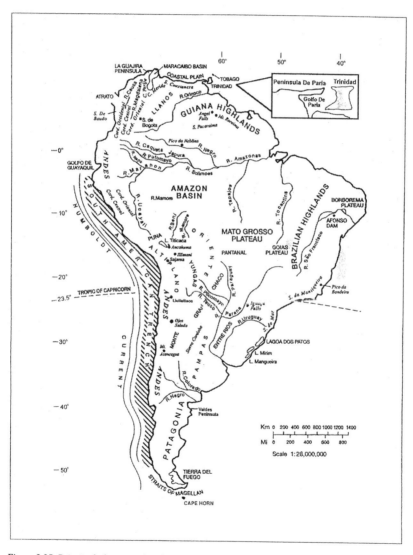

Figure 2.35 Principal physiographic features of South America.

Figure 2.36 Political subdivisions of South America.

lowlands of the Paraguay-Uruguay-Paraná rivers (Cordani 2000; Veblen et al. 2007). Andean countries include Venezuela, Colombia, Ecuador, Peru, Bolivia, Chile, and Argentina. (PBS's *Nature* broadcast a series on the region— "Andes: The Dragon's Back"—now available as a DVD.) Basin countries are Guyana, Suriname, and French Guiana in the north, Paraguay and Uruguay in the south, and Brazil centered in the expanse of the Amazon Basin.

Figure 2.37 Sign across road SP-255 at the Tropic of Capricorn marking the southern limits of the tropics at 23.5°S, Brazil (cf. figs. 2.15 and 2.35). Photograph courtesy of Taciana Cavalcanti.

Figure 2.38 Valdes Peninsula, Argentina. Photograph by J. Arthur Herrick.

The Southern Andes and Lowlands

The crustal extension (stretching) that preceded separation of South America from Africa was accompanied by extensive volcanism that deposited flood basalts 2000 m thick in the Paranà Basin of southeastern Brazil and westward. The weathering of ancient crystalline rocks has created structures such as the granitic Sugar Loaf near Rio de Janeiro. The actual separation of South America from Africa beginning in the south at about 120 Ma created lowlands along the passive eastern coast called pull-apart basins, and they have been accumulating plant and animal remains from the Cretaceous onward. One of these lowlands is the Magellanes Basin of southern Argentina, with Cretaceous floras preserved in the Springhill Formation. Another is the San Jorge Basin, which is the principal source of Argentina's extensive oil reserves. Moreover, as South America was forced against the Pacific plate, the movement exerted compression forces along the active western margin. A subduction zone called the Peru-Chile Trench developed that, along with movement at the triple junction of the South America, Nazca, and Antarctic plates, caused uplift of the Southern Andes. The glaciated landscape produced by these forces is spectacular and inspiring (fig. 2.39). As Charles Darwin wrote in his *Journal of Researches*, now known

Figure 2.39 Torres del Paine National Park, north of Punta Arenas, Chile. Photograph by J. Arthur Herrick.

as the *Voyage of the Beagle*: "I can only add raptures to the former raptures . . . each new valley is more beautiful than the last. It is not possible to give an adequate idea of the high feeling of wonder, admiration and devotion which fill the mind" (Darwin's works are widely available; readers can refer to the Penguin edition or to such sites as the Complete Works of Darwin Online, http://darwin-online.org.uk/).

These mountains are slightly older and much lower than the Central and Northern Andes, averaging about 1 km in height. The two principal intervals of rapid plate movement and mountain uplift occurred between 50–42 Ma and 25–10 Ma. The movements also resulted in accretion of island arcs that form part of the coastal mountains. The continental margin was pulled down along the subduction zone, and more recently sea levels have risen from glacial melting to further inundate the coast. As a result, the landscape is a series of islands extending from Chiloé Island at 42°S (just southwest of Puerto Montt) to Tierra del Fuego at 55°S (fig. 2.40).

Tierra del Fuego, Land of Fire, is separated from mainland South America by the 334-mile-long Strait of Magellan. Any number of things could have suggested the region's name: the native people signal with fire, they use torches to hunt at night, and there is often a smokelike mist over the land. The strait is characterized by darkness, dense fog, extreme cold, and fierce westerly winds called the Roaring Forties. The setting is vividly captured in Samuel Taylor Coleridge's *The Rime of the Ancient Mariner*: "The ice was here, the ice was there, / The ice was all around: / It crack'd and growl'd and roar'd and howl'd, / Like voices in a swoon!" Magellan and his crew spent thirty-eight days navigating the uncharted labyrinth of passages between the masses of ice:

> We ate biscuit, which was no longer biscuit, but powder of biscuits swarming with worms, for they had eaten the good. It stank strongly of the urine of rats. We drank yellow water that had been putrid for many days. We also ate some ox hides that covered the top of the mainyard to prevent the yard from chafing the shrouds, which had become exceedingly hard because of the sun, rain, and wind. We left them in the sea for four or five days, and then placed them for a few moments on top of the embers, and so ate them; and often we ate sawdust from boards. Rats were sold for one-half ducado [about $1.16 in gold] apiece, and even then we could not get them. The gums of both the lower and upper teeth of some of our men swelled, so they could not eat under any circumstances and therefore died. Nineteen men died from that sickness and the [Patagonian] giant together with an Indian from the country of Verzin. (Antio Pigafeta, quoted in Boorstin 1983, 265)

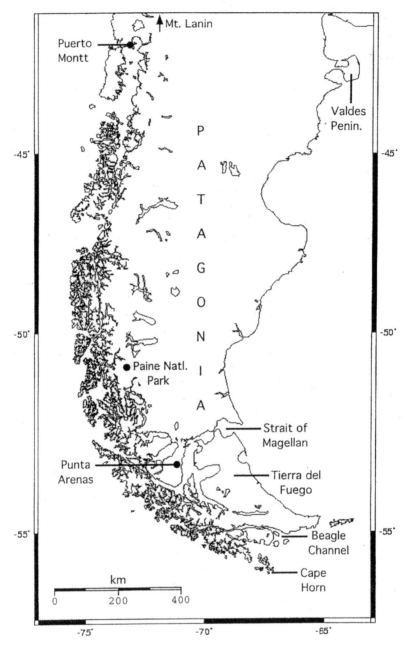

Figure 2.40 Key points of southern South America. Adapted from Paul Wessel and Walter H. F. Smith, Generic Mapping Tools, http://gmt.soest.hawaii.edu/.

When Magellan and his crew exited from the strait in 1520, they sailed for three months and twenty days in calm waters, for which they named the ocean "Pacific." In 1830, when Captain James FitzRoy visited the island with Charles Darwin aboard as naturalist, he took back to England four natives they named Jemmy Button, York Minster, Fuegia Basket, and Boat Memory (who died in England). The Tierra de Fuegans were dressed in proper British attire, presented to the royal court, and educated with the intention of returning them to the island to spread Christianity. It was a bizarre experiment, as insensitive as it was a failure, and the natives quickly returned to their familiar ways. To the south of Tierra del Fuego in the Beagle Channel are the several islands that constitute Cape Horn. Named by the Dutch for their home port of Hoorn, it is the southern terminus of the South American continent.

The vegetation is mostly bog and moorland, with scattered forests of *Araucaria* (Ruiz et al. 2007) and *Nothofagus* (southern beech; McEwan et al. 1997; Veblen 2007). Grasslands and dry shrublands or steppe are characteristic of the Río de la Plata plains and Patagonia. Rainfall is 2600 mm a year along the Pacific coast, 4000 mm on the western mainland, 10,000 mm on the upper slopes of the Southern Andes, and 430 mm in the frigid south and east at Punta Arenas on the Strait of Magellan. MAT ranges from about 14°C in the south to 19°C in the north, and temperatures are relatively mild in most of the lowlands and at midelevations because of the maritime climate. The climate and vegetation regime of this far southern and fragmented landscape is delicately balanced, however, and slight changes in climate-forcing mechanisms produce significant alterations in the grassland-steppe-forested communities. Glaciation was already underway by the Pliocene (3.5 Ma), and spores and pollen from the bogs have been a source of considerable information on Quaternary vegetation and environmental history through studies by Calvin Heusser, M. E. Quattrocchio, M. L. Salgado-Labouriau, and others (chap. 8).

The Central Andes

As the Atlantic Ocean continued to open northward and subduction of the Pacific plate intensified, uplift of the Central Andes accelerated. They extend from the Golfo de Penas at 46°S in southern Chile to near a major megashear zone called the Amotape Cross at 5°S in southern Ecuador at the Golfo de Guayaquil. This is the longest, highest, widest, and most complex segment of the Andean cordillera. There are large deposits of copper, gold, iron, lead, silver, tin, and zinc in Chile, Bolivia, and Peru, and, partly because of this, there have been numerous studies on the geology of the Central Andes. Field experiences associated with these studies are vividly recounted by Simon Lamb in *Devil in the Mountain* (2004).

Names of the physiographic provinces differ according to country, but they consist generally of a coastal plain rising to about 1000 m elevation; the Andes Mountains with the Cordillera Oriental, Cordillera Occidental, and an intervening valley or a high plateau called the Puna in Peru and the Altiplano in Bolivia. The eastern slopes are called the Yungas, and below is the Oriente, or Amazon lowlands. It is a tectonically active region, and there are more volcanoes in Ecuador than in any other country in South America. Among the sixty-five active volcanoes, Pichincha has erupted five times in the last five hundred years, and in four of these eruptions the capital city of Quito was nearly destroyed. Humboldt, traveling through the region between 1799 and 1804, named the two cordilleras the "avenue of volcanoes." The most recent eruption of Pichincha was in October and November 1999.

About 950 km to the east in the Pacific Ocean are the Galápagos Islands, administered by Ecuador. The Humboldt Current is diverted westward by the southeast trade winds and by the coastal configuration of northern Peru and Ecuador; thus, the climate of the Galápagos Islands is relatively cool (21°C–26°C May through February, 28°C March and April). Darwin made observations on the finches and other organisms that showed the effect of geographic (reproductive) isolation, and this was a key component in his formulating the theory of evolution. The Galápagos Islands are presently the site of a beleaguered biological preserve. The islands arise over a hot spot between the Cocos and Nazca plates (Hoernle et al. 2002), and the outpourings of basalt may have formed the floor of the Caribbean Basin as it opened in the Late Cretaceous and early Tertiary. Also offshore is a zone of high rainfall and weather instability called the Intertropical Convergence Zone (ITCZ), and its annual meandering north to south is a major factor in determining the equatorial climates of the New World. In January and February it moves south from near the equator to 2°S–3°S, and in June and July it moves to about 9°N. Its passage over the different regions brings the characteristic rainy seasons. In Peru, the Central Andes reach a height of 6768 m at Huascarán, where in 1970 ice from the mountain loosened by an earthquake killed nearly 40,000 people in the nearby villages.

Important contributions to the study of tropical paleoclimates have been made in Peru by Lonnie Thompson based on cores taken through the annually layered ice of the Quelccaya ice cap (13°S, 5670 m elevation). The cores reveal significant fluctuations in temperature and precipitation during the late Pleistocene even at this near-equatorial latitude (chap. 8). One consequence of global warming not widely covered in the media, but mentioned in Al Gore's *An Inconvenient Truth*, is that this valuable source of information on recent climate change is rapidly melting away. Fortunately, cores from glacial ice no longer in existence have been stored by Thompson and his colleagues for future study. The mountains are crossed by several arid valleys that geographically isolate páramo species at the peaks and montane forests on the slopes. The lowest point in the Central Andes is the Huancabamba Valley of Peru at 2145 m. The Huallaga Valley is the principal coca-growing region of South America.

Bolivia and the Altiplano

> Just over a century ago a diplomatic crisis was brewing in La Paz over a glass of chicha. The new British ambassador to Bolivia had made the mistake of sneering at this local drink when served it by the incumbent dictator. As a punishment, he was forced to drink a barrelful of chocolate and be led through the streets of the capital strapped to a donkey. When news reached London, Queen Victoria was not amused. She demanded a map of South America, drew a cross through the country and declared, "Bolivia does not exist!"
>
> —ALAN MURPHY, *Bolivia Handbook*, 1997

In Bolivia, the Central Andes reach 700 km in width and the average height is 4000 m. The highest peaks are in the Cordillera Oriental and include Nevado Ancohuma (Illampu) near La Paz at 6550 m, Huayna north of La Paz Potosí at 6088 m, and Illimani to the southwest at 6485 m. Atmospheric pressure is one-half that at sea level, making fieldwork and longer-term stays by lowlanders dangerous because of HACE (high altitude cerebral edema), HAPE (high altitude pulmonary edema), and decompression sickness (bubbles of nitrogen forming in the blood). La Paz is the world's highest capital at 3600 m, Lake Titicaca is the second largest in South America, after Lake Maracaibo in Venezuela, and it is the world's highest navigable lake at 4000 m. Fieldwork is made even more difficult by the fact that only about 4 percent of the roads are paved; during the rainy season from November through February, 80 percent are impassable.

The Altiplano is a high, cold, windswept plateau in the rain shadow of the two cordillera. It is 800 km long, about 100 km wide, extending from southern Peru to Argentina, and at 3700 m, it is the second largest and highest plateau on land. The largest and highest is the Tibetan Plateau at 4690 m, and the largest on Earth is the submerged Ontung-Java Plateau. The MAT of the Altiplano is 7°C–10°C and temperatures may drop to −20°C. It is dry with a MAP of 758 mm (30 in). Winds from the east provide moisture off the Amazon lowlands; winds from the west off the coastal desert are dry; and during strong El Niños, conditions are even drier. About 70 percent of the Bolivian people live on the Altiplano in La Paz, around Lake Titicaca, or in mining towns scattered on the plateau. A digital geologic map of the Altiplano and the Cordillera Occidental can be found in a catalog of earth science data made available by the U.S. Geological Service (http://geo-nsdi.er.usgs.gov/).

There are several Tertiary macrofossil floras known from the Altiplano near Corocoro, site of one of the world's largest deposits of copper, and at Cerro Rico with its extensive silver mines. There is a plant microfossil assemblage from north of Cochabamba at Pislepampa that together with the macrofossils provide an estimate of the time and extent of uplift for the Central Andes (about half the present elevation was attained after 10 Ma; chap. 7). The Central Andes were extensively glaciated in the Pleistocene, leaving many lakes on the Altiplano. Water evaporates during dry interglacial times, leaving highly saline lake beds called salars, and the Salar de Uyuni is the largest salt lake in the world. The study of lake levels, spores, pollen, diatoms, rodent middens, and water chemistry by Paul Baker, Sherilyn Fritz, M. Grosjean, Christa Placzek, Jay Qudae, and their colleagues has provided a wealth of information on the Quaternary vegetation and environments of the Altiplano. It indicates considerable change in climate and vegetation since 120 kyr (chap. 8).

The Yungas—the name is an Aymara word for warm lands—extends from the highlands of the eastern cordillera to the lowlands. Climates are dry in the upper part and moist toward the Amazon Basin. The Aymara-speaking people are known for the distinctive bowler-style hats worn by the women, widely used in tourism advertisements. The Aymara and the Quechua are descendents of the Inca. The Yungas has a distinctive corrugated topography of ridges and valleys running parallel to the long axis of the mountains. It is an isolated, sparsely populated region, and it was here in 1966 that Che Guevara began his revolutionary activities in South America. He was executed in a schoolhouse in La Higuera, Bolivia, on 9 October 1967.

The Central Andes are folded mountains augmented by volcanics and accreted terranes. Uplift was especially rapid in the Miocene, in part owing to the steep-angle subduction of the descending Nazca plate at 25°–30°, and decreased after 5 Ma because of lower-angle subduction (Orme 2007a). The Central Andes continue to rise 10–25 mm each year from crustal shortening (crumpling)—hence the devastating instability of the landscape. Charles Darwin witnessed an earthquake at Valdiva, Chile, on 20 February 1835 that elevated the coast 3 m in a matter of seconds; and on 19 January 1835, he saw a "great glare of red light" that was the eruption of Corcovado Volcano near the island of Chiloé in southern Chile. On 22 May 1960, the largest earthquake ever recorded in the New World occurred near Concepción with a magnitude of 9.5 on the Richter scale. Such events result from three forces driving the uplift of the Central Andes. First, subduction of the Nazca plate into the Peru-Chile Trench at 50–85 mm each year accounts for the volcanic activity in the region and about one-fifth the present height of the mountains. Second, compression generated from the Atlantic spreading center to the east forces South America against the Pacific/Nazca plate at an annual rate of about 25–35 mm, causing shortening of the crust along the western coast. The Nazca and Cocos plates originated from breakup of the older Farallon plate in the late Oligocene around 28 Ma. Third, lithosphere thinning (Garzione et al. 2008), for example, beneath the Ojos del Salado Mountains in the southern part of the Central Andes, has contributed to regional instability. The heat coming through the thin lithosphere weakens the crust and allows it to be buckled to especially high elevations, as at Cerro Aconcagua (6960 m).

Once great heights are attained, other processes are required to sustain them to avoid widespread collapse. High elevations usually result when two continental plates collide. Thus, the collision of the Indian plate with the Asian plate caused the uplift of the Himalaya Mountains and the Tibetan Plateau. Where one plate subducts under another, elevations are usually not so high. If some sustaining mechanism were not operating in the region of the Central Andes, it is estimated they would average about 2 km in elevation rather than the present 4 km.

The dryness of the coasts of Peru and Chili due to the cold offshore Humboldt Current and the rain shadow of the Andes is one such mechanism. It reduces outwash and river flow that would normally bring lubricating sediments into the subduction zone. As a result, the downward slide of the Nazca plate is impeded, friction builds up, compression forces increase, and high elevations are sustained (Lamb and Davis 2003).

Several other features of the Central Andes are explained by ongoing geologic processes. The corrugated aspect of the Yungas is due to the Cordillera Oriental being compressed against the subterranean extension of the Brazilian Shield. The mountains are being held in place along their western margin by movement of the Pacific/Nazca plate into the Peru-Chile Trench, and they are being pushed from the east by spreading of the Mid-Atlantic Ridge. The crustal shortening of the Cordillera Oriental was at least 100 km between 30 and 20 Ma, and at least another 100 km from 10 Ma to the present. The eastern margin is now riding up and over the subterranean western edge of the shield, producing the distinctive ridge and valley topography of the Yungas.

Another key feature of the Andes Mountains is their westward bend (the "elbow of the Andes," or the Bolivian Orocline) near Santa Cruz. This is a consequence of two geologic processes operating in the Central Andes. One is the lithosphere thinning already noted, which produces a more heated flexible crust there than along segments to the north and south. Subduction since 10 Ma has forced this pliable part of the Andes to the east, while the more rigid segments on either side have remained relatively stationary, or have been pushed westward, producing the distinctive curvature. In addition to being a signature feature of the Andes Mountains, the orocline affects the biology of the adjacent Amazon Basin. As winds blowing from the northeast to the southwest across the lowland basin rise along the slopes of the Andes Mountains, they lose moisture and create zones of high rainfall. The orocline provides a slope that faces more directly at right angles to these winds, and the region below receives an even greater amount of rain. It may have been a refugium for rain forest during cold dry periods of the Quaternary (chap. 9). Farther north the east-facing concave configuration of the Andes between about 15°S to 5°N has created another regional concentration of moisture, augmenting precipitation from winds off the Amazon Basin. (In addition to the 1997 *Oxford Atlas of the World*, 150, good regional maps can be found online at the Google Earth and Worldmapper sites.)

Timing and sequence of events in the Central Andes

At about 100 Ma, the western coast of central South America was swampland and lakes without bordering highlands, as shown by the horizontal bedding of the limestone strata and the absence of eroded terrestrial sediments. Preserved in the limestone is one of the most extensive displays of dinosaur footprints in the world. To the west in the Pacific Ocean there was an arc of volcanoes. Beginning at 40 Ma, eroded sandstone and siltstone covered the limestone, indicating the early presence of highlands. These highlands are slightly younger than those to the south (Southern Andes) and older than those to the north (Northern Andes). In the latest Eocene and early Oligocene, sedimentation increased and deposited the Potoco Formation, which contains plant macro- and microfossils. Thus, 40 Ma marked an important early period of uplift in the Central Andes resulting from a slowing of convergence between the South American and Nazca plates from 15 to 5 cm a year, possibly in compensation for increased collision of India with Eurasia. As the crust cooled and hardened along the subduction zone, friction increased and the highlands were uplifted.

The eroding sandstones and siltstones came from a low ridge in the former swampland, and by 30 Ma similar sediments were also being deposited to the east that would form part of the Altiplano. The arc of volcanoes had now coalesced into a continuous upland to constitute the proto–Cordillera Occidental. The Altiplano was a river basin near sea level between this mountain system and the mostly nonvolcanic Cordillera Oriental that was rising to the east from compression against the subterranean extension of the Brazilian Highlands. By 15 Ma, half the present altitudes had been attained, and this was sufficient to create arid conditions to the west, as shown by sediments of that age in northern Chile. This was the beginning of the dry zone along the coasts of Peru and Chile that would later become the hyperarid Atacama Desert. As noted previously, the remaining altitude of the Central Andes Mountains has been attained since 10–6 Ma.

The Juan Fernandez Islands lie offshore from the southern Central Andes at 33°S. They were the setting for Daniel Defoe's novel, *Robinson Crusoe*, based on the life of Alexander Selkirk, who spent four years on one of the islands, from September 1704 to February 1709, before being rescued. Selkirk was a buccaneer on board the *Cinque Ports* captained by the famed pirate William Dampier, whose own life was chronicled by Diana and Michael Preston in *A Pirate of Exquisite Mind* (2004). Selkirk left the ship because he was convinced it was not seaworthy (it later sank). In a bit of rare coincidence, he was rescued on 2 February 1709 by the ship *Duke*, captained by the same William Dampier.

The Northern Andes

The northern section of the Andes Mountains extends from a megashear zone called the Amotape Cross in southern Ecuador at the Golfo de Guayaquil northward to the Caribbean plate. It consists of two subprovinces or blocks. The southern block, from Guayaquil to southern Colombia, was uplifted by subduction of the Nazca plate in the Late Cretaceous through the Eocene. In Ecuador the Cordillera Occidental, the Central Valley (or Interandean Depression), and the Cordillera Oriental were elevated at this time (fig. 2.35). In the Central Valley there is a rich assemblage of Pleistocene

Figure 2.41 Northern Andes of Colombia and western Venezuela showing, left to right, accreted oceanic terranes of the Pacific Coast, Cordillera Occidental, Cauco Valley, Cordillera Central, Magdalena Valley, and Cordillera Oriental dividing northward into the Sierra de Perija and Mérida Andes surrounding the Maracaibo Basin. From Orme 2007a. Used with permission from Oxford University Press, Oxford.

mammals from the Cangahua Formation that records the disappearance of mastodonts from Ecuador, along with later fluctuations in the abundance of mylodonts (ground sloths) and equids (hooved animals, horses and related genera) with the cold, dry climates of the approaching LGM (21–18 kyr). As late as the early Miocene (23 Ma), the mountains were still relatively low, and there were passages through them because the Pacific Ocean extended eastward to the Oriente Province of Ecuador, and both the proto–Orinoco River and the proto–Amazon River were draining westward into the Pacific Ocean.

In southern Colombia, the Northern Andes branch to form the Cordillera Occidental, Central, and Oriental (fig. 2.41). The Cordillera Oriental further branch into the western Sierra Nevada de Santa Marta / Sierra de Perija and the eastern Cordillera Mérida of northern Colombia and Venezuela, enclosing Lake Maracaibo, nearly filled with 7000 m of sediment (fig. 2.35). There are several features of the region important to the history and economic development of South America. The city of Cumaná was established on the northeast coast of Venezuela by Dominican friars in 1513, almost a hundred years before Jamestown, Virginia, and it is the oldest Spanish settlement in South America. The Maracaibo Basin is an inlet between the Sierra Nevada de Santa Marta and the Cordillera de Mérida, and it contains large petroleum reserves. The discovery of oil in the New World by Europeans dates from the 1589 explorations by the Spaniard Juan de Castellanos on Cubagua off the coast of Venezuela. However, three decades earlier, Humboldt had noted a strong smell of petroleum near Cape de La Brea where the petroleum covers the surface of the sea a thousand feet from the coast ([1852] 1971). Studies by Jan Muller of the Dutch Shell Oil Company, and by Estella de Di Giacomo, Maria Lorente, and others of the Petróleos de Venezuela, have provided valuable information on the Tertiary vegetation and environments of the region, and the plant microfossil record is known in considerable detail because of its use in the petroleum industry for correlation, zonation, and paleodeposition reconstructions.

Within the mountain highlands, basins have accumulated fossil spore and pollen-bearing sediments thousands of feet thick dating back to the late Pliocene (4–3 Ma). The early studies of Thomas van der Hammen, and current investigations by Henry Hooghiemstra, Antoine Cleef, and their colleagues have made this one of the paleobotanical "hot spots" in Latin America.

In the Late Cretaceous, the volcanic Cordillera Occidental was accreted onto northwestern South America, as indicated by a line of ophiolites along the Romeral fault zone. It began its principal rise in the Oligocene and early Miocene. Such knowledge of volcanoes, vegetation, and other features can easily be imparted, but this belies the hazards involved in getting the information. In November 1985, Nevado del Ruiz (5300 m) in southern Colombia suddenly erupted, causing the death of 23,000 people; and on 14 January 1993, nine geologists were killed in another explosive eruption of Galeras in the same region. Slightly different accounts of the Galeras disaster are given by Victoria Bruce in *No Apparent Danger* (2001) and by Stanley Williams and Fen Montaigne in *Surviving Galeras* (2001).

The volcanic Cordillera Central and the essentially nonvolcanic Oriental were uplifted by subduction of the Nazca plate in the Miocene and reached their present heights at around 6–3 Ma. Uplift is estimated at an annual rate of 3 mm during the Pliocene. Other parts of northwestern South America were added on at about the same time through movement of the Caribbean plate. These include the Guajira Peninsula, the Sierra Nevada de Santa Marta block, and the offshore Caribbean Mountain system. The peaks of the Caribbean Mountains constitute the present-day Netherlands Antilles, or the ABC islands of Aruba, Bonaire, and Curaçao, which emerged in the

Miocene. There is a broad floodplain along the Orinoco River between the Cordillera de Mérida and the Guiana highlands called the Llanos Orientales. It is a low rolling landscape about 1.1 million km² in area with more than 2300 shallow lakes. Forests grow along the rivers and lake margins, and Latin America's largest savanna, the Gran Sabana, is located in the adjacent Guiana highlands.

Páramo

> Biologists say climate change may already be affecting high-mountain ecosystems around the world, where plants and animals adapted to cold, barren conditions now face higher temperatures and a surge of predators and competitors.
> —KEVIN KRAJICK, "All Downhill from Here?" 2004

A distinctive plant formation of the Andean highlands is the treeless vegetation between 3500 m and the permanent snow line at 5000 m (fig. 2.42). Temperatures during the day may range from −3°C to 12°C, and the MAP from a dry 500 mm to over 3000 mm. In the drier parts a prominent bunch-grass is *Calamagrostis effusa*, and in the humid páramo it is the bamboo grass *Chusquea tessellata*. There are also giant rosettes of *Puya* (fig. 2.43). The páramo consists of 4000 species of plants and, surprising to many temperate-trained biologists, includes 22 genera of orchids. The rosaceous shrubs *Acaena* and *Polylepis* are common as fragmented populations at high altitudes from Ecuador to Argentina, and their pollen is found in Quaternary deposits in the Andes Mountains, preserving a record of presettlement fires, human activity (the wood is gathered for fuel and for making charcoal), and shifts in tree line with changing climates (chap. 8).

The páramo has been highly disturbed through fire, grazing, cultivation, and lumbering of the adjacent high-altitude forests. This will become increasingly significant in the future because, compared to most mountainous countries of the Earth where cities are located in the lowlands and at midelevations, in the Northern Andes most of the population lives in the highlands—50 percent in Ecuador, 70 percent in Venezuela, and 75 percent in Colombia. The páramo is the principal water catchment, filtration, and regulation system for cities such as Bogotá and its 7 million people. Understanding the páramo as an ecosystem is essential for formulating sustainable management practices. Part of that information is preserved in the plant fossil record, which provides insight into the pace and extent of Quaternary climatic changes and their effect on the vegetation of the High Andes. At the LGM, the lower limit of the páramo was at about 2000 m, a downward

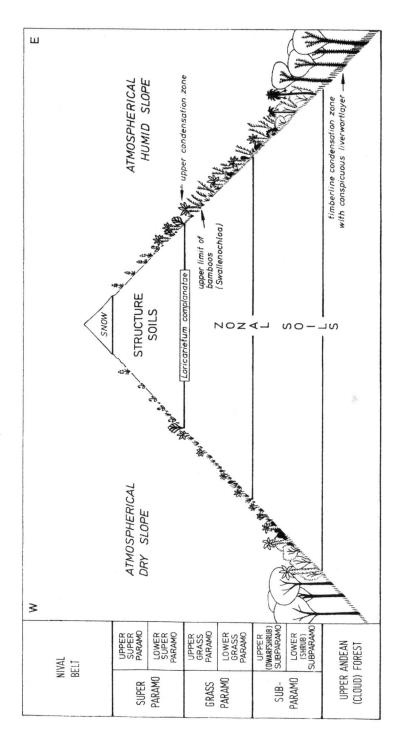

Figure 2.42 Zones of páramo vegetation, Cordillera Oriental, Colombia. From van der Hammen and Cleef 1986. Used with permission from Oxford University Press, Oxford.

Figure 2.43 Puya raimondii in the Peruvian puna, or high-altitude zone (páramo). From Weber 1969. Used with permission from Springer Science and Business Media, Dordrecht.

shift of 1500 m. Recalling that since about 2.6 Ma there have been between eighteen and twenty glacial maxima, the dynamic nature of the biota is evident. If the current trend in global warming persists, the páramo will continue to move upward until it first becomes limited and then is forced off the mountains. If the adjacent forests continue to be lumbered from the slopes, the consequences for erosion control, filtration of drinking water, and aquifer levels and dependability are obvious.

The Amazon Basin

As northeastern South America separated from Africa around 90 Ma, pull-apart basins like the Pernambuco Basin formed along the coastal region. They border the lowlands of Amazonia to the east, which occupies a depression of about 6 million km^2. The floor of the Amazon Basin follows the broad contour of the ancient cratonic platform and the later Cretaceous surface. It is covered by the world's largest extent of lowland rain forest (Richards 1996; Morley 2000; Furley 2007; Galindo-Leal and Gusmão-Câmara 2003). The Amazon forest is separated from the similar Atlantic forest by the Brazilian Highlands region and its extensions to the north (Borborema Plateau) and to the south (Mato Grosso Plateau). The basin is now filled with over 4000 m of sediment. The oldest was eroded from the bordering Brazilian and Guiana shields and deposited as the 1.6 Ga Roraima Group, which in turn was eroded to form the distinctive tepuis of Venezuela (fig. 2.44). These flat-topped mesas rise to a height of a mile or more from the lowland mist, and they provided the setting for Sir Arthur Conan Doyle's novel *The Lost World*. On the summit there are meadows, some low forests in depressions, and occasional aquatic vegetation in pools of standing water.

The basin is traversed by a labyrinth of rivers on which small boats can navigate 100,000 miles of waterway. The Amazon River (Hoorn 2006) is the second longest in the world at 6450 km. (The Nile is the longest at 6670 km, and the Yangtze is third at 6380 km, although the Amazon may be counted first at 6800 km if one identifies its origin in southern rather than northern Peru.) Large sections of land called cahidas (fallen lands) become detached along its banks, leaving thousands of furos, or small bays. The river discharges 175,000 m^3 of water per second, deposits 1 billion tons of silt into the ocean each year, and during floods the river can rise 10 m in some places. Among the well-known features of the Amazon Basin are its blackwater and whitewater rivers. Waters that flow from the north cross the hard Precambrian Shield enter as sediment-poor, organic-rich black waters. Those that drain the Andes Mountains are sediment-rich white waters:

Figure 2.44 Cerro Autana, a tepui in northwestern Amazonas, Venezuela, about 1300 m in elevation. From O. Huber 1995. Used with permission from the Missouri Botanical Garden Press, St. Louis.

"Don Francisco de Orellana departed from Quito . . . in February 1541 as a member of an expedition led by Gonçalo Pizarro in search of cinnamon and the land of 'El Dorado,' the legendary golden kingdom. When the expedition ran out of food the situation became perilous, and Pizarro and Orellana decided to separate. Driven by famine, the latter continued with his men in an eastern direction in search of food. Finally, no longer able to proceed on land, he ordered that a boat be built . . . at the confluence of the 'Rio de los Omáguas,' now the Rio Napo, with the Rio Aguárico. By the end of 1541 he had embarked on it, together with 55 Spanish soldiers and two Fratres, one of whom, Frey Gaspar de Carvajal, became the chronicler of the voyage. Instead of finding either El Dorado or cinnamon, Orellana discovered the mightiest river on earth, the 'Rio de las Amazonas,' first given the name 'Rio de Orellana.' The boat followed the current of the river down into the unknown until it came out at the mouth into the Atlantic on August 26, 1542.

Only a few sections of Father Carvajal's report are interesting from a natural sciences' standpoint. One of the most worthwhile is the paragraph, "On the same day . . . we saw the mouth of another big river, on the left side, which entered that one in which we navigated, and of water as black as ink, and therefore we gave it the name Rio Negro." In these laconic words the greatest tropical blackwater river was made known for the first time in history to European civilization." (Sioli 1984, 1–2)

A popular misconception about the Amazon Basin is that it has a uniform topography and climate that supports a continuous cover of rain forest. It is true that the relief is mostly less than 900 m, and rain forest is the prominent vegetation, but over the expanse of the basin there is considerable variation. A moderate dry season extends from June to August when "popcorn" clouds form over the forest but not the rivers (see fig. 2.47 below). Salt and brackish water is carried 650 km upstream by rapidly flowing tides. Its tumultuous meeting with the onrushing fresh water of the river is called the pororoca, and it produces waves up to 6 m high. There are lake and river communities, inundated floodplains called várzeas, and lands beyond the floodplains called terra firme. Different microhabitats exist on the trunks of the giant forest trees that extend from damp, dimly lit ground cover to full sunlight 60–70 m above ground. Along the way there are various species of bromeliads, which store water in overlapping, funnel-shaped leaves in which live bacteria, algae, aquatic fungi, protozoans, small crustaceans, snails, spiders, and mosquito larvae, on which frogs live, that are in turn eaten by a variety of small snakes—altogether 250 species may live on the trunk of a single rain forest tree. There are topographic and edaphic variations within the basin that support patches of grassland, savanna, and dry forest on promontories and sandy soil, and the surrounding cerrado and caatingas extend into the basin varying distances from the surrounding margins (fig. 2.45). These different habitats and their biota are important

Figure 2.45 Ecological subdivisions (floristic regions and vegetation types) of South America. From Young and others 2007 (see also references cited therein). Used with permission from Oxford University Press, Oxford.

because they allow for rapid response of the vegetation to environmental change. Also, they figure prominently in models proposed to explain the high levels of biodiversity based on periodic fragmentation and reuniting of habitats and the biota due to changes in climate (chap. 9, Refugia).

There are three key geologic questions relating to the biotic history of northern and central South America: (a) What was the early environment like in the Amazon Basin during the Cretaceous and Tertiary? (b) When did the mountains reach sufficient heights to reverse the flow of the rivers from west into the Pacific Ocean to east into the Atlantic Ocean? (c) When did the mountains begin to cast a rain shadow to the west that contributed to the formation of the Atacama Desert?

In the Late Cretaceous, marine deposits from the Pacific Ocean were being laid down in the upper Amazon Basin and formed units like the Río Acre Formation of western Brazil and eastern Peru. By 90 Ma, the Atlantic Ocean Basin had opened to the latitude of northeastern South America, and Cretaceous sediments from the Atlantic were deposited in the lower Amazon Basin to form units like the Marajó Formation. At 65 Ma, tsunami deposits near Recife, Brazil, associated with an iridium anomaly nearly seventy times the concentration in adjacent strata, record the asteroid impact at Chicxulub, Mexico. By the beginning of the middle Miocene, the basin was also receiving sediments eroded from the rising Andes Mountains. Cores from the basin reveal that the landscape consisted of lowland swamps, lagoons, and meandering rivers (Hoorn and Vonhof 2006). There were periodic incursions from the Caribbean Sea through the Maracaibo Basin, as shown by fossil faunas containing marine fish, and by freshwater fish derived from marine ancestors (Albert et al. 2006; Lovejoy et al. 2006). By the middle Miocene, about 15 Ma, the Andes Mountains had reached sufficient heights to tilt the lowlands eastward, and the Orinoco River began depositing sediments along the Venezuelan coast at Delta Amacuro, and the Amazon River deposited its first sediments into the Atlantic Ocean at Ilha de Marajó in Brazil. In the late Miocene, the Maracaibo River was dammed by uplift of the Cordillera Mérida extension of the Cordillera Oriental to form Lake Maracaibo. There are Amazonian paleofaunas from Peru dated at 3 Ma, and these mark the last marine incursion into the basin. After that time, the Amazon Basin continued a more subdued physical evolution through deposition of sediments from the highlands, some further tilting of the basin floor, and changes in secondary drainage patterns. The vegetation was affected by these processes, and especially by the intensifying fluctuations in climate as the ice ages approached and altered precipitation, temperature, and water tables in response to sea level changes. These changes ushered in the latter phases of rain forest history that are revealed in the fossil record to be dynamic even in the lowland basins at these tropical latitudes. This is in contrast to the view held only a few decades ago that the climates and biota of the tropics were stable and unchanging (chap. 9).

The Northern Andes by about 15 Ma had reached sufficient heights that as winds off the Pacific Ocean begin to rise along the western slopes they lost moisture, creating one of the wettest regions of the world—the Colombian Chocó. The mountains also block moisture from the east, and their uplift is one in a series of events that created on the coast farther to the south one of the driest regions on Earth—the Atacama Desert. Thus, in addition to influencing global patterns of climate through an effect on atmospheric circulation, other consequences of uplift were changes in the configuration of the Amazon Basin and its present-day drainage patterns and topography: the Colombian Chocó with its exceptionally high rainfall, and the Atacama Desert with its almost unique lack of rainfall. The biotas supported in each region are all in significant measure a result of the evolution of the Andes Mountains (Antonelli et al. 2009).

The Lowland Neotropical Rain Forest

[The rain forest is] the most remarkable expression of nature to grace the surface of the planet in four billion years of life's history.
                              —NORMAN MYERS, "Four for the Forest's Future," 1991

The rain forest often makes a deep impression on first-time visitors when they are given a few moments of solitude in its vast silence. The sensory experience is profound and unfamiliar: gigantic trees dripping with moisture and festooned with ferns, lichens, bryophytes, and lianas—"twiners entwining twiners," as Charles Darwin put it—grasses the size of bamboos, ferns as large as small trees, leaves one can stand under, water lilies one can stand on, enormous water snakes (*Boa aquatica*), and butterflies and tarantulas seven inches across; the oddities of strangler figs, sloths, and the nasty little piranhas; the beauty of passion flowers (*Passiflora*) and 750 species of the spectacular jacaranda family (Bignoniaceae); the estimated 10,000 New World orchids (of more than 22,000 species worldwide); and large flowers white in color, scented, and open at night to be pollinated by bats. There are flowers of the orchid *Ophrys* and others that mimic the female species of wasps and bees, complete with pheromones, and pollinated by males of the species. The forest is mostly evergreen, without periods of noticeable leaf fall, and regeneration occurs unobtrusively throughout the year. This subdued change from season to season may have contributed to an impression in the minds of earlier visitors that the rain forest was equally unchanging from epoch to epoch. Flowering is often in brilliant but widely scattered patches (fig. 2.46).

In a somewhat more sober and quantitative characterization, the lowland neotropical rain forest is typically found on level terrane between 10°N

Figure 2.46 Aerial view of the lowland neotropical rain forest, Ecuador. Photograph by
J. Arthur Herrick.

and 10°S, where MATs are a uniform and relatively high of 24°C in the
warmest month and at least 18°C in the coldest month. In other words,
there is no pronounced cold period. The minimum MAP is uniform and
relatively high at 1800–5000 mm, that is, variable but with no pronounced
dry season (fig. 2.47). If MAP drops below 1800–1500 mm, the vegetation
trends toward tropical dry forest or savanna. Soils are thin, sterile, often
red clay laterites, and there is rapid mineral recycling, so root systems are
shallow and there is frequent buttressing of the trunks. Angiosperms are
the prominent vascular plant group, especially the Fabaceae/Leguminosae
(legumes), Moraceae (fig family), Annonaceae, Euphorbiaceae, Lauraceae,
Sapotaceae, Myristicaceae, and Arecaceae (palms), pollinated mostly by
animals, particularly insects, as opposed to the more common wind pollination in drier, open, upland, and temperate habitats. Leaf margins are
mostly entire, there is often a drip tip, and growth rings are usually absent
or poorly developed (see fig. 5.5 below and compare with fig. 2.7). These
features of leaves and wood are important to vegetation history because
they are evident in fossil assemblages, and as noted previously, they allow
recognition of tropical environments without taxonomic identification of
the fossils. In earlier days identifications, even for tropical paleofloras in the

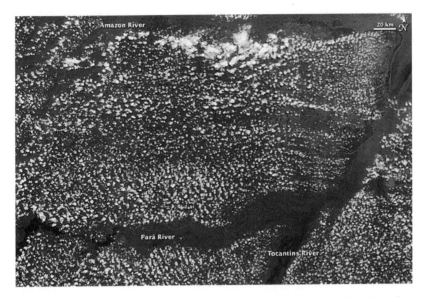

Figure 2.47 "Popcorn" clouds over the land of Amazonia during the dry season (June–August) caused by release of moisture higher into the atmosphere above tall trees than from over the rivers. From the Moderate Resolution Imaging Spectroradiometer (MODIS) on NASA's Aqua Satellite; image available on the NASA Earth Observatory site, http://earthobservatory.nasa.gov/.

Eocene of the Mississippi Embayment region, were notoriously unreliable (Dilcher 1973), and it can only be guessed what the figure is in the older literature for fossil floras in the tropics.

The exact number of species in the neotropical rain forest is not known, but for tree species that reach a size of more than 10 cm dbh, the number is estimated at 11,120 (Hubbell et al. 2008). Of these, about 3248 species have more than 1 million individuals, and 5308 species have less than 10,000 individuals. Among the latter, the extinction rate is estimated at 37–50 percent, and over the next several decades the mean total extinction rate of tree species in the Brazilian Amazon is estimated at between 20 percent (optimistic scenario) and 33 percent (nonoptimistic scenario). Humid forest clearing between 2000 and 2005 was about 1.39 percent, making for a 2.36 percent reduction in the area of humid tropical forest. About 55 percent of total biome clearing occurs within only 6 percent of the rain forest area (Latin America), with Brazil accounting for 47.8 percent, or four times that of Indonesia, the next most-cleared country at 12.8 percent (Hansen et al. 2008). As Richard Black put it in a 2008 BBC News report, "The global economy is losing more money from the disappearance of forests than through the current banking crisis, according to an EU-commissioned study. It puts the annual cost of forest loss at between $2 trillion and $5 trillion."

A sense of the diversity in the lowland neotropical rain forest, which covers 3 percent of the Earth's surface and contains 50 percent of its species, is conveyed by comparing the number of vascular tree species in plots 1000 m² from the modern northern deciduous forest to the rain forest: Ozark Mountains of Missouri, 23 species; tropical dry forest (Venezuela and Costa Rica), 63; moist tropical forest (Brazil and Panama), 109; wet tropical forest (Ecuador and Panama), 143; and superwet tropical forest (Colombia), 258 (Gentry 1982). One factor involved in the tenfold difference in diversity is predation that can reach epidemic levels under the favorable year-round growing conditions. Wide spacing (fig. 2.46) is one means whereby these levels are kept low enough that some individuals and their propagules survive.

People have occupied Brazilian Amazonia in relatively large numbers and in a sustainably balanced system for at least 1500 years. The archeological site of Upper Xingu has twenty-eight settlements, each with a population of about 2500, for an estimated regional population of 50,000 people (Heckenberger 2009; Heckenberger et al. 2008).

The Atlantic Forest

> The young Darwin had his first experience there with the tropical fauna and flora. In 1836 on his way home on the "Beagle" Darwin visited Bahia for the second time and had a strange premonition when he wrote: "In my last walk, I stopped again and again . . . and endeavoured to fix in my mind for ever, an impression which I knew sooner or later must fail . . . the thousand beauties which unite these into one perfect scene must fade away; yet they will leave, like a tale heard in childhood, a picture full of indistinct, but most beautiful figures." Not even the smallest remainder of a forest survives where Darwin stood.
>
> —FRANCIS DOV POR, *Sooretama*, 1992

The dry vegetation on the crests of the Río São Francisco running the length of the Brazilian highlands separates the Amazon from the Atlantic rain forest (Morellato and Haddad 2000). The southern boundary of the forest once included the magnificent *Araucaria angustifolia* (fig. 2.48), but conditions are ideal for growing *Hevea brasiliensis* (rubber), and *Theobroma cacao* (cocoa), the harvesting of commercially valuable timbers for lumber (*Astronium concinnum, Cedrela odorata, Dalbergia nigra, Plathymeia foliosa*) and

Figure 2.48 Paraná pine (*Araucaria angustifolia*) in Aparados de Serra Park, Rio Grande do Sul, Brazil. From Por 1992. Used with permission from Backhuys Publishers, Leiden.

for making bows of musical instruments (*Caesalpinia echinata*, pernambuco or pau-brazil, from which the country takes its name). A hypercaffeinated drink called guaraná is produced from *Paullinia cupania* that in the unrefined form imparts to the unprepared a shocking blood-red color to the urine. Another factor in the virtual destruction of the forest and its reservoir of medicinal, drug, and agricultural genetic stock, is the chaotic growth of megapopulation centers like São Paulo (more than 10 million people; 15 million in greater São Paulo) and Rio de Janeiro (11 million people; 12 million in greater Rio de Janeiro). Less than 5 percent of the forest remains.

The Colombian Chocó

The Pacific lowland from Buenaventura north to Panama along the western slopes of the Cordillera Occidental (figs. 2.35, 2.36) is one of the wettest regions on Earth. It gets a maximum of about 11,770 mm of rainfall in some years and the relative humidity is more than 85 percent. The wettest twelve months ever recorded was in Cherrapunji, India (annual average rainfall 11,420 mm) where 26,470 mm (1040 in) fell in 1980–81. The twenty-four-hour record is 1870 mm (73 in) on Réunion Island in the Indian Ocean in March 1952. The chocó is unusually rich in plants, with current tabulations of 262 tree species at least 2.5 cm dbh in only 0.1 ha. This suggests that within a warm-temperate to subtropical-tropical environment, exceptionally high rainfall and reduced evapotranspiration (two components of energy flow), along with edaphic and elevational variations (spatial heterogeneity), are important in accounting for high biodiversity. Other components now being incorporated into biodiversity models are historical and biological/evolutionary factors, such as the stochastic presence or absence in a region of rapidly speciating / highly specious lineages.

Alwyn Gentry (1986) noted the association between high rainfall and diversity. He was the world's foremost authority on the important tropical plant family Bignoniaceae, and among many other contributions, he developed the standardized sampling method of recording all vascular plants at least 2.5 cm dbh from ten plots 0.1 ha in size, now called Gentry plots. Thus, when it is stated that forest diversity is 258 species/ha at a rich tropical site, compared to 23 species/ha at a temperate locality, it is with the confidence that the same sampling method was employed. The procedure is used by organizations such as Conservation International for rapid assessment in determining conservation priorities, and by the International Union for the Conservation of Nature and Natural Resources for establishing the red list of endangered species. Gentry developed keys for identifying plants from sterile vegetative material (i.e., lacking flowers), especially valuable in the study of tropical biology; and with the luxury of flowering material, it is said he could recognize on sight the majority of neotropical trees and shrubs.

Gentry was killed in a plane crash on 3 August 1993 while surveying the Ecuadorian rain forest. Also killed were Conservation International ornithologist Theodore Parker III and three others. Parker could recognize 4000 species of birds by their calls alone. When one recalls the deaths of the five marine biologists in the Bay of California, and the nine geologists at Galeras, it is clear that field research can carry a very heavy price. The loss of such charismatic individualists, uniquely focused, capable, and determined, if sometimes difficult, creates a huge void in the study of tropical biology—hence the respectful dedication of this text to them. They are reminiscent of the admirable eccentric Erik Ekman mentioned earlier, and the remarkable Isabela de Grandmaison y Bruno. Isabela was not a scientist, but her story of survival in the Amazon as recounted by Victor Wolfgang

Von Hagen in *South America Called Them* (1945) recalls the hardships encountered by many field naturalists in the tropics, and it merits retelling.

Isabela was the daughter of Don Pedro Manuel de Grandmaison, corregidor (territorial governor) of Otavalo in Ecuador. She was very intelligent and by the age of thirteen, in addition to her native Spanish, she spoke fluent French and Quechua. Jean Godin des Odonais was a member of the Charles-Marie de La Condamine expedition, the first scientific expedition to South America. Godin married Isabela in 1742, and in 1749 he left on a four-month voyage to Cayenne, French Guiana. The correspondence between Godin and La Condamine reveals the men's stoicism in anticipation of the hellish journey. As Von Hagen notes: "[Godin] did not go into the suffering of his 3000-mile trip down the stygian Amazon. He did not have to. La Condamine was thoroughly acquainted with its horrors. To people used to living in a country of relative civilization such an action would seem on the point of lunacy . . . for the mere purpose of preparing accommodations for a second" (1945, 69)

By 1749, Godin had already spent fifteen years exploring Amazonia and was showing signs of the strain. Fifteen more years passed while he waited in Cayenne for permission from the Portuguese government to begin an expedition into the Amazon Basin. He tried to expedite matters by drafting a plan whereby the French government could take over Amazonia. When no reply was received, he became convinced the letter had fallen into the hands of the Portuguese; and when a ship from Portugal finally arrived to take him on the voyage, he refused to go, dispatching in his place his friend Tristan d'Oreasaval. Tristan carried letters for Isabela, waiting at her home in Riobamba, detailing how she was to meet up with Godin. After an eight-month voyage, d'Oreasaval delivered the letters to the wrong person. Nonetheless, rumors had reached Isabela that Godin was still in Cayenne, and after five more years with no word, she disposed of her property and in October 1769 prepared to leave:

> Isabela de Godin was a remarkable woman. She had been cloistered all her life in the cold, remote citadel of the Inca and Spaniard. Since Jean Godin had left . . . all four children died from malaria, yellow fever, and dysentery. Now in middle age she was called upon to take a journey that no one, even in the fullness of youth [would attempt], a journey down the whole of the Amazon. (Von Hagen 1945, 72)

Her father, Don Pedro, made such arrangements as he could, advising her to keep the number of people in her party to a minimum. Even so, when three Frenchmen (one who claimed to be a doctor) pleaded to go along, she agreed, and by the time they eventually left, the group consisted of herself, a twelve-year-old nephew, two brothers, four servants, the three Frenchmen, and a cadre of Indian guides. The walk through the rain and mud to the first river settlement was far more difficult than expected, and when at last they arrived exhausted at Canelos, they found it had been burned to the ground by the local people, who believed the recent smallpox epidemic that had ravaged the site was a curse by evil spirits. Unsettled by this news, the Indian guides deserted during the night. None of the rest had any experience with canoes, the rapid flow of the river made it impossible to go back, and the next settlement was days away. During one attempt to reach the next village one of the Frenchmen drowned, and during another the canoe capsized and most of the food was lost. Another of the Frenchmen suggested that he and Joachim, the only male servant, take the boat to the village and return later for the others. Without other alternatives, it was agreed, and they departed, leaving Isabela, four women servants, Isabela's two brothers, the third Frenchman (who remained as a kind of hostage), and the boy, who was now ill with fever. After a month of waiting under miserable conditions of unrelenting rain, insects, infection, and near starvation, the situation deteriorated entirely. The Frenchman, awaking to a vampire bat feeding on his foot, went mad. Isabela decided they had to leave, and another canoe was built; but it too was wrecked and the few remaining provisions lost. The boy, her two brothers, one maidservant died, and another walked into the jungle in a delirium and never returned. They all prepared for death. But then Isabela, envisioning Godin still waiting for her in French Guiana "cut off the shoes of her dead brothers, fashioned these into crude sandals, picked up a machete . . . and stumbled into the jungle" (Von Hagen 1945, 78).

She often heard imaginary voices in the days ahead, but one turned out to be real—the servant Joachim, returning with a canoe and supplies. They never made contact. He saw the camp strewn with the remains of dead bodies and assumed she must have perished. The rumor of her death reached Godin in Cayenne, and even circulated later in magazines and in the salons of Paris. But Isabela did not die. After nine days of wandering alone, she came upon three Indians sitting on the bank of the river. By this time she was completely white-haired, naked, and frighteningly gaunt, but she managed to address them in fluent Quechua, asked to go to Andoas, and then collapsed. They took her to the mission at Andoas, arriving in early January 1770.

Even then her ordeal was not over because when she gave the Indians gold necklaces as a reward, the padre "in my very presence" took the necklaces for himself and gave them in return only a few yards of cloth. As Isabela recounts, she was outraged, but being too weak to walk, she demanded to be put back in the canoe and taken to the next settlement at Loreto. A Portuguese ship intercepted the canoe and took Isabela the nearly 2000 miles to French Guiana, where she was at last reunited with Jean Godin after more than twenty years. They remained there for two years longer, then with her father and the servant Joachim, they returned to La Rochelle, France. At the dock Charles-Marie de La Condamine was waiting to greet them after Isabela's heroic sojourn and Jean Godin's thirty-eight years in Amazonia.

The Atacama Desert

The 300-km-long coastal region between southern Peru (5°S) and Chile (30°S) is hyperarid except in El Niño years, when much of it becomes a lake-dotted flowerland. MAP is 2–15 mm in southern Peru, less than 1 mm in northern Chile, and reaches 100 mm to the south. Fogs provide some moisture along the coast. MAP at places of 5 to 45 mm, and 400 to 1500 mm elsewhere, reach the higher values only during strong El Niños. Many parts of the Peruvian and Atacama deserts are nearly devoid of plants; and in other places, it is such a barren rocky landscape of exposed lava and shifting sand dunes that it was used to test the robot Nomad used as a prototype for the lunar rover. The dry soils were also used to establish protocols for detecting the possible biological content of Martian samples from the Viking missions. Personnel of the BBC, filming at Paranal Observatory, noted that in five years it had rained three times. The Atacama Desert extends about 100 km inland; and at the upper elevations, in the region of the southern Altiplano, there are mineral deposits of nitrate, gypsum, and other evaporites documenting that dry conditions periodically extended far into the geologic past. In central Chile from about 30°S to 36°S, there is a transition zone of Mediterranean-type vegetation and climate (winter-wet, April to September; MAP 200–700 mm) between the Atacama Desert to the north and temperate forests to the south (Armesto et al. 2007).

When environments change, ecosystems that are novel in extent and composition may come about through several alterations in the biota: (1) a gradual change (evolution) in the ecological requirements of the species, although this is probably rare; (2) an expression of already existing broader ecological amplitudes as conditions change or barriers are removed; (3) immigration; and (4) the almost universal presence of preadapted individual species and restricted communities within a more prominent vegetation type. The latter three are probably the most common; that is, "it may often be easier for lineages to move than it is for them to evolve" (Donoghue 2008, 11549). Moreover, there is "the tendency for species to retain their ancient ecology . . . over tens of millions of years and across continents" (Crisp et al. 2009, 754). For example, gallery forests and their associated fauna provide mesic assemblages within desert environments. Such mosaics of vegetation mean that when conditions change, in whatever direction, some groups of species will likely be favored and thus be available to move and coalesce into plant formations suitable to the new environment. After each reshuffling, the exact composition of the associations may not be the same, but the formation will likely be recognizable except after the most drastic and enduring of environmental changes. In the region of the Atacama Desert, slope, exposure, and coarse soils supported dry elements within the mesic environment that preceded desert conditions.

The first event to favor desert formation along the coast is indicated by a sharp drop in global sea level around 30 Ma. This was the result of several factors, including drift of Antarctica toward a more polar position, the cooling trend in global temperatures after the Late Paleocene Thermal Maximum / Early Eocene Climatic Optimum (LPTM/EECL), and the separation of South America from Antarctica through widening of the Drake Passage (currently 800 km across) beginning just prior to about 30 Ma. The opening of the passage allowed a west-to-east-flowing circumpolar current to develop that thermally isolated Antarctica from warmer tropical waters and served as a positive feedback to southward drift and cooling temperatures. The combination of these events led to the appearance of significant continental ice in the Gamburtsev Mountains of Antarctica about 34 Ma (Bo et al. 2009). At that time, atmospheric CO_2 concentration had declined from approximately 1600 ppmv in the mid-Cretaceous and more than 1125 ppmv at the LPTM/EECL to about 800 ppmv, that is, to about three times that of pre–industrial revolution values (280 ppmv) and roughly twice current values (380 ppmv). At 30 Ma, continental ice was significant in extent, as reflected in the drop in sea level, and it reached present-day volumes at about 15 Ma. It should be recalled that the formation and melting of sea ice, such as the extensive ice sheet covering the Arctic Ocean, has little effect on ocean levels, much like the melting of an ice cube in a glass of water. When extensive amounts of ice are anchored on land, however, their coming and going has a great impact. It is estimated that if all the Greenland, Iceland, valley glaciers, and Antarctic ice were to melt, sea level would rise by about 80.5 m (Greenland and Antarctica would account for

80.0 m). As the Antarctic ice sheet increased, causing the drop in sea level at 30 Ma and contributing cold meltwater to the world's oceans, it further cooled the Humboldt Current flowing northward along the west coast of South America. Westerly winds blowing across the cold water arrived on the coastal lands of Peru and Chile as dry desiccating air, and that began the process of aridification.

The second event was that cooling of ocean waters worldwide reduced the amount of moisture evaporating into the atmosphere to a point that drier and more seasonal climates began to expand. The beginning of enhanced seasonality is shown by spore and pollen evidence from the Norwegian-Greenland Sea that indicates a drop in cold-month temperature of about 5°C just prior to the Eocene-Oligocene transition (Eldrett et al. 2009). The trend continued, and by about 15 Ma, it had produced a noticeable reduction in mesic forests and an increase in grasslands and dry forests. Also by about 15 Ma, the Andes Mountains had reached sufficient heights to begin blocking moisture from the Amazon lowlands. Thus, by the middle Miocene, the combination of these events was (1) favoring organisms adapted to arid environments, (2) causing coalescence of elements preadapted to dry habitats, and (3) allowing expansion of formerly restricted dry-habitat species and communities into a biotic formation recognizable as the Atacama Desert. After that, arid conditions periodically intensified during the cold dry intervals of the Pleistocene, fluctuating with the El Niños and La Niñas that have affected the coast since at least 15 kyr. Prior to that time, evidence for El Niño events in the form of flood debris becomes difficult to distinguish from landslides caused by other factors, such as earthquakes.

The Monte, the Gran Chaco, and the Pampas

In Argentina, as winds rise from the lowlands up the eastern slopes of the Andes Mountains, moisture is lost at the intermediate elevations so the higher slopes are dry. Continentality, slope, and coarse soils like those on alluvial fans play a role, and in addition, rainfall from the west is blocked by the mountains. The result is a near-desert plain with a MAP of 30–350 mm, a MAT of 13°C–15° C, and a vegetation similar to that in the basins of the Basin and Range Province of the western United States and on the Central Plateau of Mexico. The region is called the Monte, and it covers about 450,000 km^2 between 24°S and 43°S . Plants similar to those in the dry areas of North America include *Acacia*, *Caesalpinia*, *Cassia*, *Cercidium*, *Larrea*, *Mimosa*, and *Prosopis*. Molecular data for *Larrea* suggest a probable South American ancestor and indicate possible arrival in North America between

8.4 and 4.2 Ma (Ickert-Bond and Wojciechowski 2004). Explanations for such disjunct distributions include long-distance dispersal and migration along a former connecting corridor. There is little paleobotanical evidence for a continuous or near-continuous arid corridor between the northern deserts and the Monte, and the most parsimonious explanation for many of the plants is long-distance dispersal by birds such as sparrows and plovers. The latter move annually between North and South America in an enormous migratory loop 2000 miles across and 8000 miles long.

The Gran Chaco, or "the green hell," is a plain at a slightly lower elevation that covers 1 million km² between the Andes Mountains and the Paraguay-Paraná rivers north to the Mato Grosso Plateau of Brazil, 17°S to 33°S (see, e.g., *Oxford Atlas of the World*, 150, or NASA's Earth Observatory site). It is a dry shrubland/chaparral-woodland-savanna with commercially valuable stands of *Schinopsis balansae* (breakaxe, or quebracho), one of the hardest woods known, used for railroad ties, fence posts, and construction timbers. Annual rainfall is lowest in the southwest (350 mm), and there is a hot, tick-infested dry season lasting six months from April to September.

The pampas is a slightly more moist plain to the southeast, situated between 30°S and 40°S (e.g., from Córdoba to Buenos Aires) and covering 650,000 km². It is bordered on the north by the Gran Chaco and on the south by Patagonia. Pampas is a Quechua word for flat, and it is a grassland, supporting one of the world's leading cattle-producing regions and tended by the picturesque gauchos. MAP is about 508 mm along the driest western margin. The cool and dusty southwesterly summer wind meets warm air from the north to produce the gales and rain storms called pamperos that characterize the pampas.

The Cerrados and Caatingas

The cerrado—the Portuguese word means "closed"—is a savanna with transitions to (1) low to medium dry shrubland or dense woodland that is nearly impenetrable, and (2) to a more open community approaching grassland (figs. 2.49 and 2.50). The MAT is 20°C–26°C, MAP is from 750 to 2000 mm, and the average elevation is 500–600 m. It is centered in Brazil bordering the Amazon Basin and covers about 2 million km², or 22 percent of the country. The cerrado is a diverse plant formation with some 10,000 plant species, of which 4400 are endemic and represent about 1.5 percent of the Earth's vascular flora (Sano et al. 2008). One kind of cerrado, found on rocky substrate, is called the campo rupestre. Among the notable features of this vegetation are the candelabra-like branching, cruciate leaf arrangement (leaves arranged in fours along the stem and appearing cross-shaped when viewed from above), an aspect reminiscent of heath, and the special beauty of the many vellosias and orchids.

To the northeast toward the Brazilian Highlands, conditions become drier (MAP 500–750 mm), and the vegetation includes a spiny component. This is the caatingas, or chaparral. The arclike configuration formed of caatingas, cerrado, and chaco vegetation from the states of Bahia and Minas Gerais in the Brazilian Highlands and the Mato Grosso Plateau of Brazil in the northeast to the Gran Chaco of northwestern Argentina is the principal dry vegetation of northern South America. Along the rivers there are me-

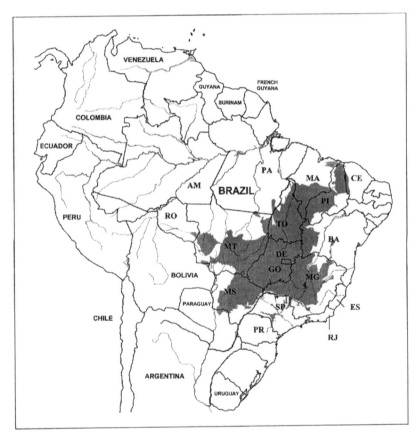

Figure 2.49 Distribution of cerrado vegetation in South America. From Oliveira and Marquis 2002. Copyright © Columbia University Press, New York. Used with permission from the publisher.

Figure 2.50 Cerrado vegetation, Brazil. Photograph courtesy of Professor Guarino R. Colli, Department of Zoology, University of Brasilia.

sic communities available for expansion if future conditions in this region of South America become more moist, just as drier elements were present in the mesic past and coalesced into the present arid plant formations. This dynamic vision of lineages, communities, and environments, extending into the past and changing with environmental fluctuations defines the ecosystem view presented here of the New World and its natural history.

Pantanal

In contrast to desert, shrubland/chaparral (caatingas)–woodland–savanna (cerrado), and Grassland, there is also in South America the world's largest wetland, the Pantanal, Portuguese for "great swamp" (Heckman 1998; Junk and Nues de Cunha 2005). It covers 170,000 km², or about fifteen times the area of the Florida Everglades, and occurs south of the Mato Grosso Plateau from about Cuiabá to the border of Mato Grosso–Mato Grosso do Sul in Brazil (fig. 2.51). It consists of a vast network of rivers, lakes, alluvial plains, and ancient raised deltas. It is flooded by the Paraguay River during October through March, leaving only a few hills standing above a vast sheet of water. Throughout its extent, there are aquatic communities, sedge marshes of *Cyperus giganteus*, swamp forests, flooded grasslands, and drier

Figure 2.51 View of the Pantanal, Nhecolândia, Mato Grosso do Sul, Brazil. From Por 1995.
Used with permission from Springer Science and Business Media, Dordrecht.

forests with chaco/cerrado elements. To the west the Pantanal merges with
the Gran Chaco, and gallery forests along the rivers connect it northward,
through the cerrado, with the Amazon and Atlantic rain forests. This mosaic of interconnected communities illustrates, once again, the potential
for recombining various preadapted elements in response to changes in
climate.

An unusual feature of the Pantanal is the murundu islands. These are
mounds 1–2 m high in parts of the drier but seasonally inundated cerrado
phase of the Pantanal. They are built by termites (*Rotuinditermes braganti-
nus*) or leaf-cutter ants (species of *Atta* and *Acromyrmex*), and the characteristic spacing is due to the territoriality of the insects. The islands' value
to the ecosystem is that they provide wood for the termites, protect root
systems from water logging, and afford dry sites for trees during extended
multiyear flooding.

There are also balseiras, or camalotes, called embalsados in Argentina and batumes in Paraguay. These are floating mats several meters in diameter made up of the floating fern *Salvinia auriculata*, the water hyacinth *Eichornia crassipes*, and the water lettuce *Pistia stratiotes* (a search for *Pistia stratiotes* on, for example, the TROPICOS Web site will yield images, habitat, distribution maps, and descriptions for this widespread tropical plant).

The balseiras may fuse into larger mats, lodging on roots or at river bends and forming even larger floating islands, eventually developing a semipermanent flora and fauna. Occasionally they become dislodged and float down the Paraguay River carrying a serenely unaware jaguar or capybara into the Paraná River and out to the Atlantic.

Also serenely unaware, at least toward the end of his life, was the redoubtable Count George Heinrich von Lanssdorff (1774–1852). The count led expeditions into the Pantanal under the patronage of the czar of Russia; but like many early explorers, he lost his collections, after which he went mad and soon died. Theodore Roosevelt, another picturesque figure, visited the Pantanal in 1913.

The environments of shallow wetlands are notoriously variable, even over short intervals. One of the biological consequences, like unstable habitats on other shifting substrates (e.g., mangroves, coastal dunes), is that they are not suitable for long-term adaptation and speciation by a wide variety of organisms. Thus, they are usually inhabited by a limited number of specialist species capable of coping with unpredictable environments. As noted by Por:

> The present conditions are too fluctuating on a yearly and a secular basis, and they were probably even more so in the longer range of the Pleistocene climatic history. Perhaps with the exception of the isolated old hills [with their caves], there has never been an environmental stability and a definition of isolated areas conductive to in situ speciation. (1995, 39)

Only two endemic plants are prominent here, *Coccoloba cujabensis* (Polygonaceae) and *Mentzelia corumbaensis* (Loasaceae).

The Pantanal plays a critical role in flood abatement and water filtration, and it provides the habitat for an estimated 656 species of birds, 263 species of fish, 162 species of reptiles that include a population of caimans (*Caiman yacare*) of over one million individuals, and 95 species of mammals (46 of which are rare or endangered). These are conservative figures and the region is regarded as a "mother lode of unrecorded life" (Swarts 2000).

It is also true that 98 percent of the land is privately owned, and there is a proposal for a waterway 3440 km long. The Hidrovia is intended to develop agriculture, ranching, mining (diamonds, gold, iron, manganese), and tourism. The area has "woefully inadequate health services, poor schools, inferior communications and transport, and lack of employment opportunities" for the pantaneiros (Swarts 2000, xii). Whether the proclaimed economic benefits of development would trickle down to the workers is open to question. The Pantanal has been the subject of two international conferences to discuss the multifaceted issues that are symbolic of the immensely complex problems of conservation, economic growth, and sustainable development facing many parts of the world. Latin Americans will hopefully have patience with those from other countries who have "developed" their own resources but counsel conservation elsewhere. A reason outside attention is often placed on sustainability in Latin America is because there is still so much of value, and so very much to lose.

: :   **REFERENCES**   : :

Ackroyd, Peter. 2005. *Shakespeare: The biography*. Nan A. Talese / Doubleday, New York.

Aide, T. M., and H. R. Grau. 2004. Globalization, migration, and Latin American ecosystems. *Science* 305:1915–16.

Albert, J. S., N. R. Lovejoy, and W. G. R. Crampton. 2006. Miocene tectonism and the separation of cis- and trans-Andean river basins: Evidence from neotropical fishes. *J. South Am. Earth Sci.* 21:14–27.

Alley, R. B. 2000. *The two-mile time machine: Ice cores, abrupt climate change, and our future*. Princeton University Press, Princeton.

Angier, N. 2009. Panama Canal project opens a tropical window. *New York Times*, 7 July 2009, http://www.nytimes.com/2009/07/07/science/07angier.html?scp=1&sq=Panama%20 Canal%20Project&st=cse.

Antonelli, A., J. A. A. Nylander, C. Persson, and I. Sanmartin. 2009. Tracing the impact of the Andean uplift on neotropical plant evolution. *Proc. Natl. Acad. Sci. U.S.A.* 106 (24), www .pnas.org/cgi/doi/10.1073/pnas.0811421106.

Armesto, J. J., M. T. K. Arroyo, and L. F. Hinojosa. 2007. The Mediterranean environment of central Chile. In *The physical geography of South America*, ed. T. T. Veblen, K. R. Young, and A. R. Orme, 184–99. Oxford University Press, Oxford.

Betancourt, J. L., T. R. Van Devender, and Paul S. Martin. 1990. *Packrat middens: The last 40,000 years of biotic change*. University of Arizona Press, Tucson.

Black, R. 2008. Nature loss "dwarfs bank crisis." BBC News, Science and Environment, 10 October 2008, http://news.bbc.co.uk/2/hi/7662565.stm.

Bo, S., et al. (eight coauthors). 2009. The Gamburtsev Mountains and the origin and early evolution of the Antarctic ice sheet. *Nature* 459:690–93, doi: 10.1038/08024. Published online 4 June 2009.

Bolay, E. 1997. *The Dominican Republic: A country between rain forest and desert*. Margraf Verlag, Weikersheim.

Boorstin, D. J. 1983. *The discovers*. Random House, New York.

Bottke, W. F., D. Vokrouhlicky, and D. Nesvorný. 2007. An asteroid breakup 160 Myr ago as the probable source of the K/T impactor. *Nature* 449:48–53.

Bruce, V. 2001. *No apparent danger: The true story of volcanic disaster at Galeras and Nevada del Ruiz*. HarperCollins Publishers, New York.

Burton, K. W., H.-F. Ling, and R. K. O'Nions. 1997. Closure of the Central American Isthmus and its effect on deep-water formation in the North Atlantic. *Nature* 386:382–85.

Carciumaru, D., and B. Ortega. 2008. Geologic structure of the northern margin of the Chihuahua Trough: Evidence for controlled deformation during Laramide orogeny. *Bol. Soc. Geol. Mex.* 60:43–69.

Chapman, F. M. 1929. *My tropical air castle*. D. Appleton and Co., New York.

———. 1938. *Life in an air castle*. D. Appleton–Century Co., New York.

Cohen, J. M., trans. 1992. *The four voyages of Christopher Columbus*. Penguin Classics, New York.

Cordani, U. G., E. J. Milani, A. T. Filho, and D. A. Campos, eds. 2000. *Tectonic evolution of South America: 31st International Geological Congress, August 6–17, Rio de Janeiro, Brazil*. Geological Society of America, Boulder.

Crisp, M. D., et al. (nine coauthors). 2009. Phylogenetic biome conservation on a global scale. *Nature* 458:754–56.

D'Arrigo, R. D., and G. C. Jacoby. 1999. Northern North American tree-ring evidence for regional temperature changes after major volcanic events. *Climatic Change* 41:1–15. [Notes short-term temperature signals in tree rings from the following volcanic eruptions: Huaynaputina, Peru, 1660; Laki, Iceland, 1783; Tambora, Indonesia, 1815; Coseguina, Nicaragua, 1935.]

De-Nova, J., and V. Sosa. 2007. Phylogeny and generic delimitation of *Adelia* (Euphorbiaceae) inferred from molecular and morphological data. *Taxon* 56:1027–36.

Diamond, J. 2005. *Collapse: How societies chose to fail or succeed*. Penguin Books, New York.

Dilcher, D. L. 1973. A paleoclimatic interpretation of the Eocene floras of southeastern North America. In *Vegetation and Vegetational History of Northern Latin America*, ed. A. Graham, 39–60. Elsevier Science Publishers, Amsterdam.

Donnelly, T. W. 1989. History of marine barriers and terrestrial connections: Caribbean paleogeographic inference from pelagic sediment analysis. In *Biogeography of the West Indies: Past, present, and future*, ed. C. A. Woods, 103–18. Sand Hill Crane Press, Gainesville.

Donnelly, T. W., G. S. Horne, R. C. Finch, and E. López-Ramos. 1990. Northern Central America: The Maya and Chortis blocks. In *The geology of North America*, vol. H, *The Caribbean region*, ed. G. Dengo and J. E. Case, 37–76. Geological Society of America, Boulder.

Donoghue, M. J. 2008. A phylogenetic perspective on the distribution of plant diversity. *Proc. Natl. Acad. Sci. U.S.A.* 105:1549–55.

Eldrett, J. S., D. R. Greenwood, I. C. Harding, and M. Huber. 2009. Increased seasonality through the Eocene to Oligocene transition in northern high latitudes. *Nature* 459:969–73, doi: 10.1038/08069. Published online 18 June 2009.

Esper, J., E. R. Cook, and F. H. Schweingruber. 2002. Low-frequency signals in long tree-ring chronologies for reconstructing past temperature variability. *Science* 295:2250–53.

Espinosa, A. F., ed. 1976. *The Guatemalan earthquake of February 4, 1976: A preliminary report.* U.S. Geological Survey Professional Paper 1002. U.S. Department of the Interior, Washington, D.C.

Fagan, B. 2000. *The Little Ice Age: How climate made history, 1300–1850.* Basic Books, New York.

Farabee, M. J. 1990. Triprojectate fossil pollen genera. *Rev. Palaeob. Palyn.* 65:341–47.

Fittkau, E. J., J. Illies, H. Klinge, G. H. Schwabe, and H. Sioli, eds. 1969. *Biogeography and ecology in South America.* 2 vols. Dr. W. Junk, N. V., Publishers, the Hague.

Flannery, T. 2005. *The weather makers: How man is changing the climate and what it means for life on Earth.* Atlantic Monthly Press, New York.

From words to action. 2007. Special report, *Nature* 445:578–79. [J. Giles commentary on the IPCC 2007 report on global warming.]

Furley, P. A. 2007. Tropical forests of the lowlands. In *The physical geography of South America*, ed. T. T. Veblen, K. R. Young, and R. Orme, 135–57. Oxford University Press, Oxford.

Galindo-Leal, C., and I. Gusmão Câmara, eds. 2003. *The Atlantic forest of South America: Biodiversity status, threats, and outlook.* Island Press, Washington, D.C.

Garzione, C. N., et al. (seven coauthors). 2008. Rise of the Andes. *Science* 320:1304–7.

Gentry, A. H. 1982. Patterns of neotropical plant species diversity. In *Evolutionary biology*, vol. 15, ed. W. Hecht, B. Wallace, and G. T. Prance, 1–84. Plenum Publishing, New York.

———. 1986. Species richness and floristic composition of Chocó region plant communities. *Caldasia* 15:71–91.

Hansen, M. C., et al. (eleven coauthors). 2008. Humid tropical forest clearing from 2000 to 2005 quantified by using multitemporal and multiresolution remotely sensed data. *Proc. Natl. Acad. Sci. U.S.A.* 105:9439–44.

Heckenberger, M. J. 2009. Lost cities of the Amazon. *Scientific American*, October, 64–71.

Heckenberger, M. J., et al. (seven coauthors). 2008. Pre-Columbian urbanism, anthropogenic landscapes, and the future of the Amazon. *Science* 321:1214–17.

Heckman, C. W. 1998. *The Pantanal of Poconé: Biota and ecology in the northern section of the world's largest pristine wetland.* Kluwer Academic Publishers, Dordrecht.

Heinicke, M. P., W. E. Duellman, and S. B. Hedges. 2007. Major Caribbean and Central American frog faunas originated by ancient oceanic dispersal. *Proc. Natl. Acad. Sci. U.S.A.* 104:10092–97.

Hoernle, K., et al. (six coauthors). 2002. Missing history (16–71 Ma) of the Galapagos hotspot: Implications for the tectonic and biological evolution of the Americas. *Geology* 30:795–98.

Hoorn, C. 2006. The birth of the mighty Amazon. *Scientific American*, May, 52–59.

Hoorn, C., and H. Vonhof, eds. 2006. New Contributions on Neogene geography and depositional environments in Amazonia. *South Am. Earth Sci.* 21:1–172.

Hubbell, S. P., et al. (five coauthors). 2008. How many tree species are there in the Amazon and how many of them will go extinct? *Proc. Natl. Acad. Sci. U.S.A.* 105:11498–504.

Huber, O. 1995. Vegetation. In *Flora of the Venezuelan Guayana*, vol. 1, *Introduction*, ed. J. A. Steyermark, P. E. Berry, and B. K. Holst. Missouri Botanical Garden Press, St. Louis / Timber Press, Portland, Ore.

Humboldt, A. von. [1852] 1971. *Personal narrative of travels to the equinoctal regions of the Americas.* Trans. Tomasina Ross. Reprint ed., Blom, New York.

Ickert-Bond, S. M., and M. F. Wojciechowski. 2004. Phylogenetic relationships in *Ephe-*

dra (Gnetales): Evidence from nuclear and chloroplast DNA sequence data. *Syst. Bot.* 29:834–49.

IPCC (Intergovernmental Panel on Climate Change). 2007. Climate Change 2007: World Meteorological Organization / United Nations Environment Programme, Geneva, Switzerland. [See Summary for policymakers.]

Junk, W. J., and C. Nues de Cunha. 2005. Pantanal: A large South American wetland at a crossroads. *Ecological Engineering* 24:391–401.

Keen, B., trans. 1992. *The Life of the Admiral Christopher Columbus by his son Ferdinand.* Rutgers University Press, New Brunswick.

Keller, G., S. Abramovich, Z. Berner, and T. Adatte. 2009. Biotic effects of the Chicxulub impact, K-T catastrophe and sea level change in Texas. *Palaeogeogr. Palaeocl. Palaeoecol.* 271:52–68.

Kier, G., et al. (seven coauthors). 2009. A global assessment of endemism and species richness across island and mainland regions. *Proc. Natl. Acad. Sci. U.S.A.*, 106 (23), www.pnas.org/cgi/doi/10.1073/pnas.0810306106.

Knoll, A. H. 2004. *Life on a young planet: The first three billion years of evolution on earth.* Princeton University Press, Princeton.

Krajick, K. 2004. All downhill from here? *Science* 303:1600–1602.

Lamb, S. 2004. *Devil in the mountain: A search for the origin of the Andes.* Princeton University Press, Princeton.

Lamb, S., and P. Davis. 2003. Cenozoic climate change as a possible cause for the rise of the Andes. *Nature* 425:792–97.

Leggitt, V. L., and R. A. Cushman, Jr. 2001. Complex caddisfly-dominated bioherms from the Eocene Green River Formation. *Sediment. Geol.* 145:377–96.

Lovejoy, N. R., J. S. Albert, and W. G. R. Crampton. 2006. Miocene marine incursions and marine/freshwater transitions: Evidence from neotropical fishes. *J. South Am. Earth Sci.* 21:5–13.

Lunt, D. J., P. J. Valdes, A. Haywood, and I. C. Rutt. 2008. Closure of the Panama seaway during the Pliocene: Implications for climate and Northern Hemisphere glaciation. *Clim. Dym.* 30:1–18.

Lyle, M., K. A. Dadey, and J. W. Farell. 1995. The late Miocene (11–8 MA) eastern Pacific carbonate crash: Evidence for reorganization of deep-water circulation by the closure of the Panama gateway. *Proc. ODP (Ocean Drilling Program), Scientific Results* 138:821–38.

Malfait, B. T., and M. G. Dinkelman. 1972. Circum-Caribbean tectonic and igneous activity and the evolution of the Caribbean plate. *Geol. Soc. Am. Bull.* 83:251–72.

Mann, P., ed. 1999. *Sedimentary basins of the world.* Vol. 4, *Caribbean basins.* Elsevier Science Publishers, Amsterdam.

McCullough, D. 1977. *The path between the seas: The creation of the Panama Canal.* Touchstone, Simon and Schuster, New York.

McEwan, C., L. A. Borrero, and A. Prieto, eds. 1997. *Patagonia: Natural history, prehistory, and ethnography at the uttermost end of the earth.* Princeton University Press, Princeton.

McPhee, J. 1981. *Basin and Range.* Farrar, Straus, and Giroux, New York.

Minter, J. E. 1948. *The Chagres: River of westward passage.* Reinhart and Co., New York.

Morellato, L. P. C., and C. F. B. Haddad, eds. 2000. *The Brazilian Atlantic forest.* Special issue, *Biotropica* 32:786–956.

Morley, R. J. 2000. *Origin and evolution of tropical rain forests.* John Wiley and Sons, New York.

Mulch, A., S. A. Graham, and C. P. Chamberlain. 2006. Hydrogen isotopes in Eocene river gravels and paleoelevation of the Sierra Nevada. *Science* 313:87–89.

Myers, N. 1991. Four for the forest's future. *BioScience* 41:511–12.

Newkirk, D. R., and E. E. Martin. 2009. Circulation through the Central American seaway during the Miocene carbonate crash. *Geology* 37:87–90.

Nisancioglu, K. H., M. F. Raymo, and P. H. Stone. 2003. Reorganization of Miocene deep water circulation in response to the shoaling of the Central American seaway. *Paleocean-ography* 18, no.1: 1006, doi:10.1029/2002PA000767. Published online 11 February 2003.

Oliveira, P. S., and R. J. Marquis, eds. 2002. *The cerrados of Brazil: Ecology and natural history of a neotropical savanna.* Columbia University Press, New York.

Oppenheimer, C. 2002. Limited global change due to the largest known Quaternary eruption, Toba ~74 kyr BP? *Quat. Sci. Rev.* 21:1593–1609.

Orme, A. R. 2007a. The tectonic framework of South America. In *The physical geography of South America*, ed. T. T. Veblen, K. R. Young, and A. R. Orme, 3–22. Oxford University Press, Oxford.

———. 2007b. Tectonism, climate, and landscape change. In *The physical geography of South America*, ed. T. T. Veblen, K. R. Young, and A. R. Orme, 23–44. Oxford University Press, Oxford.

Oxford atlas of the world. 1997. 5th ed. Oxford University Press, Oxford.

Pindell, J. L. 1994. Evolution of the Gulf of Mexico and the Caribbean. In *Caribbean geology: An introduction*, ed. S. K. Donovan and T. A. Jackson, 13–39. The University of the West Indies Publishers' Association (UWIPA), Mona, Jamaica.

Pindell, J. L., and S. F. Barrett. 1990. Geological evolution of the Caribbean region: A plate-tectonic perspective. In *The geology of North America*, vol. H, *The Caribbean region*, ed. G. Dengo and J. E. Case, 405–32. Geological Society of America, Boulder.

Por, F. D. 1992. *Sooretama: The Atlantic rain forest of Brazil.* SPB Academic Publishing, The Hague.

———. 1995. *The Pantanal of Mato Grosso (Brazil): World's largest wetlands.* Kluwer Academic Publishers, Dordrecht.

Richards, P. W. 1996. *The tropical rain forest: An ecological study.* With contributions by R. P. D. Walsh, I. C. Baillie, and P. Greig-Smith. Cambridge University Press, Cambridge.

Rosenberg, T. 1991. *Children of Cain: Violence and the violent in Latin America.* William Morrow and Co., New York.

Roth, J. M., A. W. Droxler, and K. Kameo. 2000. The Caribbean carbonate crash at the middle to late Miocene transition: Linkage to the establishment of the Miocene global ocean conveyor. *Proc. ODP, Scientific Results* 165:249–73.

Ruiz, E., et al. (eight coauthors). 2007. Genetic diversity and differentiation within and among Chilean populations of *Araucaria araucaria* (Araucariaceae) based on allozyme variability. *Taxon* 56:1221–28.

Rzedowski, J. 1978. *Vegetación de Mexico.* Editorial Limusa, Mexico City.

Sano, S. M., S. Pedrosa de Almeida, and J. F. Ribeiro, eds. 2008. *Cerrado: Ecologia e flora.* 2 vols. Embrapa, Brasilia.

Santiago-Valentín, E., and J. Francisco-Ortega, eds. 2008. *Caribbean Biodiversity.* Special issue, *Bot. Rev.* 74:1–207.

Schneider, B., and A. Schmittner. 2006. Simulating the impact of the Panamanian seaway closure on ocean circulation, marine productivity, and nutrient cycling. *Earth Planet. Sci. Lett.* 246:367–80.

Sioli, H. 1984. Introduction: History of the discovery of the Amazon and of research of Amazonian waters and landscapes. In *The Amazon: Limnology and landscape ecology of a mighty tropical river and its basin*, ed. H. Sioli, 1–13. Dr. W. Junk Publishers, Dordrecht.

Steyermark, J. A., P. E. Berry, and B. K. Holst, eds. 1995. *Flora of the Venezuelan Guayana*. Vol. 1, *Introduction*. Missouri Botanical Garden Press, St. Louis / Timber Press, Portland, Ore.

Swarts, F. A., ed. 2000. *The Pantanal: Understanding and preserving the world's largest wetland*. Paragon House, St. Paul.

Tremlin, T. 2006. *Minds and gods: The cognitive foundations of religion*. Oxford University Press, Oxford.

Tryon, A. F., and B. Lugardon. 1991. *Spores of the Pteridophyta: Surface, wall structure, and diversity based on electron microscope studies*. Springer Verlag, Berlin.

Tryon, R. M., and A. F. Tryon. 1982. *Ferns and allied plants with special reference to tropical America*. Springer Verlag, Berlin.

Tschudy, R. H. 1975. *Normapolles pollen from the Mississippi embayment*. U.S. Geological Survey Professional Paper 865. U.S. Department of the Interior, Washington, D.C.

Tyler, S. A., and E. S. Barghoorn. 1954. Occurrence of structurally preserved plants in Pre-Cambrian rocks of the Canadian Shield. *Science* 119:606–8.

Van der Hammen, T., and A. M. Cleef. 1986. Development of the High Andean páramo flora and vegetation. In *High altitude tropical biogeography*, ed. F. Vuilleumier and M. Monasterio, 153–201. Oxford University Press, Oxford.

Veblen, T. T. 2007. Temperate forests of the southern Andean region. In *The physical geography of South America*, ed. T. T. Veblen, K. R. Young, and A. R. Orme, 217–31. Oxford University Press, Oxford.

Veblen, T. T., K. R. Young, and A. R. Orme, eds. 2007. *The Physical geography of South America*. Oxford University Press, Oxford.

Von Hagen, V. W. 1945. *South America called them*. Alfred A. Knopf, New York.

Vuilleumier, F., and M. Monasterio, eds. 1986. *High altitude tropical biogeography*. Oxford University Press, Oxford.

Wagner, P. L. 1964. Natural vegetation of Middle America. In *Handbook of Middle American Indians*, vol. 1, *Natural environment and early cultures*, ed. R. C. West, 216–64. University of Texas Press, Austin.

Weber, H. 1969. Zur natürlichen Vegetations-gliederung von Südamerika. In *Biogeography and ecology in South America*, ed. E. J. Fittkau, J. Illies, H. Klinge, G. H. Schwabe, and H. Sioli, 475–518. Dr. W. Junk, Publishers, The Hague.

Williams, S., and F. Montaigne. 2001. *Surviving Galeras*. Houghton Mifflin, Boston.

Woodring, W. P. 1966. The Panama land bridge as a sea barrier. *Proc. Am. Phil. Soc.* 110:425–33.

Young, K. R., P. E. Berry, and T. T. Veblen. 2007. Flora and vegetation. In *The physical geography of South America*, ed. T. T. Veblen, K. R. Young, and A. R. Orme, 91–100. Oxford University Press, Oxford.

Zebrowski, E. Jr. 2002. *The last days of St. Pierre: The volcanic disaster that claimed thirty thousand lives*. Rutgers University Press, New Brunswick.

Additional Readings and Updates

Allaby, A., and M. Allaby. 1999. *A Dictionary of earth sciences*. 2nd ed. Oxford University Press, Oxford.

Amazon garden city. 2008. *Economist*, 28 August 2008. ["The world's favourite rainforest once had (pre-Columbian) towns in it." Available to subscribers on the online edition, http://www.economist.com/.]

Avise, J. C. 2009. Phylogeography: Retrospect and prospect. J. *Biogeogr.* 36:3–15.

Barbour, M. G., and W. D. Billings, eds. 2000. *North American terrestrial vegetation*. 2nd ed. Cambridge University Press, Cambridge.

Barrionuevo, A. 2008. With guns and fines, Brazil takes on loggers. *New York Times*, 19 April 2008.

Bermingham, E., and A. P. Martin. 1998. Comparative mtDNA phylogeography of neotropical fresh water fishes: Testing shared history to infer the evolutionary landscape of lower Central America. *Mol. Ecol.* 7:499–518.

Blakey, Ron. Paleogeography and Geologic Evolution of North America. http://jan.ucc.nau.edu/~rcb7/nam.html. [Website for paleogeographic maps.]

Bortolus, A. 2008. Error cascades in the biological sciences: The unwanted consequences of using bad taxonomy in ecology. *Ambio* (Royal Swedish Academy of Sciences) 37:114–18.

Botanic Gardens Conservation International. Climate Change Information Center. http://www.bgci.org/climate/.

Broecker, W. S., and R. Kunzig. 2008. *Fixing climate: What past climate changes reveal about the current threat—and how to counter it*. Hill and Wang, New York. [See review by Chris Turney, *Nature* 453 (2008): 158.]

Campbell, D. 2008. Don't forget people and specimens that make the database. *Nature* 455:590.

Cantú, C., R. G. Wright, J. M. Scott, and E. Strand. 2004. Assessment of current and proposed nature reserves of Mexico based on their capacity of protect geophysical features and biodiversity. *Biol. Conserv.* 115:411–17.

Carmona, A. 2007. A conversation on conservation: Contemplating the impact of climate change in the Latin America-Caribbean region. Council on Hemispheric Affairs. http://www.coha.org. [Web site can be browsed by country and by year.]

Chapman, A. D. 2005. *Numbers of living species in Australia and the world*. Report for the Australian Government, Department of the Environment and Heritage, Canberra.

Chazdon, R. L., et al. (ten coauthors). 2008. Beyond reserves: A research agenda for conserving biodiversity in human-modified tropical landscapes. *Biotropica* 41:142–53.

Cole, S. 2007. Long-term increase in rainfall seen in tropics. NASA Goddard Space Flight Center. http://www. nasa.gov/centers/goddard/news/topstory/2007/rainfall increase .html.

Coll-Hurtado, A., coord. 2007. *Nuevo Atlas Nacional de México*. Universidad Nacional Autónoma de México, Instituto de Geografía, Mexico City.

Cooley, S. R., V. J. Coles, A. Subramaniam, and P. L. Yager. 2007. Seasonal variations in the Amazon plume-related atmospheric carbon sink. *Global Biogeochem. Cycles* 21, GB3014, doi: 10.1029/2006gb002831.

Craeynest, L. 2007. 7 July 2007—Half way to the Millennium Development Goals. World Wildlife Fund. http://www.wwf.org.uk/filelibrary/pdf/mdg7.pdf.

Cressey, D. 2007. Arctic melt opens Northwest Passage. *Nature* 449:267.

Crisp, M. D., et al. (nine coauthors). 2009. Phylogenetic biome conservation on a global scale. *Nature* 458:754–56.

Cyranoski, D. 2007. Ocean drilling: In the zone. *Nature* 449:278–80. ["The world's biggest, best equipped research drilling vessel is about to set off on its first scientific voyage . . . to catch a formidable earthquake in the act."]

Dahdoug-Guebas, F., and N. Koedam, eds. 2008. *Mangrove ecology—applications in forestry and coastal zone management.* Special issue, *Aquat. Bot.* 89:77–274.

Dalton, R. 2007. Time traps. *Nature* 449:20–21. ["The whole world felt the effects of the dinosaur-killing mass extinction 65 million years ago. But a spot in Colorado may have the best record of it."]

———. 2009. NOAA chief ready to tackle climate. *Nature* 458:396.

Daly, D. C., and J. D. Mitchell. 2000. Lowland vegetation of tropical South America—an overview. In: Imperfect balance: Landscape transformations in the pre-Colombian Americas, ed. D. Lentz, 391–454. Columbia University Press, New York.

De Aguiar, M. A. M., M. Baranger, E. M. Baptestini, L. Kaufman, and Y. Bar-yam. 2009. Global patterns of speciation and diversity. *Nature* 460:384–87.

Denk, T., and G. W. Grimm. 2009. Significance of pollen characteristics for infrageneric classification and phylogeny in *Quercus* (Fagaceae). *Int. J. Plant Sci.* 170:926–40.

De Terra, H. 1955. *Humboldt.* Alfred A. Knopf, New York.

Diversitas: An International Programme of Biodiversity Science. http://www.diversitas-international.org.

Droxler, A. W., et al. (seven coauthors). 1998. Caribbean constraints on circulation between Atlantic and Pacific oceans over the past 40 million years. In *Tectonic boundary conditions for climate reconstructions,* ed. T. J. Crowley and K. C. Burke, 160–91. Oxford University Press, Oxford.

Duque-Caro, H. 1990. Neogene stratigraphy, paleoceanography and paleobiogeography in northwest South America and the evolution of the Panama seaway. *Palaeogeogr. Palaeocl. Palaeoecol.* 77:203–34.

Eberhart-Phillips, D., et al. (twenty-eight coauthors). 2003. The 2002 Denali Fault earthquake, Alaska: A large magnitude, slip-partitioned event. *Science* 300:1113–18. ["The 7.9 event on 3 November 2002 was the largest strike-slip earthquake in North America in almost 150 years."]

Encyclopedia of Life. http://www.eol.org.

Fagan, B. 1999. *Floods, famines, and emperors: El Niño and the fate of civilizations.* Basic Books, New York.

———. 2004. *The long summer: How climate changed civilization.* Basic Books, New York.

FAPESP (The State of São Paulo Research Foundation). The BIOTA-FAPESP Program. http://www.fapesp.br/en/materia/4662. [The aim of BIOTA-FAPESP Program is to study the biodiversity of the state of São Paulo. Among the projects is to put Martius's forty-volume *Flora Brasiliensis* online.]

Farr, K. 2003. *The forests of Canada.* Photographs by J. D. Andrews. Fizhenry and Whiteside, Markham, Ont.

Ferraro, P. J., and A. Kiss. 2002. Direct payments to conserve biodiversity. *Science* 298:1718–19.

Finer, M., C. N. Jenkins, S. L. Pimm, B. Keane, and C. Ross. 2008. Oil and gas projects in the western Amazon: Threats to wilderness, biodiversity, and indigenous peoples. *PLoS ONE* 3 (8): e2932, doi: 101371/journal.pone.0002932.

First antenna switches on in the Atacama. 2009. *Nature* 457:16.

Fortey, R. 2004. *Earth: An intimate history*. Alfred A. Knopf, New York.

Foster, D. R., and J. D. Aber, eds. 2004. *Forests in time: The environmental consequences of 1,000 years of change in New England*. Yale University Press, New Haven.

Francisco-Ortega, J., et al. (six coauthors). 2007. Seed plant genera endemic to the Caribbean island biodiversity hotspot: A review and a molecular phylogenetic perspective. *Bot. Rev.* 73:183–234.

Gaston, K. J., S. F. Jackson, L. Cantú-Salazar, and G. Cruz-Piñon. 2008. The ecological performance of protected areas. *Annu. Rev. Ecol. Evol. System.* 39:93–113.

Gottsberger, G., and I. Silberbauer-Gottsberger. 2006. *Life in the cerrado: A South American tropical seasonal ecosystem*. 2 vols. Reta Verlag, Ulm.

Goulding, M., R. Berthem, and E. Ferreira. 2003. *The Smithsonian Atlas of the Amazon*. With cartography by R. Duenas. Smithsonian Books, Washington, D.C.

Governments fail to reduce global biodiversity decline. 2009. *Nature* 460:163.

Gut, B. J. 2008. *Trees in Patagonia*. Birkhäuser, Basel.

Harvey, C. A., and J. C. Saénz, eds. 2008. *Evaluación y conservación de biodiversidad en paisajes fragmentados de Mesoamerica*. INBio, Heredia, Costa Rica.

Harvard University Library Open Collections Program. Expeditions and Discoveries: Sponsored Exploration and Scientific Discovery in the Modern Age. http://ocp.hul.harvard .edu/expeditions. [Web site includes digitized copies of more than 250,000 pages from 700 books and serials, 50,000 pages from Harvard's manuscript collections, more than 1200 photographs, 200 maps, 21 atlases, and numerous drawings and prints.]

Havstad, K. M., L. F. Huenneke, and W. H. Schlesinger, eds. 2006. *Structure and function of a Chihuahuan desert ecosystem: The Jornada Basin Long-Term Ecological Research Site*. Oxford University Press, Oxford.

Hildebrand, A. R., M. Pilkington, M. Connors, C. Ortiz-Aleman, and R. E. Chavez. 1995. Size and structure of the Chicxulub crater revealed by horizontal gravity gradients and cenotes. *Nature* 376:415–17.

Hopkin, M. 2008. Biodiversity: Frozen futures. *Nature* 452:404–5. House of Lords, Science and Technology Committee. 2008. *Systematics and taxonomy: Follow-up*. 5th Report of Session 2007–08. The Stationary Office, London.

The International Compositae Alliance. http://www.compositae.org.

Huber, M., and R. Caballero. 2003. Eocene El Nino: Evidence for robust tropical dynamics in the "hothouse." *Science* 299:877–81.

Jarvis, C. 2007. *Order out of chaos: Linnaean plant names and their types*. The Linnean Society of London, London. [See the review by P. S. Soltis, *Nature* 448 (2007): 868–60.]

Joppa, L. N., S. R. Loarie, and S. L. Pimm. 2008. On the protection of "protected areas." *Proc. Natl. Acad. Sci. U.S.A.* 105:6673–78.

Kaiser, J. 2003. Panel suggests a different shade of NEON. *Science* 301:828. [National Ecological Observatory Network.]

Keaten, J. 2007. Data show Arctic ice melt-off. Associated Press report, 16 September.

Kerr, R. A. 2007. Global warming coming home to roost in the American West. *Science* 318:1859. ["Assigning blame for regional climate disasters is hard, but scientists have finally implicated the greenhouse in a looming water crisis."]

———. 2008. The Andes popped up by losing their deep load. *Science* 320:1275.

Kohler, T. A., M. D. Varien, A. M. Wright, and K. A. Kuckelman. 2008. Mesa Verde migrations. *Am. Sci.* 96:146–53. [New archaeological research and computer simulations suggest why ancient Puebloans deserted the northern part of the southwest United States.]

Kolbert, E. 2009. The catastrophist. *New Yorker*, 29 June 2009. ["NASA's climate expert (James Hansen) delivers the news no one wants to hear."]

Körner, C. 2007. The use of "altitude" in ecological research. *Trends Ecol. Evol.* 22:569–574.

Loarie, S. R., L. N. Joppa, and S. L. Pimm. 2007. Letters: Satellites miss environmental priorities. *Trends Ecol. Evol.* 22:630–32.

Loarie, S. R., et al. (six coauthors). 2008. Climate change and the future of California's endemic flora. *PLoS ONE* 3 (6): e2502, doi: 10.1371/journal.pone.0002502.

Lozier, J. D., P. Aniello, and M. J. Hickerson. 2009. Predicting the distribution of Sasquatch in western North America: Anything goes with ecological niche modelling. *J. Biogeogr.* 36:1623–27, doi: 10.1111/j.1365-2699.2009.02152.x.

Lyall, S. 2007. Warming revives flora and fauna in Greenland. *New York Times*, 28 October 2007.

Malhi, Y., R. Betts, and T. Roberts, comp. 2008. *Climate change and the fate of the Amazon.* Special issue, *Phil. Trans. R. Soc. B* 363.

Manchester, S. R., and B. H. Tiffney. 2001. Integration of paleobotanical and neobotanical data in the assessment of phytogeographic history of Holarctic angiosperm clades. *Int. J. Plant Sci.* 162 (suppl.): 519–27.

Mann, C. C. 2008. Ancient earthmovers of the Amazon. *Science* 321:1148–52.

Markets can save forests. 2008. Editorial. *Nature* 452:127–28.

Meltdown. 2007. *Time*, vol. 170, October 15. [M. Jarraud, Secretary General, World Meteorological Organization: "It's very, very alarming. It's a warning signal."]

Message in a bubble. 2008. *Nature* 253. [Topic of issue; articles on global warming, effects of deforestation on climate, records from the Antarctic ice core, Quaternary CO_2 and methane concentrations, Eocene methane emission.]

Minobe, S., A. Kuwano-Yoshida, N. Komori, S.-P. Xie, and J. Small. 2008. Influence of the Gulf Stream on the trophosphere. *Nature* 452:206–209.

Morawetz, W., and C. Raedig. 2007. Angiosperm biodiversity, endemism and conservation in the neotropics. *Taxon* 56:1245–54.

Morehead, A. 1969. *Darwin and the Beagle.* Harper and Row, New York.

Mummenhoff, K., and A. Franzke. 2007. Gone with the bird: Late Tertiary and Quaternary intercontinental long-distance dispersal and allopolidization in plants. *Systematics and Biodiversity* 5:255–60.

Murphy, J. B., et al. (seven coauthors, the Rheic Ocean Team). 2008. Tectonic plates come apart at the seams. *Am. Sci.* 96:129–37.

Myster, R. W. 2009. Plant communities of western Amazonia. *Bot. Rev.* 75:271–91, doi: 10.1007/s12229-009-9032-1. Published online 23 June 2009.

National Botanic Garden of Cuba. Bissea. http://groups.google.com/group/bissea/files?hl= en&pli=1. [Plant conservation newsletter.]

Nature Conservancy. 2009. Colombia. *Nature Conservancy*, summer 2009, 22–23 ["In October, the Nature Conservancy purchased more than 3,000 acres in Colombia's Sierra Nevada de Santa Marta, a landscape that includes tropical forests, high-altitude grasslands called *páramos*, and the snow-capped peak of the tallest coastal mountains in the world.

The Conservancy donated the land to indigenous communities to expand their ancestral territory."]

Nevle, R. J. and D. K. Bird. 2008. Effects of syn-pandemic fire reduction and reforestation in the tropical Americas on atmospheric CO_2 during European conquest. *Palaeogeogr. Palaeocl. Palaeoecol.* 264:25–38.

New, M., M. Hulme, and P. Jones. 1999. Representing twentieth-century space-time climate variability. Part 1, Development of a 1961–90 mean monthly terrestrial climatology. *J. Climate* 12:829–56.

Nogués-Bravo, D., M. B. Araújo, T. Romdai, and C. Rahbek. 2008. Scale effects and human impact on the elevational species richness gradients. *Nature* 453:216–19.

Novack, D. 2008. *Burning the future: Coal in America*. Sundance Channel film, released on DVD 13 May 2008. [See review by Emma Marris, *Nature* 453 (2008): 158.]

O'Connell, M. 2008. *Mountain top removal*. Film Arts Theater, Ashville, North Carolina, 15 May 2008; released on DVD December. [See review by Emma Marris, *Nature* 453 (2008): 158.]

Ontario acts to protect its boreal forests. 2008. *Nature* 454:381.

Orr, D. W. 2009. *Down to the wire: Confronting climate collapse*. Oxford University Press, Oxford.

Pagli, C., and F. Sigmundsson. 2008. Will present day glacial retreat increase volcanic activity? Stress induced by recent glacial retreat and its effect on magmatism at the Vatnajökull ice cap, Iceland. *Geophys. Res. Lett.* 35:L09304.

Parker, T. 2003. *Manual of dendrology, Jamaica*. Tree Press, Antigua, Guatemala.

———. 2009. *The trees of Guatemala*. Tree Press, Antigua, Guatemala.

Pennington, R. T., G. P. Lewis, and J. A. Ratter. 2006. *Neotropical savannas and seasonally dry forests: Plant diversity, biogeography, and conservation*. CRC Press, Taylor and Francis Group, Boca Raton.

Perret, M., A. Chautems, R. Spichiger, T. G. Barraclough, and V. Savolainen. 2007. The geographical pattern of speciation and floral diversification in the neotropics: The Tribe Sinningieae (Gesneriaceae) as a case study. *Evolution* 61:1641–60.

Petit, R. J., F. S. Hu, and C. W. Dick. 2008. Forests of the past: A window to future changes. *Science* 320:1450–52.

The Phylodiversity Network. http://www.phylodiversity.net/. [A permanent home for biodiversity, systematics, phylogenetics, and bioinformatics research.]

Powell, J. L. 2005. *Grand Canyon: Solving Earth's grandest puzzle*. Pi Press, New York. [See review by John Schmidt, *Science* 309 (2005): 1818–19.]

Prairie in the city. 2009. *Economist*, 8 April 2009. ["An unexpected survival in St. Louis." Available to subscribers on the online edition, http://www.economist.com/.]

Prange, M., and M. Schulz, 2004. A coastal upwelling seesaw in the Atlantic Ocean as a result of the closure of the Central American seaway. *Geophys. Res. Lett.* 31:L17207, doi: 10.1029/2004GL020073.

Ranker, T. A., and C. H. Haufler, eds. 2008. *Biology and evolution of ferns and lycophytes*. Cambridge University Press, Cambridge.

Raven, P. H. 2006. Honoring the past–building for the future. Remarks of Peter Raven during the BSA [Botanical Society of America] Strategic Planning Discussion. *Plant Sci. Bull.* 52:120–23.

Ricklefs, R. E. 2006. Evolutionary diversification and the origin of the diversity-environment relationship. *Ecology* 87 (suppl.): S3–S13.

Robertson, D. S. , M. C. McKenna, O. B. Toon, S. Hope, and J. A. Lillegraven. 2004. Survival in the first hours of the Cenozoic. *Geol. Soc. Am. Bull.* 116:760–68.

Rudwick, M. J. S. 2005. *Bursting the limits of time: The reconstruction of geohistory in the age of revolution.* University of Chicago Press, Chicago. IL. [See review by Naomi Oreskes, *Science* 314 (2006): 596–97.]

Santos, C., C. Jaramillo, G. Bayona, M. Rueda, and V. Torres. 2008. Late Eocene marine incursion in north-western South America. *Palaeogeogr. Palaeocl. Palaeoecol.* 264:140–46.

Schiermeier, Q. 2007. The new face of the Arctic. *Nature* 446:133–35.

———. 2007. No solar hiding place for greenhouse skeptics. *Nature* 448:8–9.

Schneider, H., A. R. Smith, and K. M. Pryer. 2009. Is morphology really at odds with molecules in estimating fern phylogeny? *Syst. Bot.* 34:455–75.

Schuettpelz, E., and K. M. Pryer. 2007. Fern phylogeny inferred from 400 leptosporangiate species and three plastid genes. *Taxon* 56:1037–50.

Schmidt, Diane. International Field Guides. University Library, University of Illinois at Urbana Champaign. http://www.library.illinois.edu/bix/fieldguides/main.htm. [From the Web site launch announcement: "Diane Schmidt, the biology librarian at the University of Illinois Library, has built and launched the most complete database of field guides to date."]

Smith, S. Y., and R. A. Stockey. 2007. Pollen morphology and ultrastructure of Saururaceae. *Grana* 46:250–67.

Smithsonian MNH, Department of Botany. Georeferencing plants of the Guiana Shield. Smithsonian National Museum of Natural History. http://botany.si.edu/bdg/georeferencing .cfm.

———. Biological diversity of the Guiana Shield. Smithsonian National Museum of Natural History. http://www.mnh.si.edu/biodiversity/bdg. [Site viewable on Google Map or Goggle Earth.]

Stork, N. E. 2007. Biodiversity: World of insects *Nature* 118:657–58. ["When it comes to understanding patterns of biodiversity, ours is a little-known planet. Large-scale sampling projects, as carried out in two investigations of insect diversity, show a way forward."]

Streck, C., R. O'Sullivan, T. Janson-Smith, and R. G. Tarasofsky, eds. 2008. *Climate change and forests: Emerging policy and market opportunities.* Chatham House, London. [See review by M. Obersteiner, *Nature* 458 (2009): 151.]

Stuessy, T. F. 2008. *Plant taxonomy.* Columbia University Press, New York.

Sugden, A., J. Smith, and E. Pennisi. 2008. Introduction to *The future of forests.* Special issue, *Science* 320:1435.

Thompson, A. 2007. Ocean science: Carbon consumers. *Nature Geoscience* 1 (1 September), doi: 10.1038/ngeo.2007.11. Published online 13 September 2007, http://www.nature .com/ngeo/reshigh/2007/0907/full/ngeo.2007.11.html.

Tollefson, J. 2008a. All eyes on the Amazon. *Nature* 452:137.

———. 2008b. Brazil goes to war against logging. *Nature* 452:134–35.

———. 2008c. Climate war games. *Nature* 454:673. ["Role-play negotiations test the outcomes of global warming."]

———. 2008d. Save the trees. *Nature* 452:8–9.

————. 2008e. Think tank reveals plan to manage tropical forests. *Nature* 454:373. ["Novel way to use carbon credits to save trees."]

Upchurch, P. 2007. Gondwanan break-up: Legacies of a lost world. *Trends Ecol. Evol.* 23:229–36.

USDA Natural Resources Conservation Service. Plants Database. U.S. Department of Agriculture. http://www.plants.usda.gov. [Provides information about and images if the plants of the U.S. and its territories.]

Vajda, V., and S. McLoughlin. 2007. Extinction and recovery patterns of the vegetation across the Cretaceous-Paleocene boundary—a tool for unraveling the causes of the end-Permian mass-extinction. *Rev. Palaeobot. Palynol.* 144:99–112.

Vega, F. E. 2008. The rise of coffee. *Am. Sci.* 96:138–45.

Virginia Tech Department of Forest Resources and Environmental Conservation. Forest Biology and Dendrology Education. http://www.cnr.vt.edu/DENDRO/. [Fact sheets covering the most common trees of North America.]

Walker, G., and D. King. 2008. The Hot Topic: How to tackle global warming and still keep the lights on. Bloomsbury Publishing, London.

Water under pressure. 2008. *Nature* 452. [Topic of issue.]

Weber, A., W. Huber, A Weissenhofer, N. Zamora, and G. Zimmermann. 2001. *An introductory field guide to the flowering plants of the Golfo Dulce rain forests, Costa Rica: Corcovado National Park and Piedras Blancas National Park.* Biologiezentrum des Oberösterreichischen Landesmuseum, Linz.

Weidensaul, S. 2007. The last stand. Photographs by Garth Lenz. *Nature Conservancy*, summer 2007, 20–33. [Boreal forest.]

Whelan, C. 2007. BP's answer to food-based ethanol. *Fortune*, 17 September. ["The oil giant believes an inedible plant called *Jatropha* can ease global fuel demands."]

Whitfield, S. M., et al. (seven coauthors). 2007. Amphibian and reptile declines over 35 years at La Selva, Costa Rica. *Proc. Natl. Acad. Sci. U.S.A.* 104:8352–56.

Who owns the Arctic? 2007. *Time*, vol. 170, November 14. ["As global warming shrinks the ice to record lows, the global battle for resources heats up."]

Williams, J. W., et al. (thirteen coauthors). 2006. *An atlas of pollen-vegetation-climate relationships for the United States and Canada.* AASP Foundation, Dallas.

Wittemyer, G., P. Elsen, W. T. Bean, A. Coleman, O. Burton, and J. S. Brashares. 2008. Accelerated human population growth at protected area edges. *Science* 321:123–26.

World population will grow fastest in poorest areas. 2009. *Nature* 460:943. [Between the present and 2050 population in the least-developed regions will grow from 828 million to 1.66 billion.]

Wortley, A. H., V. A. Funk, and J. J. Skvarla. 2008. Pollen and the evolution of Arctoideae (Compositae). *Bot. Rev.* 74:438–66.

Zuloaga, F. O., O. Morrone, and M. J. Belgrano, eds. 2008. *Catálogo de las plantas vasculares del cono sur (Argentina, Sur de Brasil, Chile, Paraguay y Uruguay).* Monographs in Systematic Botany 107, Missouri Botanical Garden Press, St. Louis.

3

Floods, Temperature, Evolution, and It's about Time

GLOBAL SEA LEVELS

One of the valuable contexts within which paleobiological information can be assessed and interpreted is the history of the Earth's sea levels. When they are high, continental margins and low-lying interiors are flooded, making for more extensive maritime climates that buffer the land from temperature extremes. High waters may disrupt or limit the extent of land bridges such as those across Beringia, the North Atlantic Ocean, Panama, and the Scotia Sea between Antarctica and South America. When the extent of land surface changes, albedo (reflectivity) is affected because water absorbs and releases heat more slowly than land. If sea levels were higher than at present by 100–300 m, as they were during the Cretaceous, this factor, if acting alone, would change the albedo index by about 0.03 and result in a temperature rise of around 3°C.

Another consequence of sea level change, together with orogeny, is the effect it has on atmospheric chemistry (e.g., CO_2 concentration). When land is reduced, less silicate rock is available for weathering and that affects the following chemical transformations:

$$CO_2 + \text{weathering of CaSiO}_3 \text{ (e.g., wollastonite)} \longrightarrow CaCO_3 + SiO_2$$

and

$$CO_2 + \text{weathering of MgSiO}_3 \text{ (e.g., enstatite)} \longrightarrow MgCO_3 + SiO_2$$

That is, during periods of lower sea levels, as during glacial intervals, increased erosion takes place and the trend is for less CO_2 and lower temperatures. With decreased weathering of silicate rocks, less CO_2 is removed, more accumulates in the atmosphere, and higher temperatures result (as in the Cretaceous).

The reconstruction of past sea levels is based on acoustic signals sent through subsurface strata and reflected back in patterns determined by the thickness, density, and porosity of the rock. If the sedimentary layer is solid, indicating an interval of deposition (= high sea levels), the reflected amplitude, frequency, waveform, and velocity is different than if the rock is unconsolidated, indicating a time of exposure to erosion. By examining seismic profiles from localities worldwide, geologists have constructed curves showing periods of major inundation by the sea followed by regressions (fig. 3.1). There were widespread inundations of at least 100 m throughout the Cretaceous, Paleocene, and early Eocene (Miller et al. 2005), a regression around 30 Ma (Oligocene) corresponding to early formation of glaciers in the Arctic and especially on Antarctica circa 33.5–34 Ma (Pearson et al. 2009), and rapid fluctuations after about 5 Ma, when late Pliocene and Quaternary glaciations were underway. Fossil floras and faunas can be plotted on the curve to give an indication of their relationship to this particular environmental factor (fig. 3.1). There have been absolute or eustatic fluctuations in the level of the sea through geologic time as a result of climate change (glaciation) and nonclimatic factors (ocean water displacement from subsurface volcanism). There have also been relative changes in sea level due to upward or downward movement of the land. Either way, the curve constitutes a useful context for viewing biotic history.

On a finer time scale, sea levels are presently rising 30 to 40 cm per century, and this portends increasing trouble for coastal areas. As noted by Pilkey and Cooper (2004), the recent rush to the shore, with "developments" on beachfronts and wetlands, and the massive population increases on deltas and coastal lowlands, requiring associated subsurface

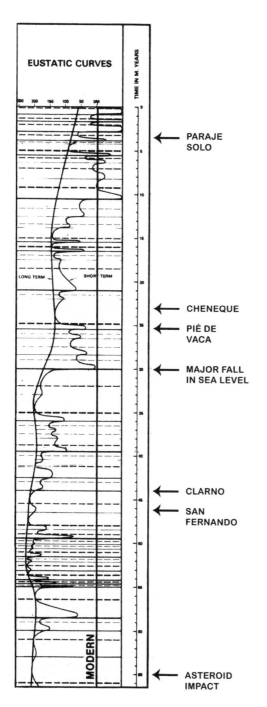

Figure 3.1 Cycles of long-term and short-term eustatic changes in sea level. Eustatic refers to an actual change in water volume in contrast to relative change that may involve movement of the land. Modified from Haq and others 1987. Reprinted with permission from the American Association for the Advancement of Science, Washington, DC. Location and age of the fossil floras: Paraje Solo, Mexico, 3.5 Ma; Chenque, Argentina, circa 23 Ma; Pié de Vaca, Mexico, circa 25 Ma; Clarno, Oregon, 44 Ma; San Fernando, California, circa 44 Ma (middle Eocene).

infrastructure and freshwater aquifers free from the invasion of salt water, has coincided with a rise in sea level. This rise is projected to increase in the centuries ahead to a maximum of about 80 m if the Antarctic ice sheet (7.7 million cubic miles of ice, currently losing about 36 cubic miles of ice each year) and the Greenland ice sheet (680,000 cubic miles of ice) melt completely with continued global warming. Already there has been a 20 percent increase in intense-level floods in the past few decades, prompting articles in the popular media asking, "Are we losing Louisiana?" and similar questions about the Antarctic icecap, the Arctic coastlines, Bangladesh, and other places. The answer from the present is yes. The answer from the past is yes—and it has happened repeatedly, and the evidence has long been available from geologic deposits, soil profiles, and historical records.

PALEOTEMPERATURES

The technique for determining temperatures of the distant past is based on measurement of the isotopes ^{16}O and ^{18}O in marine waters. Both are part of the CO_2 incorporated into the calcite and silicate shells of marine invertebrates, for example, foraminifera, marine gastropods, radiolarians (fig. 3.2). The ratio of the isotopes in the shells changes with temperature (more ^{18}O is taken up as the water cools). Curves have been constructed from Deep Sea Drilling Project (DSDP) cores drilled in all the world's oceans by research vessels such as the Glomar *Challenger* (fig. 3.3) as recounted by Kenneth Hsü in *Challenger at Sea* (1992). Calibration of the ratios to specific temperatures is based on ecological studies of modern species similar to those in the core, and on aquaria experiments. The calcium carbonate–oxygen isotope geothermometer is now the most widely used tool for estimating ancient ocean temperatures. Fossil floras and faunas and geologic events can be plotted on the curve to provide a temperature provenance for both marine and adjacent terrestrial biota (fig. 3.4).

A complication arose in calculating temperatures from the isotopes because as water evaporates from the ocean surface, more of the lighter ^{16}O is distilled, potentially affecting the $^{16}O/^{18}O$ ratio in the basin. During nonglacial times, the water is rapidly returned by river flow and groundwater seepage, circulated by marine currents, and the ratio remains relatively constant. In glacial times, however, the disproportionate amount of ^{16}O in the evaporated water is retained in the ice for long periods of time, and the ratio in marine basins is altered by a factor other than temperature (i.e., ice volume). If there is agreement about the presence of glaciers during a given

Figure 3.2 Marine gastropod *Turritella* from surface outcrop of the Gatún Formation, in the canal region of Panama.

period, adjustments can be made in the calculations, and the technique can be used to estimate both temperature and ice volume. If there is disagreement, as there has been until recently for the early Tertiary, then different estimates of the paleotemperature will result.

Part of the problem was that through the 1980s, there was little reason to believe glaciers had been present in the Arctic and Antarctic during the early Tertiary when climates were warm or warmer than at anytime since 100 Ma. Then Birkenmajer (1988) found evidence for mountain glaciers in Antarctica, now recognized as beginning about 43 Ma (Sluijs et al. 2006; Stoll 2006). Ice was present in the Arctic as early as 45 Ma (DeCanto et al. 2008; Jahren et al. 2009; but see Edgar et al. 2007); Spielhagen and Tripati (2009) believe temperatures periodically reached near freezing in the Arctic in the early Tertiary; Eldrett and colleagues (2009) note possible alpine outlet glaciers in Greenland at 33.5 Ma; and Davies, Kemp, and Pike (2009) suggest intermittent winter sea ice was even present in the Late Cretaceous. In considering these dates, it is important to note that some are concerned with the earliest onset of glaciation, others with the presence of seasonal

Figure 3.3 Drilling pipe on the Glomar Challenger used in the DSDP (Deep Sea Drilling Project). From Hsü 1992. © 1992 Princeton University Press, Princeton, NJ. Used with permission from the publisher.

sea ice (Stickely et al. 2009; 47.5 Ma) or temporary alpine glaciers, and others primarily with determining whether the glaciations were bipolar and synchronous. Some uncertainty continues, so two temperature scales are given on the curves, one for ice-free and one for glacial conditions. With new data becoming available on continental ice volume, it is possible now

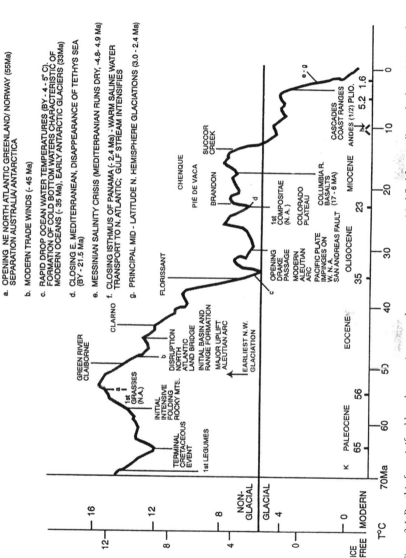

Figure 3.4 Benthic foraminiferal-based oxygen isotope paleotemperature curve. Modified from Miller and others 1987. Used with permission from the American Geophysical Union, Washington, DC.

to better calculate temperature values and provide an improving curve for the past 100 million years.

Temperatures were warm in the Middle Cretaceous and were beginning to cool in the Late Cretaceous when the asteroid hit, causing a brief rebound, then they continued to decline. Temperatures suddenly rebounded dramatically in the early Tertiary, in part, because of an explosive emission of methane (a greenhouse gas) from the Norwegian Sea. Another factor was the continuing albeit more gradual release of CO_2 and additional methane from volcanism in East Greenland (Storey et al. 2007) associated with activity along the Mid-Atlantic Ridge. In total, some 1500 gigatons of CO_2 were emitted, temperatures rose by 5°C–6°C, and ocean acidity increased, resulting in the extinction of 30–50 percent of benthic (deepwater) foraminifera species. The warming lasted for about 220,000 years, with the most rapid increase in the first 20,000 years, and temperatures ultimately reached maximum values at about 55 Ma (the EECL/LPTM). Tropical biotas attained their most extensive geographic distribution during this interval, while temperate organisms, such as deciduous trees and shrubs, were mostly restricted to high latitudes and high altitudes. If ocean waters in tropical regions also experienced substantial warming, reaching 35°C–40°C for sea surface temperatures (see review by Huber 2008), this would sustain strong ocean circulation via thermal gradients. It would also cause some marine organisms of surface and mid-depth habitats to shift their ranges northward and southward to escape physiologically incompatible equatorial temperatures. To the extent that low- to midelevation terrestrial temperatures in tropical regions were higher than the approximate present temperature of 28°C, this would be a factor in forcing the extinction of some tropical species, in addition to a shift in range of others to the north and south (migration) because of unsuitably warm climates in the lower latitudes. These species would be in addition to those expanding their range in response to warming temperatures in the higher latitudes. There is now some evidence that temperatures in the tropics may have been substantially warmer in the Paleocene than at present. As noted, the discovery of a spectacular giant boid snake thirteen meters long and weighing over a ton from the Cerrejón Formation of Colombia at 60 Ma (Head et al. 2009), prior to the even warmer temperatures at 55 Ma, is interpreted to indicate MATs of 30°C–34°C, or 6°C–8°C warmer than at present and higher than previous estimates based on paleobotanical data.

As the input of CO_2 and the effects of the methane excursion waned (Pearson et al. 2009), temperatures dropped until about 35 Ma, then

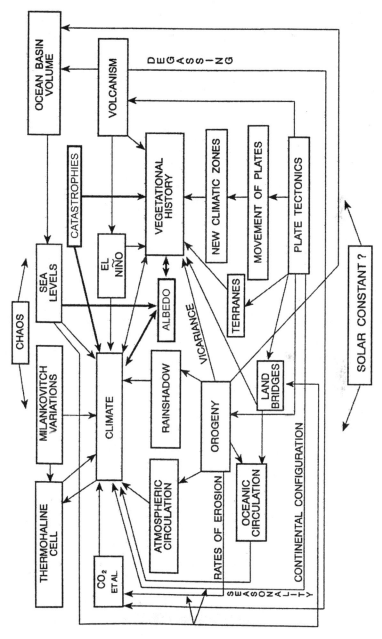

Figure 3.5 Some interacting factors determining vegetation change over time. Modified from Graham 1999. Used with permission from Oxford University Press, Oxford.

they remained in a transitional glacial/nonglacial range until about 15 Ma (fig. 3.4). By that time, ocean waters and land surfaces had cooled significantly, and temperate deciduous elements had spread widely. Shortly afterward, many New World mountain systems attained their highest elevations, providing additional temperate habitats. Less moisture was evaporating from the cooler oceans into the atmosphere, so wet/dry seasonality was also intensifying and spreading. At about 2.6 Ma, another tipping point in global temperatures was reached that ushered in the Quaternary glaciations. During all this time, and documented as far back as at least the Late Triassic (Vollmer et al. 2008), the extraterrestrial Milankovitch variations were operating; but by the Quaternary, the overriding influence of CO_2, although still important, had waned to the point that these variations were unmasked and became more evident as regulators of climate.

Climate-forcing mechanisms interact in a complex system of positive and negative feedbacks that can reinforce, cancel out, or reverse the many individual factors—and they have done so throughout geologic time. To make the point, some of the interactions are sketched in figure 3.5, including the still unquantified role of solar variations (Foukal et al. 2004). Among the conclusions reinforced by this diagram are (1) climate is an immensely complex system, and considering its essential role in sustaining life on the planet, common sense would dictate it should be pertubated with caution; (2) a systems (multidisciplinary) approach incorporating data from many independent lines of inquiry is necessary for understanding the ecosystems as they respond to the varied factors; and (3) context information is necessary to assess conclusions about ecosystems as a basis for intelligently reading the past, managing the present, and preparing for the future.

PHYLOGENY

Knowledge about the evolutionary relationships among extant plants is important for determining meaningful patterns of biogeographic distribution, as opposed to ones based on artifacts of taxonomy. For example, a different explanation would be required to explain the disjunct occurrence of a species in two widely separated localities than if taxonomic study revealed the taxon represented two distantly or unrelated species. Another value of phylogeny is that it can serve as a basis for assessing identifications of fossils from particularly important strata or localities. Fossils preserve in rocks of different ages, organisms evolve over time, and there should be some general correspondence between them. Primitive (basal) groups (e.g., Magnoliids) would be expected in older strata, advanced (derived) groups (e.g., Euasterids) in younger strata, and an explanation (or an acknowledged realization) offered for any significant difference. For example, the Asteraceae, with multiple flowers on a single head (asters, chrysanthemums), are among the most advanced of the flowering plants, so reports or assumptions about their presence in the Cretaceous or early Tertiary should be made with caution and received with a healthy skepticism. The developing phylogeny for the angiosperms is shown in figure 3.6. Similar phylograms are actively being developed for other plant and animal groups, for example, for the bryophytes at Duke University (www.biology.duke.edu/bryology), for the ferns by Schuettpelz and Pryer (2007), and at the Arnold Arboretum at Harvard University for gymnosperms (see the site Gymnosperms on the Tree of Life, or Gymnosperm AToL, http://www.huh.harvard.edu/research/mathews-lab/).

Each of these contexts—sea level, temperature, phylogeny—can serve as a guideline for assessing identifications and for accurately interpreting the paleontological record. Other contexts mentioned previously are faunal history, orogeny, plate tectonics, geomorphic features, and sedimentology. These frameworks should not constrain novel reports or innovative thinking, however, because new paradigms often arise from the unexpected. The value of context is that during a time of increasing responsibility for providing correct identifications, it gives cause for reflection before placing in the literature information that is now being used for such a wide variety of important purposes. Fossils are used to add a time dimension to evolutionary scenarios, propose relationships between organisms, and reconstruct environments within which new groups of plants, animals, and early humans appeared, migrated, and evolved. The fossil record is also being used, together with models of energy flow, spatial heterogeneity, regional (historical) effects, and biological/evolutionary factors (e.g., the stochastic distribution of highly speciose plant groups), to explain and predict zones of unusual species richness in underexplored regions so as to prioritize conservation efforts (Distler et al. 2009; Graham 2009). The modern rain forest was once depicted as ancient, unchanging, and enduring, with the implicit rationalization that it could be lumbered, burned, farmed, and ravaged with impunity because it had always recovered or had persisted unaltered for millions of years. Subsequent paleobotanical studies have shown that concept to be wrong on all counts. There is a lot riding these days on a proper reading of fossils as a basis for detecting past lineages and for reconstructing past environments, and context is one way to improve the chances of getting it right.

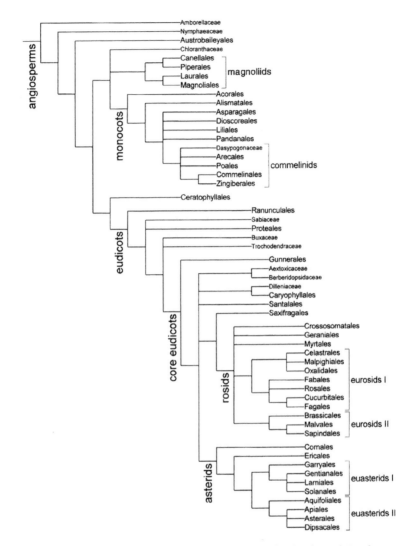

Figure 3.6 Interrelationships of angiosperm orders (-ales) and families (-aceae). Basal groups (e.g., magnoliids), anticipated in older strata, are oriented toward the top in this diagram, and advanced groups (e.g., euasterids), anticipated in younger strata, are shown toward the bottom. From the Angiosperm Phylogeny Group 2003. Used with permission from the Linnean Society of London and Wiley-Blackwell Publishing, Oxford.

TIME

Another essential context for properly reading Earth history is the accurate measurement of geologic time, for which many methods are now available. It has long been recognized that the center of the Earth is composed primarily of molten iron in a constant state of motion that generates a magnetic force or geomagnetic energy. This energy flows to the surface near the geographic South Pole, envelopes the Earth in a magnetic field, and enters back into the Earth near the North Pole. The process is of fundamental importance to life on the planet because it provides a shield against the huge amounts of lethal radiation emanating from the sun and transported through the universe by the solar winds. The interaction between solar radiation and the shield are manifested in the aurora borealis and aurora australis— the northern and southern lights. What has not been recognized until recently is that the Earth's magnetic field has weakened by about 15 percent over the past 150 years, and at that pace it would be expected to collapse in about 1000–1200 years.

The consequences of this weakening, if continued, would initially be a disruption of radio- and satellite-based systems, a gradual increase in skin cancers and genetic mutations, eventual extinctions, and ultimately a barren and lifeless planet. Such a scenario is given credence by preliminary results of voyages to Mars indicating that millions of years ago it had a molten core, a geomagnetic field, liquid water, and possibly some forms of life. It is speculated that the present bleak and arid planet came about through the weakening and then loss of the magnetic envelope due to cooling of the molten core. Needless to say, this has generated interest beyond the field of astrophysics. However, there is encouraging new information that indicates this worst-case scenario will probably not occur for quite a while. Earth is larger than Mars, so its rate of cooling is slower, and any devastating effects are likely to be millions of years in the future. Also, it is now known that periodic weakening of the magnetic field is part of a normal cycle, and that these intervals are prelude not to imminent disaster but to another event of considerable scientific importance.

Geomagnetic Reversals

On a highly variable time scale ranging from tens of thousands to tens of millions of years, the Earth's magnetic field reverses, with geomagnetic energy flowing outward from near the North Pole and inward near the South Pole. In fact, the magnetic North Pole is currently drifting to the south, and since the last reversal it has moved from northernmost Canada toward Siberian Russia. The cause of the reversals is unknown, but they are preceded by a period of several hundred to a few thousand years when the magnetic field weakens, reverses, and then returns to normal strength. The last switch from normal (that is, present-day) to mostly reversed polarity was about 780,000 years ago, and it is called the Brunhes-Matuyama reversal. The pattern was discovered when magnometers were being towed by the U.S. Navy to map the topography of the ocean floor, and later during studies of plate tectonics. Long parallel stripes of lava pour out along either side of the Mid-Atlantic Ridge, and these were found to have ferric (iron) inclusions alternately oriented according to the present poles and in reverse alignment. The different intervals of normal and reversed polarity, called chrons, have been spliced together from sedimentary and igneous rocks from different parts of the world into a sequence that extends into the very remote geologic past. Among the many values of magnetostratigraphy, in addition to revealing pole reversals and for correlating lava and basaltic rocks, is their use in determining the past position of continents. As hematite or magnetite settles out during deposition, it orients toward the magnetic poles (inclination), and the tilt (declimation) reveals the position of the site on the curved surface of the Earth. The two vectors measured through a section track changes in poles and continents through relative time. The patterns of geomagnetic reversals for an early interval is shown in figure 3.7.

Other geohistoric and biohistoric information that can be read in chronological order comes from the sediments that have accumulated on the sea floor. A time limit of about 180 million years is imposed by the disappearance or subduction of the ocean floor into the great trenches of the world. For example, the western edge of the Pacific basin is subducting beneath Asia along the Mariana Trench, and the northern edge is entering the Aleutian Trench bordering the Bering region between North America and Russia. The remaining unsubducted sediments reveal whether the water was deep or shallow, the proximity of land masses, pole and continent positions from the magnetite inclusions, volcanism along midocean ridges, and changes in these features over time.

The value of magnetic and sedimentary sequences is that they afford ways of placing strata, structures, processes, and events in Earth history in proper chronological order. This constitutes the science of stratigraphy, which began with the early work of British geologist William Smith (1769–1839) as recounted by Simon Winchester in *The Map that Changed the World* (2001). Smith widely advocated that strata, and the fossils and

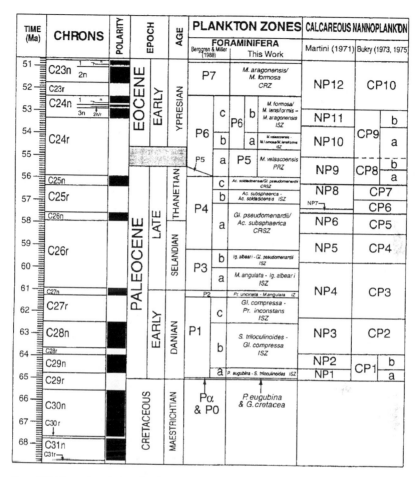

Figure 3.7 Magnetic polarities for the interval between about 51 and 68 Ma, with plankton (floating) foraminifera and calcareous nanoplankton (microscopic) zones. Black indicates normal (i.e., present) polarity, and white indicates reversed polarity. From Berggren and others 1995, which also includes diagrams from other intervals (reproduced in Graham 1999, 171, 218, 224, 243, and 244). Used with permission from the Society for Sedimentary Geology (SEPM), Tulsa, OK.

other geologic and biologic information they contain, are laid down in layers from oldest at the bottom to younger toward the top. They remain so unless disturbed by tectonic or orogenic forces, and these disturbances usually leave compelling evidence in the form of faulting or reversed sequences compared to those elsewhere. Although intuitively obvious from the present perspective, the superposition of strata was not as clear in the eighteenth century. Especially controversial, vis-à-vis some readings of biblical scripture, were the implications, first, that immense amounts of time must be involved, based on observations of processes currently taking place on Earth (erosion of deep canyons, imperceptible accumulation of sediments) that in geologic sections measure tens of thousands of feet in thickness; and second, that the sequence of life generally has proceeded from simple and few to more complex and diverse. Early on, Smith's views were discredited, he was mostly excluded from the religious-dominated academic community of eighteenth- and nineteenth-century Europe, and he actually spent time in debtor's prison. The eventual acceptance of the superposition of strata based on uniformitarian principles was a new paradigm, comparable in importance to the later recognition of catastrophes and plate tectonics in revolutionizing the geological sciences, and it facilitated one of the conceptual breakthroughs in the thinking of Charles Darwin on evolution.

Biostratigraphy

If the sequence of reversals in the Earth's magnetic field preserved in ocean basins can confidently be read from old to young based on the principals of stratigraphy, then the history of marine life preserved in those rocks can be read in the same way and correlated with the terrestrial life during, for example, the North American Land Mammal Ages, or NALMAs (fig. 3.8). As noted, one manifestation of this reading is that life has proceeded in general from morphological, anatomical, cytological, and biochemical simplicity toward greater complexity. Also, there has been a net increase in biodiversity, albeit episodically (chap. 5), from comparatively few kinds or species of organisms in older strata, to different kinds and greater numbers of species in younger strata. This understanding ushered in a new way of comparing discontinuous strata in widely separated parts of the Earth through their fossil content (correlation), and detecting small differences within strata (zonation). Such studies constitute the field of biostratigraphy, and it is based on the vast array of fossils preserved worldwide throughout the geologic column in sediments representing a variety of environments. The fossils include diatoms and other algae; foraminifera, radiolarians, and allied protozoans; corals, mollusks, gastropods, crinoids, and ammonites from the marine realm; pollen grains, spores, seeds, fruits, leaves, wood, phytoliths, and starch grains of plants from ancient bog, lake, swamp, delta, ocean, and volcanic ash deposits; birds, bats, rodents, insects, fungi, bacteria and other plant and animal fossils preserved in sediments including amber and stalactites; aquatic mammals and vertebrates from whales to

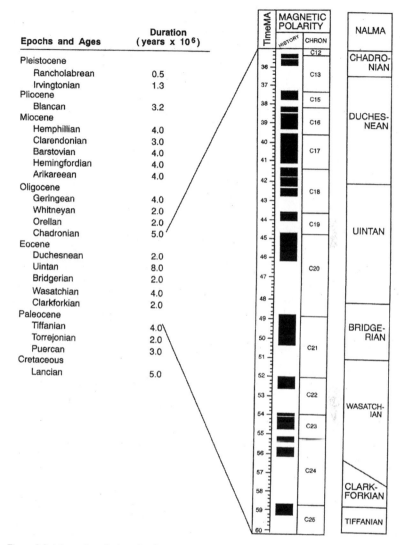

Epochs and Ages	Duration (years x 10⁶)
Pleistocene	
Rancholabrean	0.5
Irvingtonian	1.3
Pliocene	
Blancan	3.2
Miocene	
Hemphillian	4.0
Clarendonian	3.0
Barstovian	4.0
Hemingfordian	4.0
Arikareean	4.0
Oligocene	
Geringean	4.0
Whitneyan	2.0
Orellan	2.0
Chadronian	5.0
Eocene	
Duchesnean	2.0
Uintan	8.0
Bridgerian	2.0
Wasatchian	4.0
Clarkforkian	2.0
Paleocene	
Tiffanian	4.0
Torrejonian	2.0
Puercan	3.0
Cretaceous	
Lancian	5.0

Figure 3.8 Magnetic polarities for the interval between 35 and 60 Ma, with North American Land Mammal Ages. Compiled from various sources (for references, see Graham 1999, 103).

manatees, and land vertebrates such as the highly publicized dinosaurs, their eggs, embryos, footprints, feathers in the avian progenitors, and coprolites, that is, the fecal remains preserving, in turn, plant remains that indicate the local vegetation and the animal's dietary preferences (fig. 3.9; Davis 2006). There are otoliths (fish ear bones; Surge and Walker 2005)

Figure 3.9 Coprolite from the
Santonian (Late Cretaceous)
of Georgia, USA, similar to
those of modern beetle and
containing *Noferinia* pollen.
From Lupia and others 2002.
Used with permission from the
University of Chicago Press,
Chicago.

and biochemical fossils (Eglinton and Eglinton 2008) such as the long-chain alkenones produced by algae and preserved in marine sediments. Fossil-based biostratigraphy, lithostratigraphy (rock based), and magneto-stratigraphy are among the multiple ways of arranging strata and placing the history of the planet and the life it supports in proper chronological order.

Seismic/Sequence Stratigraphy and Facies

Different kinds of sediments are deposited in a fluctuating environment. For example, during periods of ingression and regression along a shoreline, silts with marine organisms, lignites with pollen of terrestrial vegetation, and sandstones and conglomerates from exposed erosion layers may alternate in repeating cycles with the coming and going of the waters. These layers, representing different depositional environments, are called facies, and their subsurface arrangement, extent, thickness as indicated by zones of separation between them (unconformities), and distortion by faulting, compression, transpression (oblique collision), extension (stretching), and overturning is revealed in seismic reflection profiles. These profiles are produced by sending sound waves, generated by explosives on land, or by air guns at sea, through the Earth's crust and reading the reflection times back to the surface with geophones or hydrophones. The time it takes for the waves to return is a function of the density and thickness of the subsurface layers. In combination with visible samples from drill cores, and study of exposures on deep canyon walls, the kind of rock composing these layers can be identified (e.g., sedimentary sandstone, limestone, shale; igneous lava intrusions; metamorphics such as anthracite coal, graphite, marble, and slate). The technique is applicable to a depth of about 5 km and it has a resolution of about 15 m. It is another method in the impressive arsenal used to establish the relative age of rocks.

The key phrase here is "relative age," because so far the methods discussed reveal the kind, sequence, and arrangement of the rocks and the biological information they contain; that is, they reveal ways of putting things in the right chronological order. Assigning an age in years, however, had to await the discovery of radioactivity and the development of absolute dating techniques.

Radiometric Dating

This broad category of techniques is based on the property of certain unstable variants (isotopes) of elements to decay (decompose) by emitting packets of energy. One form is a particle consisting of two protons and two neutrons. This emission is called alpha radiation. It was discovered by the nuclear physicist Ernest Rutherford (1871–1937), winner of the 1908 Nobel Prize in Chemistry. Among elements emitting alpha particles are plutonium, uranium, thorium, radium, and radon. Another form of radiation consists of electrons (beta radiation), also discovered by Rutherford, and emitted, for example, by isotopes of potassium. When these particles are lost, the nucleus is often in an "excited state." That is, there is an imbalance of energy between the nucleus and the electron shells, and this energy is given off, for example, in cobalt, in the form of electromagnetic radiation, or gamma rays. In 1900 gamma radiation was discovered by physicist Paul Villard (1860–1934), working in France at the time of Marie and Pierre Curie.

Many isotopes are familiar because they have practical applications in medicine (e.g., radium). Others are less well known, but among alpha emitters are polonium, prominent recently in the Russian spy news and used to remove static charges that develop during the manufacture of paper, and americium, which generates the current in some smoke detectors. This form of radiation travels relatively slowly (one-twentieth the speed of light), dissipates in a few centimeters, and has little penetrating power. It is stopped, for example, by the outer layer of dead skin and by materials the thickness of a sheet of paper. If inhaled or taken internally, however, it causes severe damage—hence the monitoring of radon levels in homes and in the workplace.

Beta radiation from strontium is used to treat bone and eye cancers, and other beta particles track thickness during the manufacture of materials such as paper. Gamma rays have great penetrating power. They can pass, for example, through a centimeter of aluminum, are used in the sterilization of medical devices and foodstuffs, as well as in CT (computed tomography) or the older CAT scans (computed axial tomography), gamma-knife surgery, scanning foreign merchant ship containers before entering domestic ports, and converting the outwardly mild-mannered but clearly unstable Bruce Banner into the Incredible Hulk.

Radioactive isotopes have other characteristics that render them immensely important in the geological and biological sciences. The decay process is irreversible once it has begun, and the rate is constant. Moreover, as the atomic structure of the isotope, for example uranium, is altered through loss of neutrons, protons, electrons, and the emission of electromagnetic rays, its physical properties are also altered, and the isotope is converted to another element, in this case, lead. No philosopher's stone is involved, and it is not an alchemic transformation. It results from the fact that the physical properties of an element are a reflection of its atomic makeup. If this makeup changes, then the altered product takes on the characteristics of some different element (i.e., uranium to lead). Other decay products are argon from radioactive potassium, and strontium from radioactive rubidium. The rate at which an isotope decays is its half-life, and this rate differs among different elements. For example, the half-life of uranium 238 is 4.5 billion years, cobalt is 5.2 years, sodium is 15 hours, and bismuth is 1.9×1019 years. The familiar carbon[14] has a half-life of 5730 years.

Because rocks and fossils from the geologic column include elements undergoing irreversible isotopic transformations at different but constant rates, it becomes possible to assign an absolute age to rocks provided other criteria are met. First, the origin of the rock containing, for example, uranium must preclude the initial presence of the decay product, for example, lead (in the case of igneous rocks any original lead is volatilized). Second, the rock must contain both uranium and lead. Third, the transformation must still be going on, for otherwise it would not be known how long ago the decay process had stopped. Theoretically, at the most precise level of application, particles one billionth of a gram can be dated. More commonly, ages are assigned to samples down to about the size of minute teeth. Because of their half-life, lead-uranium ratios are mostly used to date rocks billions to hundreds of millions of years old, potassium-argon for rocks hundreds to a few millions of years, and many other radioactive elements fall in intervening and younger ranges. In the later part of the geologic column there, are marine sediments several thousand years old deposited and lying undisturbed, as in the Cariaco Basin off the coast of Venezuela, that have been dated with the resolution of a decade or less. Sediments that are varved (annually layered), such as lake deposits, ice cores from Greenland, Antarctica, the Andean and Himalayan glaciers, and some stalactites; organic material produced in annual cycles (e.g., tree rings); and many historical records can be dated to the year. These and the other methods discussed below are used both independently and collectively to check results from different procedures in a process called cross-dating. Absolute chronologies are based on the following techniques.

Fission-track dating

When decay products of the radioactive isotope uranium 238 move through an enclosing mineral, a track is left in the form of destroyed or modified atomic latticeworks. The decay or fission occurs at a constant rate, so the number of tracks per area, called the spontaneous density (number of tracks per cm^2), is an indication of the age of the rock. It is applicable to apatite, natural glass (obsidian), zircon, and other minerals. The tracks are destroyed by heat, so the measurements indicate the last time the sample cooled. It is used to date tiles, and to assess an area for tectonic hazards such as earthquakes and the "suitability" of the site for storing nuclear waste. It has also been used to cross-check the potassium-argon ages of sediments from the anthropologically important Ulduvai Gorge in Africa. Depending on the uranium concentration, fission-track dating is applicable to material from essentially modern historical and archeological records to rocks billions of years old.

Thermoluminescent dating

The technique known as thermoluminescent dating is based on the occurrence of minute imperfections in minerals such as calcite, diamond, feldspar, or quartz, as well as in ceramics, bricks, and tiles, that trap electrons emitted in low-level radioactivity from, for example, inclusions of thorium, potassium, or uranium. If the sample is heated to about 500°C or exposed to a laser beam, this energy is released as sparks of light proportional in intensity to the energy stored, which, in turn, is a function of the time it has been accumulating. Since lava originates in the molten state, and since pottery, brick, and tile are fired in the manufacturing process, the thermoluminescent clock, like the number of fission tracks, is reset back to zero and dates the time of the last cooling. Standards are set up by bombarding material of known age and measuring the amount of emission. The method is applicable to samples mostly several thousand years old and, depending on the extent of the traps, as old as a few hundred thousand years.

Radiocarbon dating

The most familiar of the absolute dating techniques is probably radiocarbon dating because it encompasses the time span of recent human history. It was developed in 1949 by J. R. Arnold and W. F. Libby at the University of Chicago, for which they won the Nobel Prize in Chemistry in 1960, and it was initially tested on a piece of *Acacia* wood from the tomb of the pharaoh Zoser (2700–2600 BCE). The technique is based on the interaction of cosmic radiation and nitrogen in the Earth's upper atmosphere to produce radioactive carbon ^{14}C, the stable ^{13}C, and the most common ^{12}C which constitutes 98.89 percent of the Earth's carbon. Production of the isotopes takes place mostly at an altitude of 9 to 15 km at the geomagnetic poles. The isotopes drift down, or are brought to Earth in the form of $^{14}CO_2$ by rain or snow. They are then distributed by storms, winds, ocean currents, and rivers and become incorporated into organic biomass through photosynthesis in plants and ingestion by animals. The isotopes are present in a wide variety of materials, such as peat, wood, seeds, bone, shell, horn, enamel, hides, and flesh preserved in bogs and ice. Archeological materials include leather, fabrics, parchment or vellum (animal skins used for writing), and plant-derived paper. The decay process is through emission of beta particles, at a rate of about fifteen per minute per gram, to form nonradioactive nitrogen. The half-life of ^{14}C is 5730 years, and the technique is applicable to material as old as about 50,000 to 60,000 years. This limit is because after 5730 years, half of the carbon isotope in a specimen is converted to nitrogen (that is, the beta emissions are cut by half); after another 5730 years, half of the remainder disappears; and after about ten cycles (or 57,300 years), the amount left is too small to measure. Application of the technique to recent peat deposits younger than 300 years, preserving records of events such as the Chernobyl accident, has been described by Turetsky and others (2004).

In the early days of radiocarbon dating, it was not known that the amount of ^{14}C in the atmosphere varies over time. This variation is due to fluctuating intensities in cosmic radiation, and to the effectiveness of the upper atmosphere and Van Allen Belts in deflecting radiation. Van Allen Belts are zones of cosmic radiation (the outer belt), some of which becomes trapped in the Earth's magnetic field (the inner belt). The auroras at the poles are part of the inner Van Allen Belt. The widely recognized depletion of the ozone layer allows more radioactive carbon to reach the Earth's surface now than was possible before the industrial revolution beginning in the 1800s. Thus, the early Libby dates must be adjusted because they are too young by about 2000 years for material older than 12,500 years, and 500–250 years too young for more recent material. In the literature, dates are designated as uncalibrated ^{14}C years or calibrated years (years BP). Unless otherwise stated, the dates used here are calibrated years. The calibrations are done from historical artifacts of known age, tree rings of bristlecone pine (*Pinus aristata*) to about 4500 years, construction timbers and other woods to about 12,000 years, and from measurements of $^{230}Th/^{234}U/^{238}U$ and ^{14}C on corals back to about 50,000 years. An innovation has been accelerator mass spectrometry (Gove 1999). The mass spectrometer is a device invented in the 1940s and applied to radiometric dating in the 1950s. It allows direct counting of residual ^{14}C isotopes, rather than those emitted in the decay process. It is more accurate and allows dating of samples as small as a few milligrams. Ages calculated from this new method are called Cambridge dates, and they are the ones most commonly used today. Another early source of error was the unrecognized contamination of samples deposited in waters highly charged with dissolved carbonates (e.g., $CaCO_3$). These carbonates usually include some radioactive carbon ($CaC^{14}O_3$) that can be of almost any age. Such material is now avoided and, like results from other absolute dating methodologies, multiple dates of duplicate samples based on carbon isotopes are further checked for proper superpositional sequence, for consistency with comparable material from other sites, and with historical records.

Non–Radioactive Based Absolute Dating Techniques

Another method of determining the absolute age of rocks, fossils, and artifacts, but independent of radioactivity, is obsidian hydration dating. As lava cools, the silica forms volcanic glass, or obsidian. This hard shiny substance can be flaked to make tools, weapons, and ornaments, and therefore it is often found at archeological sites. The technique was introduced by American geologists Irving Friedman and Robert Smith (1960) of the United States Geological Survey. When fresh surfaces are exposed, water is gradually absorbed into the mineral lattice to form a hydration layer that increases in thickness with time. The thickness can be measured in thin sections, and the figures, when used alone, reveal the relative age of different artifacts. To establish an absolute age, it is necessary to know the rate at which the water has been absorbed. Initially, this was done by comparison to standards from sites of known age or by radiocarbon dating of organic material at or near the site (e.g., charcoal). There were a number of uncertainties in the early version of the method. For example, the rate of absorption is influenced by temperature and humidity, which vary over time. Also, the exact thickness of the hydration layer was difficult to determine with the optical microscope because it appears to vary somewhat with the focal level. An ion beam is now used to measure thickness because the reflection pattern of the beam changes at the boundary between hydrated and nonhydrated layers. Old sequences are being recalculated, and new ones are being established by using obsidian directly associated with the site (rather than from the general vicinity). These techniques are showing greater consistency within the stratigraphic sequence and with the ages of artifacts from comparable levels from multiple sites.

Once radiometric dating techniques were developed, it was possible to date and correlate sequences based on relative ages with those having absolute ages. The zones of planktonic foraminifera, calcareous nannoplankton communities, NALMAs, and magnetic reversals (figs. 3.7, 3.8) could be viewed in real time. Ages are being determined with ever-increasing precision, and the results are being coordinated through projects such as EARTHTIME (2007). Slowly, geologic, biologic, and climatic histories are emerging that show consistency across methodologies, with patterns of environmental change in distant regions, and in coevolutionary patterns between plant and animal groups. One example of the latter is the recent discovery of a fossil bee in amber from the Cretaceous of Myanmar (Poinar and Danforth 2006). One of the defining features of the angiosperms that give them an advantage over the wind and water-pollinated ferns, as well as the wind-pollinated gymnosperms they replaced as the Earth's dominant vegetation, is a more efficient system of insect pollination. Bees are prominent among these pollinators, and even though flowering plants originated about 135 Ma, and soon after were successfully radiating and differentiating, the oldest known bees were seemingly late at 35 to 45 Ma. Crepet and Nixon (1998) anticipated bee pollination for some Cretaceous Clusiaceae based on floral structure, and the 100 Ma Burmese fossil bee significantly closes the gap between the angiosperms and one of their principal pollinators. Other examples include correspondence between Cretaceous paleotemperature estimates using $^{16}O/^{18}O$ ratios and otoliths, the rings in fossilized calcite spheres from the inner ear of fish (chap. 5), and between glacial events and temperature changes based on preserved amino acids (allioisoleucine/isoleucin, A/I ratios) in fossilized gastropod shells from loess (chap. 8). There are numerous examples of consistencies gradually emerging in lineage and ecosystem histories with new technologies and that are improving the chronological framework of Earth history.

So What's the Problem?

Would there not also be some glory for man to know how to burst the limits of time, and, by observations, to recover the history of this world, and the succession of events that preceded the birth of the human species?

—GEORGES CUVIER, 1812, quoted in Martin Rudwick,
Bursting the Limits of Time, 2005

Seldom has a single discovery in chemistry had such an impact on the thinking of so many fields of human endeavor. Seldom has a single discovery generated such wide public interest.

THOMAS HIGHAM, quoted in Royal Ervin Taylor, *Radiocarbon Dating*, 1987

Challenges to the reliability of absolute dates were almost a foregone conclusion exactly because the results do impinge on human endeavors and public interest. The origin of the universe around 15 billion years ago was based initially on a theoretical and now on a documented expanding universe—the expansion documented, for example, by 1992 data from the Cosmic Background Explorer Satellite launched in 1989 and by observations and analysis of the density and distribution of ionized particles from the Hubble Telescope. Lead-uranium ratios and fission-track dating both indicate the Earth formed about 4.5 billion years ago and that life appeared about 3.5 billion years ago (Knoll 2004; Szostak 2009; Powner et al. 2009; Whitfield 2009). The latter is deduced from the presence of structurally preserved fossils, biogenetic deposits (e.g., stromatolites), and from optically active (birefringent) compounds in radiometrically dated rocks. Living organisms can produce compounds either in the dextrorotatory form that rotates polarized light to the right or in the levorotatory form. Inorganic synthesis produces a 50:50 mixture and solutions of the synthesized compounds are optically neutral. Thus, birefringent compounds can provide evidence of life in the absence of structurally preserved fossils and in rocks from other planets and extraterrestrial bodies (so far, negative), and they provide a means of assessing the organic nature of morphologically unique microscopic structures found in ancient Precambrian rocks on Earth.

When the Earth's strata were placed in proper sequence following the principles of stratigraphy and superposition, the geologic record revealed that the complexity and diversity of life increased in a timeframe also measured in eons of time. Furthermore, new forms appeared, while others like trilobites, saber-toothed tigers, giant ground sloths, Irish elks, dinosaurs, tree-sized lycopods and horsetails, seed ferns, and the cycadlike Bennettitales disappeared. This pattern did not suggest a near-simultaneous beginning for all life, the perfection of a divine creation, or an intelligent design. Rather, it showed an episodic increase in life-forms involving the appearance and the extinction of species, and explainable by evolutionary processes under genetic control extending over millions of years influenced by environmental change. Humans appeared latest in this sequence, and some of their relicts, such as the Dead Sea Scrolls and the Shroud of Turin, have been dated by ^{14}C. This further brings the measurement of time into the sphere of interest of both science and theology, and when the results are not conflicting, as in the dating of the scrolls at 21 BCE and 61 CE, they are generally accepted. When the dates are in conflict—three dates of the shroud are between 1260 and 1390 CE—these dates and the methodology used to obtain them have been challenged.

Even so, some of the species and ecosystems discussed in this text had existed since the Cretaceous or earlier until severely damaged or exterminated by human activity (Kolbert 2009). Even if it is argued they are only 6000 years old, the lack of outrage universally by theologians at the rampant abuse and destruction of creations, especially when attributed to the direct will of God, must surely be one of the great hypocrisies of modern time. Efforts being made by conservationists and paleontologists in the field, by evolutionary biologists in the laboratory, and with vocal support from the pulpit would seem to be moral and ethical common ground. As noted by Tim Flannery in reference to the extinction of the golden toad:

> It's always devastating when you witness a species' extinction, for what you are seeing is the dismantling of ecosystems and irreparable genetic loss. The golden toad's extinction, however, was not in vain, for when the explanation of its demise was published in *Nature*, the scientists could make their point without equivocation. The golden toad was the first documented victim of global warming. We had killed it with our profligate use of coal-fired electricity and our oversize cars just as surely as if we had flattened its forest with bulldozers. (2005, 118–19)

There has been ample opportunity for debating the age of the Earth, and the history of organisms preserved in its rocks. It seems apparent that unless the discoveries of science are evaluated from a rational and informed base, intellectual polarization will likely continue to the detriment of all. An accurate reckoning of time and an appreciation of the true nature of fossils is central to these evaluations.

: : REFERENCES : :

The Angiosperm Phylogeny Group. 2003. An update of the Angiosperm Phylogeny Group classification for the orders and families of flowering plants; APG II. *Bot. J. Linn. Soc.* 141:399–436. [Updated version in preparation 2009.]

Arnold, J. R., and W. F. Libby. 1949. Age determinations by radiocarbon content: Checks with samples of known age. *Science* 110:678–80.

Berggren, W. A., D. V. Kent, M.-P. Aubry, and J. Hardenbol, eds. 1995. *Geochronology, timescales and global stratigraphic correlation.* SEPM Special Publication 54.Society for Sedimentary Geology, Tulsa.

Birkenmajer, K. 1988. Tertiary glacial and interglacial deposits, South Shetland Islands, Antarctica: Geochronology versus biostratigraphy, a progress report. *Bull. Polish Acad. Sci., Earth Sci.* 36:133–45.

Crepet, W., and K. Nixon. 1998. Fossil Clusiaceae from the Late Cretaceous (Turonian) of New Jersey and implications regarding the history of bee pollination. *Am. J. Bot.* 85:1122–33.

Davies, A., A. E. S. Kemp, and J. Pike. 2009. Late Cretaceous seasonal ocean variability from the Arctic. *Nature* 460:254–58.

Davis, O. K., ed. 2006. Advances in the Interpretation of Pollen and Spores in Coprolites. *Palaeogeogr. Palaeocl. Palaeoecol.* 237:1–118.

DeCanto, R. M., et al. (five coauthors). 2008. Thresholds for Cenozoic bipolar glaciation. *Nature* 455:652–56.

Distler, T., P. M. Jørgensen, A. Graham, G. Davidse, and I. Jiménez. 2009. Determinants and prediction of broad-scale plant richness across the western neotropics. *Ann. Mo. Bot. Gard.* 96:470–91.

EARTHTIME: Calibrating Earth History through Teamwork and Cooperation. 2007. http://www.earth-time.org. [See, e.g., Timescale.]

Edgar, K. M., P. A. Wilson, P. F. Sexton, and Y. Suganuma. 2007. No extreme bipolar glaciation during the main Eocene calcite compensation shift. *Nature* 448:908–11.

Eglinton, T. I., and G. Eglinton. 2008. Molecular proxies for palcoclimatology. *Earth Planet. Sci. Lett.* 275:1–16.

Eldrett, J. S., D. R. Greenwood, I. C. Harding, and M. Huber. 2009. Increased seasonality through the Eocene to Oligocene transition in northern high latitudes. *Nature* 459:969–73, doi: 10.1038/nature 08069. Published online 18 June 2009.

Flannery, T. 2005. *The weather makers: How man is changing the climate and what it means for life on earth.* Atlantic Monthly Press, New York.

Foukal, P., G. North, and T. Wigley. 2004. A stellar view on solar variations and climate. *Science* 306:68–69.

Friedman, I., and R. L. Smith. 1960. A new dating method using obsidian. Part 1, The development of the method. *Am. Antiquity* 25:476–522.

Graham, A. 1999. *Late Cretaceous and Cenozoic history of North American vegetation (north of Mexico).* Oxford University Press, Oxford.

———. 2009. The Andes: A geological overview from a biological perspective. *Ann. Mo. Bot. Gard.* 96:371–85.

Gove, H. E. 1999. *From Hiroshima to the Iceman: The development and application of accelerator mass spectrometry.* Bristol Institute of Physics Publishing, Bristol.

Haq, B. U., J. Hardenbol, and P. R. Vail. 1987. Chronology of fluctuating sea levels since the Triassic. *Science* 235:1156–67.

Head, J. J., et al. (seven coauthors). 2009. Giant boid snake from the Palaeocene neotropics reveals hotter past equatorial temperatures. *Nature* 457:715–17.

Hsü, K. J. 1992. *Challenger at sea: A ship that revolutionized earth science*. Princeton University Press, Princeton.

Huber, M. 2008. A hotter greenhouse? *Science* 321:353–54.

Jahren, A. H., M. C. Byrne, H. V. Graham, L. S. R. Sternberg, and R. E. Summons. 2009. The environmental water of the middle Eocene Arctic: Evidence from δD, δ^{18}O and δ^{13}C within specific compounds. *Palaeogeogr. Palaeocl. Palaeoecol.* 271:96–103.

Knoll, A. H. 2004. *Life on a young planet: The first three billion years of evolution on Earth*. Princeton University Press, Princeton.

Kolbert, E. 2009. The sixth extinction? *New Yorker*, 25 May 2009.

Lupia, R., P. S. Herendeen, and J. A. Keller. 2002. A new fossil flower and associated coprolites: Evidence for angiosperm-insect interactions in the Santonian (Late Cretaceous) of Georgia, U.S.A. *Int. J. Plant Sci.* 163:675–86.

Miller, K. G., R. G. Fairbanks, and G. S. Mountain. 1987. Tertiary oxygen isotope synthesis, sea level history, and continental margin erosion. *Paleoceanography* 2:1–19.

Miller, K. G., et al. (nine coauthors). 2005. The Phanerozoic record of global sea-level change. *Science* 310:1293–98.

Pearson, P. N., G. L. Foster, and B. S. Wade. 2009. Atmospheric carbon dioxide through the Eocene-Oligocene climate transition. *Nature* 461:1110–13, doi: 10.1038/nature8447. Published online 13 September 2009.

Pilkey, O. H., and J. A. G. Cooper. 2004. Society and sea level rise. *Science* 303:1781–82.

Poinar, G. O. Jr., and B. N. Danforth. 2006. A fossil bee from Early Cretaceous Burmese amber. *Science* 314:614.

Powner, M. W., B. Gerland, and J. D. Sutherland. 2009. Synthesis of activated pyrimidine ribonucleotides in prebiotically plausible conditions. *Nature* 459:239–42.

Rudwick, M. J. S. 2005. *Bursting the limits of time: The reconstruction of geohistory in the age of revolution*. University of Chicago Press, Chicago. [See review by Naomi Oreskes, *Science* 314 (2006): 596–97.]

Schuettpelz, E., and K. M. Pryer. 2007. Fern phylogeny inferred from 400 leptosporangiate species and three plastid genes. *Taxon* 56:1037–50.

Sluijs, A., et al. (fourteen coauthors and the Expedition 302 Scientists). 2006. Subtropical Arctic Ocean temperatures during the Palaeocene/Eocene thermal maximum. *Nature* 441:610–13.

Spielhagen, R. F., and A. Tripati. 2009. Evidence from Svalbard for near-freezing temperatures and climate oscillations in the Arctic during the Paleocene and Eocene. *Palaeogeogr. Palaeocl. Palaeoecol.* 278:48–56.

Stickley, C. E., et al. (six coauthors). 2009. Evidence for middle Eocene Arctic sea ice from diatoms and ice-rafted debris. *Nature* 460:376–79.

Stoll, H. M. 2006. The Arctic tells its story. *Nature* 441:579–81.

Storey, M., R. A. Duncan, and C. C. Swisher III. 2007. Paleocene-Eocene thermal maximum and the opening of the northeast Atlantic. *Science* 316:587–89.

Surge, D., and K. J. Walker. 2005. Oxygen isotope composition of modern and archaeological otoliths from the estuarine hardhead catfish (*Ariopsis felis*) and their potential to record low-latitude climate change. *Palaeogeogr. Palaeocl. Palaeoecol.* 228:179–91.

Szostak, J. W. 2009. Systems chemistry on early Earth. *Nature* 459:239–42.

Taylor, R. E. 1987. *Radiocarbon dating: An archaeological perspective.* Academic Press, New York.

Turetsky, M. R., S. W. Manning, and R. K. Wieder. 2004. Dating recent peat deposits. *Wetlands* 24:324–56.

Vollmer, T., et al. (five coauthors). 2008. Orbital control on Upper Triassic playa cycles of the Steinmergel-Keuper (Norian): A new concept for ancient playa cycles. *Palaeogeogr. Palaeocl. Palaeoecol.* 267:1–16.

Whitfield, J. 2009. Origin of life: Nascence man. *Nature* 459:316–19.

Winchester, S. 2001. *The map that changed the world: William Smith and the birth of modern geology.* HarperCollins, New York.

Additional Readings and Updates

Alexander, D. R. 2008. Religion and science: A guide for the "perplexed." *Nature* 455:590.

Begley, S. 2007. The evolution revolution: The new science of the brain and DNA is rewriting the story of human origins. *Newsweek*, 19 March 2007.

Creation and classrooms. 2008. Editorial. *Nature* 455:431–32. ["Better to confront superstition with science than to disregard the superstitious."]

Dalton, R. 2006a. Ethiopian plan for Lucy tour splits museums. *Nature* 444:8.

———. 2006b. Telling the time. *Nature* 444:134–35.

———. 2008a. Fears for oldest human footprints. *Nature* 451:118. ["Threats to the world's oldest hominid footprints in Tanzania are again stirring debate over how to best protect the 3.7-million-year-old tracks."]

———. 2008b. Palaeontology: The new mother lode. *Nature* 455:153–55. [Palaeontologists in Argentina are exploring a trove of fossils that is rewriting evolutionary history.]

Dixon, T. H., et al. (nine coauthors). 2006. Subsidence and flooding in New Orleans. *Nature* 441:587–88.

Friedman, T.␣T. 2007. In the age of Noah. *New York Times*, Op-Ed, 23 December.

Gradstein, F. M., et al. (six coauthors). 1995. A Triassic, Jurassic and Cretaceous time scale. In *Geochronology, time-scales and global stratigraphic correlation*, ed. W. A. Berggren, D. V. Kent, M.-P. Aubry, and J. Hardenbol, SEPM Special Publication 54, 95–126. Society for Sedimentary Geology, Tulsa.

Gubbins, D. 2008. Earth Science: Geomagnetic reversals. *Nature* 452:165–67.

Harris, S. 2007. Correspondence: Scientists should unite against threat from religion. *Nature* 448:864.

Huber, M. 2009. Snakes tell a torrid tale. *Nature* 457:669–71.

Latest satellite launches to monitor sea level. 2008. *Nature* 453:1156.

Lessios, H. A. 2007. Correspondence: Admission that intelligent design is a religious view. *Nature* 448:22.

Lieberman, D. E. 2007. Homing in on early *Homo*. *Nature* 449:291–92. [Newly described fossils from Georgia in Eurasia and from Kenya shed more light on the earliest members of the genus *Homo*: "This species (*Homo erectus*) first appeared in Africa about 1.9 million years ago."]

Linnean Society celebrates seminal evolution papers. 2008. *Nature* 454:14–15. ["This week 150 years ago, papers by British naturalists Charles Darwin and Alfred Russel Wallace were read at the Linnean Society of London. . . . On Tuesday, society members recreated the 1 July 1858 reading at the Royal Academy on Piccadilly."]

Miller, K. R. 2008. *Only a theory: Evolution and the battle for America's soul.* Viking, New York. [See review by P. Z. Myers, *Nature* 454 (2008): 581–82.]

Neanderthal genomics. 2006. Special issue, *Nature* 444.

Noonan, J. P., et al. (ten coauthors). 2006. Sequencing and analysis of Neanderthal genomic DNA. *Science* 314:1113–18.

Pagani, M., K. Caldeira, R. Berner, and D. J. Beerling. 2009. The role of terrestrial plants in limiting atmospheric CO_2 decline over the past 24 million years. *Nature* 460:85–88.

Pennisi, E. 2006. The dawn of Stone Age genomics. *Science* 314:1068–71. [See inset by Michael Balter.]

The Politics of Jesus. 2006. *Newsweek,* 13 November.

Refitted drilling ship sets sail. 2009. *Nature* 457:648. [The JOIDES Resolution revamped at a cost of $130 million departed from Singapore on 25 January 2009 to drill cores in the equatorial Pacific for study of extreme climate changes.]

Schulman, S. 2006. *Undermining science: Suppression and distortion in the Bush administration.* University of California Press, Berkeley.

Silver, L. M. 2006. *Challenging nature: The clash of science and spirituality at the new frontiers of life.* HarperCollins, New York. [See reviews by J. T. Bradley, *Nature* 444 (2006): 271–72, and M. A. Goldman, *Science* 314 (2006): 423].

Spread the word. 2008. Editorial. *Nature* 451:108. ["Evolution is a scientific fact, and every organization whose research depends on it should explain why."]

Steig, E. J., et al. (five coauthors). 2009. Warming of the Antarctic ice-sheet surface since the 1957 International Geophysical Year. *Nature* 457:459–62.

Stocker, T. F., and C. C. Raible. 2005. Water cycle shifts gear. *Nature* 434:830–33. ["Various studies indicate that the hydrological cycle is speeding up at high northern latitudes. The resulting increase in freshwater flow into the Arctic Ocean is predicted to have long-range effects."]

Tzedakis, P. C., K. A. Hughen, I. Cacho, and K. Harvati. 2007. Placing late Neanderthals in a climatic context. *Nature* 449:206–8.

Wigley, P. 2008. Religion and science: Separated by an unbridgeable chasm. *Nature* 455:590.

Wynn, T., and F. L. Coolidge. 2008. A stone-age meeting of minds. *Am. Sci.* 96:44–51.

4

How Do They Do That?

The nature of fossils and their meaning has had a turbulent history. Among the early savants, some were religious and others were not; and among theologians, who often taught, practiced, and administered science, there was a wide range of views held with varying conviction about the accounts of creation in the Old Testament. It is not unexpected that this mélange produced a variety of opinions about fossils, expressed for a variety of purposes. Fossils were said to grow in rocks from propagules transported through the air (the Spermatick theory); or they were carried by light, which was composed of miniature solar systems, capable of absorbing whatever it passed over and depositing it elsewhere; or they were works of the devil placed on Earth to lead God's people astray. The gigantic bones of Irish Elk, dinosaurs, and other "monsters" being found in quarries and in farmer's fields did, indeed, complicate literal interpretation of the Bible. Extinction was discredited by many because it would be to admit imperfections in the divine creation. An alternative view was that the bizarre plants and animals still lived in unexplored parts of the world. In an 1818 letter to Francis van der Kamp, Thomas Jefferson wrote, "It may be doubted whether any particular species of animals or vegetables which ever did exist, has ceased to exist." He had expressed the same view more than twenty years earlier in a letter to John Stuart: "I cannot however help believing that this animal [megalonyx] as well as the Mammoth are still existing" (both letters cited on the Thomas Jefferson Encyclopedia site, http://wiki.monticello.org/mediawiki/index.php/Species_Extinction). Jefferson charged Lewis and Clark to be on the lookout for such life-forms during their exploration of the American west. Ordinary citizens coping with daily survival and hardships were probably either unaware of these debates or gave them little thought. For those who did concern themselves, the uncertainty represented by fossils likely emphasized the potential dangers of the present and the precariousness of the future, furthering hopes for protection by supernatural forces and creating dreams of an easier and just afterlife. It also afforded opportunity for the unscrupulous to exploit these fears by threats of punishments on Earth, visions of an eternity in hell, promises of an eternal life, and the practice widely used by church leaders and their representatives of selling dispensations (salvation) to the frightened and gullible.

The "works of the devil" hypothesis became increasingly popular as Lyell and other geologists provided uncomfortably convincing evidence for the great antiquity of the Earth and its life. Contemporaneously, Darwin and other biologists were offering plausible arguments for evolution, also involving eons of time, through geographic (reproductive) isolation, competition, and natural selection acting on the variation observable in sexually reproducing populations.

Proponents of a biblical flood and believers in God's limitless creativity claimed to have found support for their views in the fossil record when in 1789 Johann Jacob Scheuchzer (1672–1733) discovered a specimen at Oehningen, Switzerland, in deposits now known to be Miocene in age. He called the fossil specimen *Homo diluvii testis* (man testifies to the flood). It was some two or three feet long with large eye sockets on the top of a flattened head and a tail. Almost no one accepted the specimen as being a fossil human, but its assignment to the genus *Homo* speaks volumes about early views on the nature of fossils, efforts to explain the ancient world, and errors that can result from interpreting evidence a priori to support strong beliefs. Cuvier in 1812 recognized it as an amphibian, and Holl in 1832 named it *Andrias scheuchzeri*, an extinct member of the "hellbenders" group of giant salamanders. The specimen was purchased by Martinus van

Figure 4.1 *Homo diluvii testis*, Teyler Museum, Haarlem, The Netherlands.

Marum for the Teyler Museum in Haarlem, The Netherlands, where it is still on display (fig. 4.1.)

The limitless creativity of the Almighty was demonstrated for Dr. Johann Bartholomew Adam Beringer, dean of the Faculty of Medicine, University of Würzburg, in his discovery of some 2000 fossils around Eivelstadt (now Eiblestadt), in Bavaria, southern Germany. The specimens were scattered on and just under the surface of rocks from the Muschelkalk Formation of Triassic age. The events surrounding discovery of the fossils, and the subsequent publication of Beringer's account *Lithographiae Wircenburgensis* in 1726, are described in Melvin Jahn and Daniel Woolf's 1963 translation, *The Lying Stones of Dr. Johann Bartholomew Adam Beringer*. Other informative accounts are provided by Gould (2000) and Taylor (2004).

The eclectic assemblage included beetles preserved in flight, leeches, lice, fish with shells, frogs in the act of copulation, spiders complete with webs—although it is true that silk threads have been reported from Lebanese Cretaceous amber (Zschokke 2003)—and birds with nests; plants with leaves, roots, stems, flowers, and pollinating insects; moons, comets and their gaseous tails; pearls, coins, and clay tablets engraved in Hebrew with letters spelling out "Jehovah." It was posited that light had transferred the Hebrew characters from a nearby Jewish cemetery to the fossils. The

Flood hypothesis, on the other hand, did not merit consideration because among the specimens were a ripe apricot and an acorn, suggesting fossilization in the fall, whereas the Flood was thought likely to have occurred in the spring. It was suggested that the figures were the remains from a pagan culture, but then it became necessary to explain why the pagans would be writing in Hebrew. Beringer himself noted that the undersides of the fossils were smooth and featureless; the sculptured upper surfaces had knife marks and looked as though they had been carved and polished with pumice. The specimens were all perfectly oriented and fitted neatly on the pieces of stone as if they were the work of a sculptor. There was no compression, and the fossils appeared to be a continuation of the stone. (Beringer [1726] 1963).

As writing of the *Lithographiae Wircenburgensis* neared completion, questions about the authenticity of the fossils were circulated by two members of the Würzburg Academic Society, J. Ignatz Roderick and Georg von Eckhart. Beringer suspected Roderick and von Eckhart of perpetrating the charade in the first place and later brought court proceedings against them, but initially he could not admit he had been misled and continued with the publication. Clearly the charade had served its purpose of discrediting the rather pompous Beringer, and Roderick and von Eckhart were probably trying to bring it to a close. Although by April 1726 Beringer had finally acknowledged that the whole episode was a hoax, it was too late to withdraw the book. Despite the farce and humiliation, however, Beringer's name did not sink into oblivion. Indeed, in 1767, after Beringer's death, a second edition of the book that had caused him such great embarrassment was published by his heirs and did quite well.

Long before the efforts of some eighteenth-century theologians and scientists to use fossils to document ancient scripture and explain the unfamiliar, others had already recognized that fossils were nothing more than the remains of past life preserved in the rocks of the Earth. Somewhat ironically, the most prominent of these was Leonardo da Vinci (1452–1519), who painted one of the greatest of religious icons, *The Last Supper*. Unlike his artistic achievements, Leonardo's extensive scientific works, numbering over 4000 pages, were not cataloged until the 1700s. Even then, they were difficult to read because the writing is in mirror image (right to left) with cryptic spellings and abbreviations. The scientific papers were brought together in 1717 as the *Codex Leicester*, after their purchase by Thomas Coke, first Earl of Leicester. Armand Hammer acquired them in 1990 and renamed them the *Codex Hammer*. They were finally purchased in 1994 for $30.8 million by Bill Gates, who restored the name *Codex Leicester*.

Leonardo considered the interpretations of fossils prevalent during his time as ludicrous—"such an opinion cannot exist in a brain of much reason." His opinion that fossils were "remains" of past life has been modified to "evidence" of past life in recognition that shell, bone, or organic material need not be present to document life, but may include indirect evidence such as worm borings, amber (fossil plant resin), phytoliths (plant crystals), biogenetic sediments, biochemical fossils, and birefringent compounds. Once this view was generally accepted, augmented by the principles of stratigraphy and increasingly accurate estimates of the age of the Earth, the study of fossils began making meaningful contributions to our knowledge of evolution, lineages, environments, and ecosystems. They provided what the poet Edward FitzGerald in 1847 called our "vision of time" (Terhune and Terhune 1980). Even so, there was a transition period through the 1800s when both biblical literalists and "Old Earth" geologists grappled with the meaning of a fossil record far less complete and more poorly understood than today. Part of the subsequent polarization in views came from the growing belief, as Edward Forbes realized as early as 1851, that scientists must present the results of their research "in plain, readable, and comprehensive language . . . for if they do not do so, others, unqualified for the task will impose a sham philosophy." Nearly a century later, John Crosse agreed, arguing that "if men competent to the task disdain to popularize science, the task will be attempted by men who are incompetent, but popularized it will be" (both Forbes and Crosse quoted in O'Conner 2007, 213–14).

PALEOBOTANY

The study of plant fossils is traditionally designated as paleobotany if it deals primarily with macrofossils (Stewart and Rothwell 1993; Taylor et al. 2009), and palynology (from Greek palÐ, fine flour or dust) if it deals with modern and fossil spores, pollen, and other plant microfossils (Traverse 2007). In those parts of the geologic column where organisms or their relatives, at least to the level of families, are often still represented in the modern biota (roughly Middle/Late Cretaceous to Recent), the basis for both approaches is comparative material obtained from field or herbarium collections (fig. 4.2). The latter include plant specimens mounted on herbarium sheets with label data giving the name of the plant, collector and collection number, place (via GIS systems), and ecology, that is, vegetation type, habitat, and altitude (fig. 4.3). As noted previously, this information is currently being databased, together with relevant literature, images, and maps showing global distributions. Collections in the major herbaria are

Figure 4.2 Compactors in the Herbarium of the Missouri Botanical Garden, St. Louis. Photograph by Shirley A. Graham.

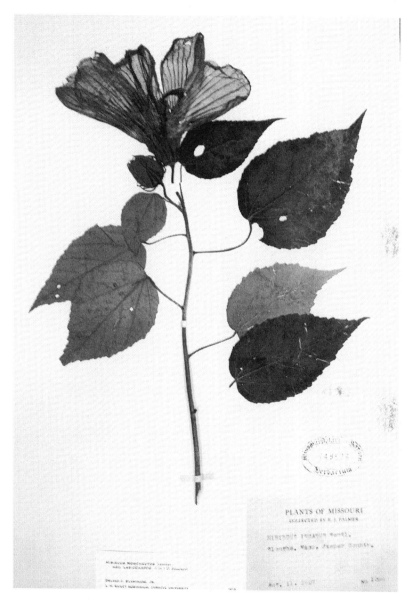

Figure 4.3 Herbarium specimen *(Hibiscus lasiocarpus)* from the collections of the Missouri Botanical Garden, St. Louis. Photograph by Shirley A. Graham.

being integrated into a searchable resource of immense value to taxonomists, biogeographers, ecologists, foresters, economic botanists, biographers, historians of science, conservationists, archeologists, and paleobotanists.

The underlying principle is that if a fossil leaf, flower, fruit, seed, or wood anatomy matches a specimen in the reference collection of modern plants, it is considered identified. The key phrase is "if it matches." In the early days of paleobotany, the characters used for matching were relatively few, comparisons were made using a hand lens or the low power of a light microscope, and the size of collections in herbaria were considerably smaller and biased toward northern temperate regions. For example, in the early 1900s, the holdings in each of three major herbaria in the New World—the Missouri Botanical Garden, the New York Botanical Gardens, and U.S. National Herbarium—were less than 1 million, whereas now they are each approaching or exceed 6 million specimens. SEM (scanning electron microscopy) and TEM (transmission electron microscopy) are standardly used to study fossils, and there are extensive collections of seeds in the U.S. Seed Herbarium of the Department of Agriculture; woods at North Carolina State University and the University of Wisconsin; and cleared leaves in the Paleobiology Division of the Smithsonian Institution, Peabody Museum of Yale University, and the Museum of Paleontology of the University of California, Berkeley. Leaf clearings (fig. 4.4) are supplemented by x-ray photomicrographs (Wing 1992), and together they reveal cellular details, venation, stomata morphology, arrangement, and density (Passalia 2009), epiphytes, parasites, and inclusions. The minute dark structures in figure 4.4, for example, are raphid crystals. Many of these features preserve in fossil leaves, and with more complete collections, modern techniques, standardized terminology, electronic imaging, databasing, and the wide array of ancillary and context information now available, plant fossils are providing an increasingly reliable inventory of ancient lineages, past vegetation, and, by extrapolation, paleoenvironments.

The interpretation of paleoenvironments and the reconstruction of vegetation history from the identified material is based on two independent approaches. One is the modern analog or NLR (nearest living relative) method. The assumption is that most Cenozoic plants (e.g., *Larrea*, creosote bush; *Opuntia*, cactus; *Nymphaea*, water lily; *Rhizophora*, mangrove) had about the same ecological requirements and occupied similar habitats as the extant plant. When this assumption is assessed against evidence such as the associates of the fossil in the assemblage, presence of morphological adaptations reflecting a particular environment, habitat diversity as revealed by the geology of the region, and established trends in climate—for

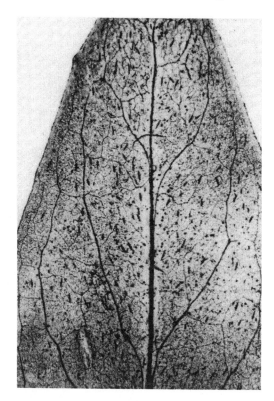

Figure 4.4 Cleared leaf of *Epilobium suffruticosum*. Illustration provided by Richard Keating. Photograph by Shirley A. Graham.

example, in the Atacama Desert (chap. 2)—the assumption appears generally valid, and overall ecological stasis at the generic level since at least 45 Ma emerges as a rather common characteristic of plants and animals.

A potential limitation to the method is that the present distribution of an organism may not necessarily reflect its potential distribution; that is, where an organism does occur is not always the same as where it can occur. After each reshuffling following major environmental changes, for example, the eighteen to twenty glacial cycles of the past 2.6 million years, and the finer-scale Holocene events of the past 11,000 years, some organisms simply may not have had time or opportunity to regain their full range. It is known that the armadillo, alligator, possum, and recent introductions like fire ants and African bees, are spreading northward partly in response to global warming but also owing to a lag time in migration. Barriers such as highways, cities, farms, ranch lands, and cut-over forests have progressively restricted the movement of many organisms over the past two centuries. The reconstruction of paleoenvironments based on the 1900 occurrence of the armadillo in North America, for example, would be quite different from one based on its present distribution.

The possibility of changes in ecological requirements over time and the incomplete range of some extant plants is widely recognized, and there are ways of mitigating both limitations. One is to base paleoenvironmental reconstructions on assemblages, rather than on the ecological and habitat characteristics of a single or limited number of "key," or "indicator," plants. The other is to consider all the fossil floras and faunas in the region to give a fuller picture of the biotic history, in other words, a context. It would be singularly improbable that all organisms from multiple assemblages would change their requirements at the same time in the same direction.

Another way of reconstructing environments, in addition to the NLR or modern analog method, is from the external morphology of leaves, or leaf physiognomy. The value of the approach is that it is independent of the NLR method, and it does not depend on identification of the fossils. Early study of extant plants showed that some leaf features correlated with rainfall and to a lesser extent with temperature. Bailey and Sinnott (1915) provided the following tabulation:

	% entire	% nonentire
Cold temperate environment (eastern U.S.)		
Trees	10	90
Shrubs	14	86
Woody	13	87
Warm tropical environment (Amazonia)		
Trees	90	10
Shrubs	87	13
Woody	88	12

Later studies showed that larger leaves, thinner texture, and drip tips occur frequently in tropical plants, while thicker, coriaceous, and smaller leaves are found most commonly in plants of dry habitats. Wolfe (1990, 1993) tabulated these features for several modern vegetation types in different parts of the world, and developed a computer program called CLAMP (Climate-Leaf Analysis Multivariate Program). Using this program, data on leaf morphology from fossil floras can be entered, generating an estimate of the paleoclimate (usually erring slightly on the cool side). The results can be cross-checked with those from the NLR method, and collectively they allow reconstructions of paleoenvironments that are increasingly accurate and mutually consistent.

PALYNOLOGY

Pollen is the male or sperm-bearing element in angiosperm and gymnosperm (seed plant) reproduction. Spores are more difficult to define because they occur in a wide range of organisms, including bacteria, fungi, algae, bryophytes, and ferns. Mostly they are microscopic asexual reproductive structures of the non–seed plants. Pollen is familiar as a principal cause of allergies, but spores can also have insidious effects, at least for about 48 percent of the world's population (Balick and Beitel 1989). Pollen has long been used in forensics to identify the place where a suspect has been recently and at what time of year. The surface of bullets is too smooth to retain fingerprints, so nanotags are being added to the surface of cartridges that transfer to the hands of the user, and the adhesive is a coating of pollen (Weintraub 2008). Ozone levels can be estimated from the amounts of DNA-protecting pigments contained in the spore wall (Barry Lomax, cited in Morton 2005).

A somewhat bizarre chapter in the annals of palynology was written by Clara S. Hires ([1965] 1978), a scion of the Hires Root Beer family. Hires had graduated from Cornell with a degree in botany, and somewhere along the way she became fixated on fern spores. She had the resources to establish Mistaire Laboratories in Milburn, N.J., and spent her life arguing that what others were calling pollen of pine, hemlock, and other plants were actually variously oriented tetrads of fern spores. Her unique perspective was known to most botanists, and when she occasionally presented papers at national meetings they were met with consideration and understanding. John Mickel of the New York Botanical Garden wrote a thoughtful review of her book and pointed out some of its pitfalls to the unwary. For example, Hires believed that many misinterpretations arose from the fact that biologists did not know how to focus a microscope or interpret three-dimensional structures. Her terminology included "scales like tiny pink roses," and there was no bibliography, although reference was made to her 1916 class notes and to a Mr. Wizard TV program.

Suggestions for improvements, and restraint, were often made to Hires privately, but unabashed she carried on. At a meeting of the American Institute of Biological Sciences in the late 1960s, she encountered a noted authority on fern taxonomy and evolution, the always effusive Warren H. Wagner Jr. of the University of Michigan. Wagner was unaware of Hires's special place in the realm of fern systematics, and after her paper, he launched into a tirade against the misconceptions and misinformation being presented. After several minutes he sat down breathing heavily while Hires stood quietly in her own world at the podium until he had finished. Then she said to him, "Mister, you come to my booth this afternoon and I'll straighten you out." Herb Wagner and everyone in the room enjoyed the moment.

The use of spores and pollen in vegetation history, like macrofossils, also relies on comparative material from herbarium collections. The method is based on several principles or generalizations. First, almost all plants produce either spores or pollen grains. This means that no major taxonomic or ecological group is a priori excluded from the fossil record. The exceptions are widely known and include some aquatics because of their thin exine or pollen wall, as well as the important tropical plant family Lauraceae—*Ocotea*, 350 species in the neotropics; *Nectandra*, 120 species; *Persea*, 100 species. The chemistry of the exine in the Lauraceae precludes preservation of the pollen, but the leaves are common as fossils. With these exceptions, pollen from most kinds of plants can be incorporated into the fossil record. It is estimated that the wind-pollinated spruce forests of southern Sweden produce 75,000 tons of pollen each year that is distributed regionally and falls as the pollen and spore rain. Insect pollinated plants are common in the tropics, and they usually produce smaller amounts of pollen that is large, heavy, sticky and less likely to be preserved. However, because of higher rainfall in the tropics, this pollen is often transported into depositional basins by outwash, and it is commonly found in small amounts in tropical microfossil floras.

Second, pollen and spores are capable of being fossilized because the spore wall and the exine are extraordinarily resistant to acid conditions. This means that sediments accumulating in stagnant, organic, low pH environments, such as peat bogs, lake bottoms, and swamps (and converted to lacustrine shales, lignite, and coal) are often rich in plant microfossils. A sample the size of a marble can yield 100,000 or more spores and pollen grains representing 350 different kinds of plants. If samples are collected from horizontal exposures extending over a long distance, and vertically every several centimeters up a section (canyon walls, river or road cuts; see fig. 6.2) or from well cores (see fig. 7.4), and compared with other fossil floras from the region, an extensive inventory is available of the vegetation and vegetation change over the landscape and through time. The resistance of the wall is reflected in the methods used to recover plant microfossils from sediments. If a hammer, bicycle chain, pair of pliers, and pollen were placed in a platinum crucible and warmed with hydrofluoric acid for a week, the metal objects would be digested or highly corroded, while the pollen walls would remain virtually unaltered.

Third, techniques are available for extracting plant microfossils from the rock. This involves placing pulverized samples in a series of acids that dissolve the mineral matrix—hydrofluoric acid (HF) for silicates, hydrochloric acid (HCL) for carbonates, and nitric acid (HNO_3) to reduce subrin, cellulose, and other organic debris. The wall chemistry is stabilized, both in modern reference material and in plant microfossils, by acetolysis which is a mixture of nine parts acetic anhydride ($[CH_3CO]_2O$) to one part concentrated sulfuric acid (H_2SO_4). The acetolysis mixture is explosive with water, so it is preceded and followed by rinses in dehydrating glacial acetic acid. The residue remaining in the centrifuge tube consists of anything originally present in the rock that is not soluble in acid, that is, mostly spores, pollen, and other plant microfossils. A small amount of mounting medium is added (glycerin jelly or silicone oil) and several slides are prepared, labeled, cataloged, and stored from each sample. The process takes about a day for a group of eight samples depending on how long they are left in each acid (typically several minutes; up to four hours in HF or HNO_3 for highly siliceous or organic samples).

Fourth, pollen grains and spores vary in morphology, so plants can be identified on the basis of pollen and spore characters (see, e.g., fig. 4.5; Moran et al. 2007; and David Roubik's site Pollen and Spores of Barro Colorado Island, http://striweb.si.edu/roubik/). Identifications are usually to the level of genus, but occasionally to species or only to family (e.g., Poaceae, grasses). One exception among the grasses is Zea mays (corn), which has pollen triple the size of most other grasses (100 vs. 30 mμ), allowing fossils of this important plant to be recognized in sediments from archeological sites.

Finally, plants differ in their ecological requirements and are restricted to varying extents to specific kinds of habitats (Larrea, desert; Nymphaea, aquatic). A corollary to this generalization is that paleoenvironments can be estimated from paleobotanical and palynological assemblages.

The goal is ultimately to be able to "sight read" (identify) at a glance a sufficient number of microfossils on a slide (e.g., fig. 2.25b, c) that one can "see" an ecosystem, that is, a mental image emerges (e.g., fig. 2.25a). The details of that image become better defined with additional identifications and with due consideration to ancillary and contextual information. A preliminary impression of ecosystem history emerges when the images are mentally superimposed like frames in a slowly moving film.

Plant macrofossils and microfossils may be preserved in the same flora or in different floras from the same region. In these cases, suggestions for

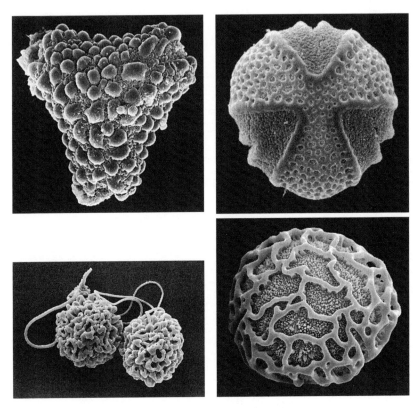

Figure 4.5 Variation among modern pollen types within a single family: Fabaceae/Legumi-
nosae. (Top left) *Eperua rubiginosa* (Surinam). (Top right) *Caesalpinia caladensis* (Mexico).
(Bottom left) *Jacqueshuberia purpurea* (Brazil); note the viscin threads connecting the grains,
allowing multiple fertilizations (each grain has two sperm) from each pollination event.
(Bottom right) *Arapatiella psilophylla* (Brazil).

the identification of unknowns in one category can be provided by the inventory of the other. Study of both kinds of fossils provides a more inclusive
list of plants because those absent as microfossils (e.g., Lauraceae) may be
present as macrofossils. Leaves of annuals typically wither on the stem and
are rarely preserved, but their pollen is often abundant in microfossil floras.
When interpreted in conjunction with faunal and geologic evidence, and
with regional trends in climate, establishing the origin and reconstructing
the history of the extant New World lineages and ecosystems is being done
over broad geographic and stratigraphic ranges and with increasing precision. The results for four intervals encompassing this history are recounted
in following chapters.

: : REFERENCES : :

Bailey, I. W., and E. W. Sinnott. 1915. The climatic distribution of certain types of angiosperm leaves. *Am. J. Bot.* 3:24–39.

Balick, M. J., and J. M. Beitel. 1989. *Lycopodium* spores used in condom manufacture: Associated health hazards. *Econ. Bot.* 43:373–77.

Beringer, J. B. A. [1726] 1963. *The Lying Stones of Dr. Johann Bartholomew Adam Beringer being his* Lithographiae Wirceburgensis. Ed. M. E. Jahn and D. J. Woolf. University of California Press, Berkeley.

Crosse, J. 1845. The Vestiges, etc. *Westminster Review* 44:152–203.

Forbes, E. 1851. A discourse on the studies of the University of Cambridge. *Literary Gazette*, 5–7.

Gould, S. J. 2000. *The Lying Stones of Marrakech*. Jonathan Cape, London.

Hires, C. S. 1965. *Spores, ferns: Microscopic illusions analyzed*. Vol. 1. Mistaire Laboratories, Millburn, N.J. [See review by J. T. Mickel, *Science* 150 (1965): 336.]

———. 1978. *Spores, ferns: Microscopic illusions analyzed*. Vol. 2. Mistaire Laboratories, Millburn, N.J. [See review by J. T. Mickel.]

Moran, R. C., J. G. Hanks, and G. Rouhan. 2007. Spore morphology in relation to phylogeny in the fern genus *Elaphoglossum* (Dryopteridaceae). *Int. J. Plant Sci.* 168:905–29.

Morton, O. 2005. Specks of evidence for ancient sunburn. *Science* 309:1321.

O'Connor, R. 2007. *The Earth on show: Fossils and the poetics of popular science, 1802–1856*. University of Chicago Press, Chicago.

Passalia, M. G. 2009. Cretaceous pCO_2 estimation from stomatal frequency analysis of gymnosperm leaves of Patagonia, Argentina. *Palaeogeogr. Palaeocl. Palaeoecol.* 273:17–24.

Stewart, W. N., and G. W. Rothwell. 1993. *Paleobotany and evolution of plants*. 2nd ed. Cambridge University Press, Cambridge.

Taylor, P. D. 2004. Beringer's iconoliths: Palaeontological fraud in the early 18th century. *The Linnean* 20:21–31.

Taylor, T. N., E. L. Taylor, and M. Krings. 2009. *The biology and evolution of fossil plants*. 2nd ed. Academic Press, New York.

Terhune, A. M., and A. B. Terhune, eds. 1980. *The letters of Edward FitzGerald*. 4 vols. Princeton University Press, Princeton.

Traverse, A. 2007. *Paleopalynology*. 2nd ed. Springer Verlag, Berlin.

Weintraub, A. 2008. Nanotech: Pollinating the crime scene. *Businessweek*, 18 August 2008.

Wing, S. L. 1992. High-resolution leaf x-radiography in systematics and paleobotany. *Am. J. Bot.* 79:1320–24.

Wolfe, J. A. 1990. Estimates of Pliocene precipitation and temperature based on multivariate analysis of leaf physiognomy. In *Pliocene climates: Scenario for global warming*, ed. L. B. Gosnell and R. Z. Poore, U.S. Geological Survey Open File Report 90-64, 39–42. U.S. Department of the Interior, Washington, D.C.

———. 1993. A method of obtaining climatic parameters from leaf assemblages. *U.S. Geol. Surv. Bull.* 2040:1–71.

Zschokke, S. 2003. Spider-web silk from the Early Cretaceous. *Nature* 424:636–37.

Additional Readings and Updates

Molecular Dating Techniques

Bandelt, H.-J. 2008. Time dependency of molecular rate estimates: Tempest in a teacup. *Heredity* 100:1–2. [In the field of human evolution, Ho et al. 2005 has suggested an exponential decay law at odds with the assumed linear relationship between genetic differences and time, contrary to Bandelt, who notes "the enormous efforts that are undertaken to treat the data with fashionable software."]

Graham, A. 1999. *Late Cretaceous and Cenozoic history of North American vegetation*, chap. 8 and references. Oxford University Press, Oxford.

Heads, M. 2005. Dating nodes on molecular phylogenies: A critique of molecular biogeography. *Cladistics* 21:62–78.

Ho, S. Y. W., M. J. Phillips, A. Cooper, and A. J. Drummond. 2005. Time dependency of molecular rate estimates and systematic overestimation of recent divergence times. *Mol. Biol. Evol.* 22:1561–68.

Rutschmann, F., T. Eriksson, K. Abu Salim, and E. Conti. 2007. Assessing calibration uncertainty in molecular dating: The assignment of fossils to alternative calibration points. *Syst. Biol.* 56:591–608.

Sanderson, M. J. 2007. Construction and annotation of large phylogentic trees. *Austral. Syst. Bot.* 20:287–301.

Beyond the origin. 2008. Editorial. *Nature* 456:281. [Darwin bicentenary, 2009.]

Burn, M. J., and F. E. Mayle. 2008. Palynological differentiation between genera of the Moraceae family and implications for Amazonian palaeoecology. *Rev. Palaeobot. Palynol.* 149:187–201.

Del Fueyo, G. M., M. A. Caccavari, and E. A. Dome. 2008. Morphology and structure of the pollen cone and pollen grain of the *Araucaria* species from Argentina. *Biocell* 32:49–60.

Haworth, M., and J. McElwain. 2008. Hot, dry, wet, cold or toxic? Revisiting the ecological significance of leaf and cuticular micromorphology. *Palaeogeogr. Palaeocl. Palaeoecol.* 262:79–90.

Hesse, M., et al. (six coauthors). 2008. *Pollen terminology: An illustrated handbook*. Springer Verlag, Berlin.

Leaf Architecture Working Group. 1999. *Manual of leaf architecture, morphological description and categorization of dicotyledonous and net-veined monocotyledonous angiosperms*. Smithsonian Institution, privately printed, Washington, D.C.

Lightman, B. 2007. *Victorian popularizers of science: Designing nature for new audiences*. University of Chicago Press, Chicago.

Lignum, J., I. Jarvis, and M. A. Pearce. 2008. A critical assessment of standard processing methods for the preparation of palynological samples. *Rev. Palaeobot. Palynol.* 149:133–49.

Moon, H.-K., S. Vinckier, E. Smets, and S. Huysmans. 2008. Comparative pollen morphology and ultrastructure of Mentheae subtribe Nepetinae (Lamiaceae). *Rev. Palaeobot. Palynol.* 149:174–86.

Punyasena, S.W. 2008. Estimating neotropical palaeotemperature and palaeoprecipitation using plant family climatic optima. *Palaeogeogr. Palaeocl. Palaeoecol.* 265:226–37.

Roubik, D. Pollen and Spores of Barro Colorado Island. http://striweb.si.edu/roubik/. [Web site posting images of the STRI spore and pollen collection for Barro Colorado Island, Panama.]

Rudwick, M. J. S. 2008. *Worlds before Adam: The reconstruction of geohistory in the age of reform.* University of Chicago Press, Chicago. [See review by V. R. Baker, *Nature* 454 (2008): 406.]

Uhl, D., et al. (six coauthors). 2007. Cenozoic paleotemperatures and leaf physiognomy—a European perspective. *Palaeogeogr. Palaeocl. Palaeoecol.* 248:24–31.

Wiemann, M. C., D. L. Dilcher, and S. R. Manchester. 2001. Estimation of mean annual temperature from leaf and wood physiognomy. *Forest Science* 47:141–49.

Willard, D. A., et al. (five coauthors). 2004. Atlas of pollen and spores of the Florida Everglades. *Palynology* 28:175–227.

Williams, J. W., et al. (thirteen coauthors). 2006. *Atlas of pollen-vegetation-climate relationships for the United States and Canada.* AASP Foundation, Dallas.

5

Early On
Middle Cretaceous through the Early Eocene

The use of contextual information to approach ecosystem history establishes several expectations about New World environments and vegetation at 100 Ma. These expectations are particularly valuable when direct paleontological information is meager or scattered geographically and stratigraphically, and when the age of the assemblage precludes assuming taxonomic or ecological equilivancy with modern forms. A hundred million years ago, landscapes were low-lying and sea levels were higher by up to 300 m (by plus of 100 m according to Miller et al. 2005), resulting in the widespread flooding of coastlines and interior lowlands (fig. 2.10). Habitats included coastal margins and deltas, stream channels, levees, swamps, marshes, lakes and lake margins, slopes, and moderately elevated uplands. High sea levels also contributed to extensive maritime climates, which are characterized by muted seasonal variations.

Atmospheric CO_2 concentration is estimated to have been at least 1620 ppmv, or 4–6 times the present 380 ppmv (Beerling et al. 2002; 2.3–4 times according to Passalia 2009, based on stomatal analysis of gymnosperm leaves from Patagonia of Aptian-Cenomanian age; see table E.1 below). If MAT increases by approximately 2°C with each doubling of CO_2, this would give a temperature warmer by at least 8°C–12°C (fig. 3.4). Oxygen isotope measurements of ocean bottom waters, indeed, do show temperatures 15°C–20°C warmer than the present 1°C–2°C. Computer simulations further predict a reduced north-south meandering of the polar jet streams and less amplitude between the troughs (low pressure systems) and ridges (high pressure systems) as the fronts moved across the continents. These physical and climatic features suggest a Cretaceous world with warm, equable conditions extending along low thermal gradients between the equator and the poles, with less seasonality and regional differentiation of climate. As a consequence of this, there were many species with extensive geographic ranges, and vegetation was abundant and diverse, as revealed, for example, by stable isotope studies of Late Cretaceous dinosaur tooth enamel and fish scales from Montana and Alberta (Fricke et al. 2008).

Still to be resolved is the possibility of the presence, extent, and location of glaciers in the Cretaceous. The evidence as reviewed in Hay 2008 (see chap. 3) consists of (1) early modeling results (e.g., Barron et al. 1981) that raised, and continue to raise, the possibility of extended periods of subfreezing temperatures over extensive polar areas; (2) fluctuations in sea level during the Cretaceous unlikely attributable solely to movement of the land or reduction in ocean basin volume; and (3) $\delta^{18}O$ ratios seeming to require the presence of large volumes of ice. The uncertainties in the estimates still allow for ice volume ranging from continental glaciation on Antarctica to limited sea ice along the coasts of Antarctica and in the Arctic. Paleobotanical evidence is most extensive for the north polar regions, and there the extent and diversity of Late Cretaceous vegetation is difficult to reconcile with widespread, sustained freezing and glaciation.

Oxygen isotope and other evidence reveals that toward the end of the Middle Cretaceous, atmospheric CO_2 concentration began to decline as plate movement and associated volcanism waned. Also, with uplift and less input of new crustal material into the ocean basins, the epicontinental seas of North and South America began to drain from the interiors. This contributed to an expansion of continental climates characterized by greater seasonal variation. Temperatures decreased toward the K/T boundary from their mid-Cretaceous highs, then began to increase in the Paleocene owing to enhanced volcanic activity along the rift between Greenland and Europe, and then more suddenly to an explosive emission of methane gas from the Norwegian Sea. At about 55 Ma, temperatures reached their maximum for all of Mesozoic and Cenozoic time, rising 5°C–6°C, and possibly averaged at least 15°C in polar regions. After that they started to decline, fluctuated in a transition interval between the late Eocene through the early Miocene, and then fell further into the ice age world of the late Pliocene and Pleistocene (fig. 3.4). The trend is often described as moving from the hothouse climates of the Paleogene to the icehouse climates of the Neogene. The expected consequence for the biota would be widespread tropical communities through about the early Eocene giving way to more temperate and seasonal communities later in the Paleogene, and to increasingly widespread dry forest and grasslands in the Neogene.

At 100 Ma, the angiosperms that define most modern terrestrial plant communities, other than the coniferous forest, were in the early stages of diversification and radiation (e.g., Wang et al. 2009). Originating at about 135 Ma, early on they probably occupied mostly aquatic (Coiffard et al. 2007), shallow-water to streamside, marsh, and other seasonally variable habitats where deciduousness would impart a competitive advantage. Taxonomically angiosperms were just emerging at the level of modern families and genera, and ecologically just differentiating into forms with broadly similar habitat requirements, so ferns and gymnosperms were still prominent components of the vegetation. From this early interval, modernization at the ecosystem level progressed from those with individual elements increasingly similar to those of the present, to coalescence of these elements into early versions, and finally greater differentiation into modern counterparts of the current communities. Except for the coniferous forest, "modern" means angiosperm-prominent communities. This tripartite progression is a convenient way to envision the development of any ecosystem through time. However, in the Cretaceous and early Paleocene, the environments and plant formations were notably different from the present, so their classification, as discussed below, must reflect the uniqueness of these ancient ecosystems.

NORTH AMERICA (NORTH OF MEXICO)

The Cretaceous climate at the high northern latitudes is described as microthermal, with some seasonality and within an overall climate distinctly warmer than at present. This temperature classification derives from following system:

Microthermal   MAT < 13°C
Mesothermal   MAT 13°C–20°C
Megathermal   MAT > 20°C

For example, on the North Slope of Alaska the MAT based on paleobio-logical evidence is estimated at 13°C for the Coniacian circa 89 Ma, with a seasonal cold month temperature of approximately 5°C–6°C (i.e., above freezing), declining to an MAT of around 8°C by the Maestrichtian circa 66 Ma, probably with occasional freezing during the dark winter months. Moderate seasonality in temperature was augmented by distinct seasonality in light regimes, contributing to the development and imparting advantage to those gymnosperms and angiosperms that were deciduous (i.e., could lose their leaves and physiologically hibernate or "chill out" in the dark and in winter).

Across the continent in western Greenland, Cretaceous floras also re-veal conditions markedly different from those of the present. At 71°N in the Early Cretaceous, the floras include an extinct group of plants called the Bennettitales (fig. 5.1; Boyd 1998a). They are morphologically similar to modern Cycadales (cycads) in having long, thick, evergreen, pinnately divided leaves, and together with the cycadophytes they constituted a prin-cipal food source for the dinosaurs (Butler et al. 2008). The closest relatives presently grow in warm-temperate regions of the world. The angiosperm fossils include thin-textured, entire-margined, heart-shaped leaves similar to the Nymphaeales (order of water lilies), the floating aquatic *Ceratophyl-lum*, *Cercidiphyllum* (a deciduous tree currently of eastern Asia), *Platano-phyllum* (a sycamore-like leaf 5.5 inches across), and *Dicotylophyllum* of unknown affinities but with leaves nearly 8.5 inches wide (Boyd, 1998b). The leaf morphology of these genera suggests warm climates, and all were growing above the Arctic Circle (66°33'N) in the Cretaceous.

From later in the Cretaceous, the Lower Atanikerluk flora of Greenland is a relatively diverse terrestrial assemblage with deciduous gymnosperms (e.g., *Metasequoia*, dawn redwood), and fifty-six angiosperm species of which roughly 77 percent have entire margined leaves (tropical) and a smaller but substantial number, about 23 percent, have toothed or lobed leaves (temperate). Identifications and the relationships with modern taxa are uncertain, but leaf types named *Proteoides*, *Dermatophyllites*, and *Chon-drophyllum* in the Late Cretaceous at 71°N are also known from the south-eastern United States at 32°N, a distance of around 4350 km (2700 mi).

In the far northwest, a palynoflora from the Campanian to the Mae-strichtian Tongue River Formation on the Arctic Slope of Alaska includes a diverse plant microfossil flora of 110 different types of spores and pollen (Frederiksen 1989). Many represent herbaceous understory vegetation rare or absent in the macrofossil record (bryophytes, ferns and allied groups, herbaceous and aquatic angiosperms). The plant microfossils are distinct

Figure 5.1 Pseudocycas thomasii (Bennettitales) from the Early Cretaceous of West Greenland. From Boyd 1998a. Used with permission from E. Schweizerbart'sche Verlags-buchhandlung Science Pub-lishers, Stuttgart (http://www .schweizerbart.de).

from most modern forms, but some are similar to cycads, the Asian de-ciduous gymnosperm tree *Ginkgo*, and angiosperm trees of the Betulaceae (birch), Myricaceae (*Myrica*, or gale), and Ulmaceae (elm) families. Fossil wood of palm is common, and other woods show poor to moderately devel-oped growth rings. On Ellesmere Island, Campanian-Maestrichtian sedi-ments and fossil assemblages from a coastal plain/mire habitat reveal active volcanism, flooding, and occasional frost (Falcon-Lang et al. 2004).

Overall, across the high northern latitudes, Cretaceous vegetation con-sisted of the relatively newly arrived angiosperms, competing with the ferns and allied groups and with the gymnosperms. There was probably some habitat differentiation into warmer and less climatically seasonal coastlands, and cooler and more climatically seasonal inland and uplands. There was also a mixture of elements that are either extinct or now found primarily in the Old World, with interchange facilitated both by the warm climates and continuity of the continents that readily allowed migration across the lands of the North Atlantic. The picture is complicated by the fact that some west coastal lands (e.g., Wrangellia) have been transported by plate movement from the south (Stamatakos et al. 2001). Jack Wolfe presents a classification of this unique Cretaceous and Paleogene vegetation of northern North America. The system is discussed below and used in figure 2.10. Wolfe calls the plant formation a polar broad-leaved deciduous forest. It is characterized as follows:

> MAT 7°C–8°C to 15°C, distinct growth rings often present; an extinct microthermal to mesothermal forest type composed of deciduous gymnosperms and angiosperms with leaves of the angiosperms large, thin-textured, and without drip tips.

This forest was part of a taxonomically and ecologically generalized vegetation widely distributed across the north polar regions of the world that in aggregate Wolfe (1975) calls a boreotropical flora. This conceptualized assemblage was not a random intermingling of ecological types, however, because, as noted, there probably was some differentiation between the tropical, warmer coastlands and the more temperate, cooler inland and moderately elevated uplands. Even so, it is a useful concept for envisioning one source and one direction of migration for some members of the proto–deciduous forest. Later these elements would move southward and eventually coalesce with others evolving in warm-temperate to tropical regions to form the deciduous forest formation. It would become better defined and more widespread with the cooling and greater seasonal variation that developed after about the middle Eocene and especially after the early Miocene.

The Cretaceous polar broad-leaved deciduous forest ecosystem had present-day Asian and New World components growing in tropical to warm-temperate environments. The forest consisted of small to moderate-sized deciduous gymnosperms and angiosperms to about 15 m, tall shrubs, and an understory of ferns and herbs. Also present was an early version of an aquatic ecosystem (Nymphaeales, *Ceratophyllum*). Clemens and Allison (1985) report that large dinosaurs lived on the Arctic Slope, and on Axel Heiberg Island (80°N) between 92 and 86 Ma there were champsosaurs (cold-blooded, long-snouted, extinct crocodilian reptiles; Vandermark et al. 2007). Over Tertiary time, the polar broad-leaved deciduous forest ecosystem gradually disappeared through extinction, elimination of present-day Asian species, migration of many deciduous angiosperms southward into the later developing deciduous forest ecosystem, and through the evolution and immigration of ecologically and taxonomically new lineages and communities. It would be replaced by warm-temperate to subtropical vegetation in the early Eocene, a more modern deciduous forest in the late Eocene through the early Miocene, then a boreal forest, and finally by tundra at the extreme high northern latitudes.

To the south, between paleolatitudes 60°N and 50°N, the MAT was 13°C–20°C, and the Late Cretaceous to Paleocene vegetation is called a notophyllous broad-leaved evergreen forest. Note the key difference here is "evergreen" rather than "deciduous." Notophyll is a size class of leaves in a classification frequently used in ecology and for vegetation mapping. Small leaves (i.e., microphyll and smaller) are commonly associated with cold, open habitats, and dry climates, and large leaves (notophyll and larger) with warmer, closed (darker) habitats, and moister climates:

Leptophyll	maximum size 0.25 cm²
Nanophyll	2.25 cm²
Microphyll	20.25 cm²
Notophyll	45 cm²
Mesophyll	182.25 cm²
Macrophyll	1640.2 cm²
Megaphyll	no maximum size

The notophyllous broad-leaved evergreen forest can be characterized as follows:

An ecotonal (transitional) vegetation between a temperate flora to the north and a subtropical flora to the south; mean of the coldest month about 1°C, MAT about 13°C; a few broad-leaved deciduous trees present; gymnosperms not common; buttressing rare, drip tips absent, mesophyll to notophyll size class; entire-margined leaves 40–60 percent.

Compared to the polar broad-leaved deciduous forest, environmental conditions in the notophyllous broad-leaved evergreen forest were warmer, moister, less seasonal, with more light, and much of the vegetation was evergreen throughout the year, except in unpredictable habitats like streamsides and along the margins of shallow lakes. Gymnosperms included the evergreen *Cunninghamia* currently of southeastern Asia (Brink et al. 2009), and members of the *Taxodium-Metasequoia-Sequoia-Sequoiadendron* (bald cypress–dawn redwood–coast redwood–big tree) complex—all deciduous and growing in relatively moist habitats. There was *Equisetum* (horsetail), *Lycopodium* (ground "pine"), ferns (Little et al. 2006; Hernández-Castillo et al. 2006; Vavrek et al. 2006; Stockey et al. 2006), tree ferns (e.g., *Rickwoodopteris hirsuta*, Cyatheaceae; late Campanian Spray Formation, Vancouver Island; Stockey and Rothwell 2004), aquatic monocots (*Cardstonia tolmanii*, Limnocharitaceae; Riley and Stockey 2004), *Cercidiphylum*-like plants, and pollen similar to *Gunnera* (at present a warm-temperate to tropical plant; Jarzen 1980; Jarzen and Dettmann 1989). Overall, however, the conspicuous feature of the forest was its prominent evergreen component.

Farther south, between paleolatitudes 50°N and 40°N, conditions were transitional between mesothermal and megathermal climates (fig. 2.10). In the northern part of this zone, a few deciduous trees with moderately developed growth rings were present. To the south evergreens became even more abundant; they were larger in size and had no or poorly developed growth rings. In Wolfe's system, this vegetation toward the southwestern United States is called a paratropical rain forest, while a similar type with some seasonal (winter) dryness in the southeast is designated a tropical forest (fig. 2.10). The distinctions are a bit fine, but essentially they are both "not quite tropical rain forests" in terms of MAT, MAP, seasonality, buttressing, lianas, drip tips, leaf texture, and entire-margined leaf percentages. As is often the case with biological systems, the parameters have to be numerically quantified more precisely than befits reality for data entry into computer programs such as CLAMP, and one feature earlier used to define a lowland tropical rain forest, a three-tiered canopy, is no longer considered as important (cf. the 1952 and 1996 editions of Richards, *The Tropical Rain Forest*). These two similar vegetation types may be defined as follows:

Paratropical rain forest may experience some frost (−1°C to −3°C), MAT from 20°C–25°C, precipitation may be seasonal but no prolonged dry season; predominantly broad-leaved evergreen with some deciduous plants, woody lianas diverse and abundant, buttressing present, some drip tips, entire margined leaves 57–75 percent.

Tropical forest has MAT of about 25°C, subhumid (slightly drier), MAP estimated at less than 1600 mm, moderate seasonality, growth rings absent to poorly developed, canopy somewhat open, mesophylls (megaphylls in the substratum), few drip tips, entire-margined leaves 39–55 percent.

Among modern vegetation types, these would be described as subtropical to warm-temperate forests. The lowland tropical rain forest had not yet developed in the modern sense, but it is similar and for future reference may be broadly defined as follows:

> MAT of the coldest month does not fall below about 18°C, generally above 25°C; no pronounced or extended dry season; broad-leaved, evergreen, drip tips, lianas, and buttressing common; leaves mostly mesophylls (megaphylls in the substratum); entire-margined leaves at least 75 percent.

In eastern North America, the Cretaceous vegetation of the Appalachian uplands is not well known, but it was likely a notophyllous broad-leaved evergreen forest with some deciduous elements. Representatives are found along the Atlantic Coastal Plain as remarkably well-preserved fossil flowers in the Raritan Formation of New Jersey (Turonian in age; 90.4–88.5 Ma). They include the Clusiaceae (fig. 5.2), Ericaceae, Iteaceae (Saxifragaceae; Hermsen et al. 2003), Magnoliidae, and others (Crepet and Nixon 1994, 1998; Crepet et al. 1992; Nixon and Crepet 1993). In the Mississippi Embayment region of Tennessee, leaves in the Late Cretaceous McNairy Sand and associated floras are 62 percent to more than 70 percent entire margined, growth rings are poorly developed, and drip tips are rare. Leaves are small and somewhat coriaceous (thick) for a typical tropical rain forest,

Figure 5.2 Flower of *Paleoclusia* (Clusiaceae) from the Late Cretaceous (Turonian) of New Jersey. Pistil with broken carpel wall and numerous ovules. From Crepet and Nixon 1998. Used with permission from the Botanical Society of America, St. Louis.

suggesting seasonal rainfall and slightly open vegetation (tropical forest; T in fig. 2.10).

In the midcontinent region, low to moderate topographic diversity provided uplands (as opposed to highlands), swampy lowlands, and aquatic habitats. Otoliths from a Maestrichtain fish (*Vorhisia vulpes*) in the Fox Hills Formation of South Dakota suggest the near-shore waters of the epicontinental sea were brackish (70–80 percent seawater), and temperatures in the open marine waters were about 18°C (Carpenter et al. 2003). The Dakota Formation (Albian, Early Cretaceous) of Kansas and Minnesota contains plants similar to the fern family Marattiaceae (Hu et al. 2006). The modern species grow along streams in shaded, wet, tropical, and subtropical habitats. Also present are Nymphaeaceae (*Aquatifolia fluitans*) and Cabombaceae (*Brasenites kansense*; Wang and Dilcher 2006), as well as *Isöetes* (quillwort, a fern-allied plant; Skog et al. 1992) and the aquatic fern *Marsilea* (Skog and Dilcher 1992). Other reports of Cretaceous aquatics include the floating ferns *Azolla* and *Salvinia* (Hall 1974), Nymphaeales/Nymphaeaceae (Friis et al. 2001; Gandolfo et al. 2004), and *Trapago*, a floating rosette of unknown affinity (Stockey and Rothwell 1997). The Nymphaeaceae is also known from the late Paleocene Almont flora of North Dakota (Taylor et al. 2006). Early and middle Eocene floras include *Isöetes*, *Salvinia*, *Decodon* (Dillhoff et al. 2005; Little and Stockey 2003), *Nelumbo*, and *Porosia*, an extinct floating aquatic angiosperm. The Nelumbonaceae and aquatic ferns (*Paleoazolla*, *Marsilea*, *Regnellidinum*) occur in the Late Cretaceous La Colonia Formation, Chubut, Argentina (Gandolfo and Cúeno 2005), so the aquatic ecosystem was widespread in the Cretaceous and Paleogene of the New World.

An interesting plant from the Late Cretaceous Lance and Hell Creek formations of Wyoming was originally described as a fig (*Ficus ceratops*), but it is most similar to the modern *Guarea* (e.g., *G. chichon*; fig. 5.3a, b). Both *Ficus* and *Guarea* grow in the American tropics and subtropics. In the earliest Paleocene the vegetation from a site near Denver was a paratropical rain forest, with MAT 22.2°C ± 2°C, and MAP 2250 mm (Johnson and Ellis 2002). The paleoclimate is based on leaf physiognomy where 69 percent of the ninety dicot species had leaves that were entire margined, and forty-eight species had drip tips.

In summary, eight ecosystems formed the plant communities of North America north of Mexico toward the end of the Cretaceous. There was a polar broad-leaved deciduous forest, notophyllous broad-leaved evergreen forest, paratropical rain forest, tropical forest, aquatic, and herbaceous bog/marsh/swamp communities. Also, there must have been versions of

Figure 5.3 Fruits of *Guarea* (Meliaceae). (Left) Fossil originally described as *Ficus ceratops* from the Late Cretaceous Lance and Hell Creek formations of Wyoming. (Right) Modern fruits of *Guarea chicon*. From Graham 1962. Used with permission from the Society for Sedimentary Geology (SEPM), Tulsa, OK.

brackish-water mangroves (without the modern *Rhizophora*), and beach/strand/dune vegetation along the sandy coasts, but conditions in this environment are not suitable for preservation and the community is poorly known. Reconstructing the history of New World ecosystems will involve tracking what happened to these eight communities and discovering reasons for the changes.

As things started to heat up toward the LPTM/EECL, with more water evaporating from the ocean surface, the continents became densely vegetated and conditions were even more tropical. Eocene vegetation from a global perspective is described by Utescher and Mosbrugger (2007). In the New World, the late Paleocene / early Eocene Thyra Ø flora is located on the north coast of Greenland at latitude 82°N (Boyd 1990, 1992). It consists of about thirty species of plants of which twenty-three are angiosperms. Of particular interest is the huge monocot leaf *Musophyllum* (*Musopsis*) *groenlandicum*, 1.5 m long and 30–40 cm wide. It belongs to the Musaceae (banana) complex of families, and it is a clear indication of the near tropical conditions existing at least locally north of the Arctic Circle. Also present were *Equisetum*, *Ginkgo*, *Metasequoia*, and angiosperm leaves similar to *Corylus* (hazel) and *Platanus* (sycamore) that constituted the warm/moist phases of a deciduous forest. The recently discovered northernmost record of the tapir lineage is from Ellesmere Island in Arctic Canada at 79°N (Eberle 2005). On the North Slope of Alaska, early to middle Eocene climates are estimated as warm temperate to subtropical, with some deciduous forest elements present such as *Alnus*, *Betula*, *Carya*, *Juglans*, and *Liquidambar* (Frederiksen et al. 2002). The estimated MAT for northern Alaska in the late Paleocene is 10°C–12°C.

In the early Eocene, temperatures rose by another 5°C–6°C, and at 55–50 Ma, global warmth reached its peak for all of latest Mesozoic and Cenozoic time. Warm temperatures persisted at maximum levels for 170,000220,000 years (Pagani et al. 2006) and remained at elevated levels for about 2–5 million years. The atmospheric CO_2 concentration is estimated at 1125 ppmv (Zachos et al. 2008) compared to the present 380 ppmv. The tropical rain forest reached its maximum extent to 45°N–50°N, and grew farther north toward the coasts. Lateritic paleosols are present at the same high latitudes. Coastal paratropical forests grew at 60°N–65°N, and a notophyllous broad-leaved evergreen forest occurs on a terrane on the Pacific Northwest coast at 70°N–75°N, although it was transported some distance from the south. In the Northwest Territories, the Eureka Sound Group preserves remains of these paratropical and notophyllous evergreen forests. The tropical tree fern *Cnemidaria* and the liana fern *Lygodium* make their first appearance at these latitudes in North America in the early Eocene.

During the Paleocene and early Eocene, a few deciduous forest trees grew in the highest lands of the eastern United States. In the U.S. Geological Survey's Oak Grove core from northern Virginia, there is pollen similar to *Alnus* (alder), *Betula* (birch), *Carya* (hickory, pecan), and *Ilex* (holly; Frederiksen 1991). Migrations across the North Atlantic from Asia and Europe were possible until about the middle Eocene, and North American floras include a number of warm-temperate species extinct there now but still found in eastern Asia (e.g., *Platycarya*, *Pterocarya* of the Juglandaceae, the pecan/hickory/walnut family). Climatically, the opportunity for interchange between the New World and the Old World, particularly of megathermal (tropical) and mesothermal (warm-temperate to subtropical) organisms, was greatest in the late Paleocene and early Eocene. Physically, land connections across the North Atlantic were nearly continuous, and a northern DeGeer route (after Baron Gerard DeGeer, Swedish geologist) and a more southerly Thulean route (from Thule, northwest Greenland) accommodated warm-temperate and tropical taxa (Tiffney 1985; Graham

1993; I, 61–64). The North American–European–Asian biotic interchanges of the early and middle Eocene (55–50 Ma) are the most extensive of the entire 20-million-year interval of the Eocene (Frederiksen 1988), and possibly for any time in the Cenozoic. Included in the interchanges were the mesic *Platycarya* (Juglandaceae) and *Eucommia* (Eucommiaceae, now of Asia); tropical cashews (*Anacardium*, now of Central and South America) also known from the middle Eocene of Messel, Germany (Manchester, Wilde, and Collinson 2007; see also Manchester, Xiang, and Xiang 2007), and an extensive associated fauna. Animal assemblages from Ellesmere Island near latitude 79°N contain alligator and an arboreal prosimian primate distantly related to modern flying foxes (lemurs). There is evidence to suggest that many animal groups (family Hyaenodontidae; orders Primates, Artiodactyla, Perissodactyla) were first present in Asia in the Paleocene and arrived in North America near the Paleocene/Eocene boundary (Beard 2002; Bowen et al. 2002). Mild conditions across the northern migration route are further documented in northwestern Europe where the MAT of the London Clay flora is estimated at 25°C (77°F) compared to the present 10°C. Spreading along the Mid-Atlantic Ridge extended progressively northward and climates started to cool from the middle Eocene. At about the same time, the North Atlantic land bridge became disrupted, and interchanges, especially of tropical elements and those with limited dispersal capacities, diminished.

Along the Gulf Coast, leaf size in Paleocene and early Eocene floras is almost always larger than in the Late Cretaceous, MAT is estimated at 27°C (Wolfe and Dilcher 2000), and about half the leaves have drip tips. The increase in moisture resulted in a replacement of the somewhat seasonally dry tropical forest by what was essentially a lowland neotropical rain forest. Fossils of plants typical of that formation increase in abundance and include the Anacardiaceae, Bombacaceae, Lauraceae, Fabaceae (Leguminosae), Rubiaceae, and Rutaceae, in addition to families with numerous lianas (Icacinaceae, Menispermaceae, and Vitaceae). The lowland neotropical rain forest ecosystem had its origin in the lower latitudes, and it dates from the middle to late Paleocene. It will be discussed further with reference to the Amazon Basin region, where it now occupies its greatest extent. In the interval of the LPTM/EECL, however, it extended far north and south into what are now the cold-temperate latitudes.

In the western part of the midcontinent region, there is a series of floras spanning the interval between the early Paleocene and the early Eocene—Fort Union, Sentinel Butte, and Golden Valley, including the Bear Den and Camels Butte members (Hickey 1977; I, fig. 5.7). These and other floras show that north of about the Colorado-Wyoming border, and in the early Paleocene, there is a good representation of temperate plants such as *Acer* (maple), *Betula*, Cornaceae (dogwood family), *Corylus* (hazel), Fagaceae (oak complex), Juglandaceae, Ulmaceae (elm family, including the Asian genus *Zelkova*), and *Viburnum*; while to the south and later in the Paleocene and early Eocene, megathermal genera such as *Cinnamomum* (cinnamon), *Menispermum*, and palms are more common.

Farther to the west, the Willwood flora in the Big Horn Basin of Wyoming is late Paleocene / early Eocene in age (52.8 Ma). It grew at the transition between microthermal and megathermal frost-free climates with an MAT of about 18°C. Entire-margined leaves increase from 41 percent to 52 percent up the section and become associated with crocodilians and land turtles (Wing 1980, 1987; Wing et al. 1995). To the far north, the Paleocene floras of Alaska contain large leaves of *Sabalites* (fig. 5.4), similar to the modern palm *Sabal* of Florida; and to the south in the Paleocene Black Peaks Formation of Big Bend National Park in Texas, fossil wood of *Paraphyllanthoxylon abbottii* lacks distinct growth rings (fig. 5.5; Wheeler 1991).

At about this time, the early Eocene Buchanan Clay Pit flora in western Tennessee was being deposited, and it contains the oldest grasses for North America. *Quercus* (oak) appears as a recognizable modern genus in the still-moderate uplands of the Sierra Nevada region, and *Pinus* had been present since the Cretaceous at several sites in North America (Millar 1998; Perry et al. 1998). These individual elements were available for assemblage with other species into grasslands, oak shrubland/chaparral–woodland–savanna, coniferous forests, and pine-oak forests when dry and more seasonal conditions expanded after the early Eocene and later in the Cenozoic.

The Paleocene and early Eocene was a time of intense deformation, and the Rocky Mountains reached one-half or more of their present elevation. Within an overall tropical to subtropical climate, topographic diversity allowed for drier habitats through slope, exposure, rain shadow, and edaphic factors. The early Eocene Wind River Formation of Wyoming (51 Ma), just to the east of the rising Rocky Mountains, includes present-day mesic Asian genera (*Alangium*, *Platycarya*, *Pterocarya*), others now found in northern Latin America (*Cedrela*, Mexican mahogany; *Oreopanax*), but only a few microthermal temperate elements (*Acer*, *Alnus*, *Populus*; Leopold and MacGinitie 1972). In the Lost Cabin Member of the Wind River Formation (50.5 Ma), there is one of the most diverse early Eocene faunas in North America. More than a hundred species have been identified, including sixty-five mammals, twenty-seven reptiles (lizards, snakes, turtles), two

Figure 5.4 Portion of leaf of *Sabalites* (Palmae) from the Paleocene of Alaska. From Wolfe 1972.

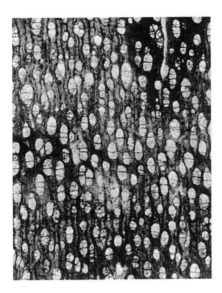

Figure 5.5 Cross-section of dicotyledonous wood, *Paraphyllanthoxylon abbottii*,
from the Paleocene Black Peaks Formation, Big Bend National Park, Texas.
Note absent or poorly developed growth
rings. From Wheeler 1991. Used with
permission from the Botanical Society of
America, St. Louis.

crocodilians, four amphibians, three fish, and two birds. The faunal evidence suggests a flood plain / swamp community in the lowlands experi-
encing occasional dry intervals, as shown by filled mud cracks; while in
the surrounding uplands, the paleobotanical evidence reveals tropical to
subtropical forests growing under moist conditions. Winter temperatures
were above 5°C and frost free. In the Green River basin in southwestern
Wyoming (Smith et al. 2008) there are also extensive floras, with some
189 species of macrofossils (Wilf 2000), and faunas that span the late Paleocene / early Eocene warm interval. The MAT rose from 12°C to an estimated high of 21°C. As will be seen in chapter 6, floras slightly younger
(middle Eocene) and to the west of the mountains (e.g., Republic, Princeton) have a significant moist microthermal component owing to the onset
of cooling temperatures, increasing elevations, and moisture-laden winds
coming unimpeded off the Pacific Ocean in the absence of any significant
elevation in the Cascade Mountains or the Coast Ranges.

Thus, enhanced topographic diversity at the end of the early Eocene
(less-eroded Appalachian Mountains in the east, rising Rocky Mountains in
the west) allowed for a mosaic of eight ecosystems in North America north
of Mexico within an overall dense, uniquely warm, and moist vegetation.
There was a now a much-restricted polar broad-leaved deciduous forest in
the inland and upland regions of the far north, a widespread notophyllous
broad-leaved evergreen forest, a paratropical rain forest, and a lowland neo-

tropical rain forest in the southeast. Locally there were continuing versions of the freshwater herbaceous bog/marsh/swamp ecosystem (*Equisetum, Lycopodium, Selaginella*; possibly some marsh grasses by the Paleocene), and versions of the aquatic ecosystem (*Isöetes, Azolla, Salvinia, Trapago* from the Cretaceous; *Nelumbo, Porosia* added in the early Eocene). As noted previously, there must have been some beach/strand/dune vegetation of fibrous-rooted plants suitable for the shifting sandy substrate because grasses were present in North America by the Eocene (Paleocene to possibly Cretaceous elsewhere). Brackish-water coastal habitats also existed, but *Rhizophora* would not appear until the middle Eocene; so mangrove-habitat plants included some ferns (like the present-day mangrove fern *Acrostichum*), possibly some grasses, and palms. The Old World mangrove palm *Nypa* was present in the Late Cretaceous and Paleocene of the New World (Maracaibo Basin/Caribbean region, Germeraad et al. 1968; Colombia, Gomez-Navarro et al. 2009). It expanded to its maximum distribution in the warm interval of the Eocene (e.g., to the southeastern United States as the pollen type *Spinozonocolpites*, Frederiksen 1988; in south Texas as *Nypa*, Westgate and Gee 1990), and disappeared from the New World in the late Eocene. Only individual elements of a coniferous forest (*Pinus*), grassland, and drier vegetation like shrubland/chaparral-woodland-savanna—possibly represented by other species of *Pinus*, grasses, and *Quercus*—were present in northern North America at the end of the early Eocene. There is no paleobotanical evidence for desert (*Ephedra* was present as an element), tundra, or alpine tundra as communities. The principal difference between the ecosystems of North America north of Mexico in the early Eocene and those in the Middle to Late Cretaceous was that in the early Eocene various modernized forms of a tropical rain forest reached their maximum extent, while, in contrast, early versions of the deciduous forest, as a community, had the most limited distribution of its entire history.

MEXICO, THE ANTILLES, CENTRAL AMERICA

Cretaceous, Paleocene, and early Eocene floras are few in Mexico, the Antilles, and Central America because the land diminishes toward the south; furthermore, the region was still being assembled from terranes or by uplift, and much of it was inundated by the high seas. In Mexico the La Misión and Rosario floras are Campanian (Late Cretaceous) in age and grew on the north coast of Baja California when that land was still part of a volcanic island arc off midcoastal Sonora. These small floras include *Araucaria*, related to the Southern Hemisphere Norfolk Island pine or monkey-puzzle tree and to *Araucarioxylon*, which makes up much of the wood in the Triassic Chinle Formation of the Petrified National Forest in Arizona. Also present was the gymnosperm *Brachyphyllum*, an extinct heterogeneous genus of small trees with flattened, mostly microphyllous, helically arranged leaves widespread in Mesozoic deposits of the New World, and the *Persea*-like angiosperm (Lauraceae, laurels). Little can be said about the floras except that the fossils and the setting are an extension of the mesothermal to megathermal tropical flora and the seasonally dry environments described for the southern United States.

Inland, the vegetation on the low proto–Sierra Madre Occidental is also poorly known, although the Maestrichtian Huepac Chert Member of the Tarahumara Formation contains fossils described as *Opispocaulis* (stem) and *Tarahumara* (inflorescence) that are part of an aquatic angiosperm of the family Haloragidaceae (Hernández-Castillo and Cevallos-Ferriz 1999). Together with forms in the Late Cretaceous of the central United States (*Isöetes, Azolla, Salvinia, Equisetum, Lycopodium, Selaginella*), this establishes the presence of freshwater aquatic and marsh habitats with some angiosperms at this time, but without present-day *Typha* (cattail), *Thalia*, and aquatic grasses that characterize the modern communities.

Farther east, in Cohuila, southwest of Saltillo, is the late Campanian Cerro del Pueblo flora (Estrada-Ruiz et al. 2009). Dating from 73.5 Ma, it contains seeds of the aquatic *Decodon tiffneyi* (Lythraceae) and *Ceratophyllum lesii* (Ceratophyllaceae). The Maestrichtian Olmos flora in the Sabinas region of Coahuila represents a flood plain or delta environment, and many of the fossils are fragmentary, suggesting transport by rivers from the foothills of the proto–Sierra Madre Oriental. They include the aquatic ferns *Salvinia* and *Dorfiella* (extinct), several extinct conifers including *Brachyphyllum*, and fossil woods of *Podocarpoxylon* (similar to *Podocarpus*, at present a warm-temperate genus of conifers) and *Taxodioxylon* (Weber 1973, 1975, 1976). There is also wood of the angiosperm families Fagaceae and Malvaceae (Estrada-Ruiz et al. 2007). The woods have poorly developed growth rings, like those from the Paleocene of the Big Bend region of Texas, indicating moderate seasonality in temperature and rainfall. This view of the climate is supported by context information from isotopic analysis of carbonate nodules in fossil soil horizons (paleosols) that suggest a MAT of at least 15°C, possibly reaching 20°C–25°C in the warmer intervals (Ferguson et al. 1991; Lehman 1990).

Still farther east is the Maestrichtian Piedras Negras microfossil flora north of Sabinas (Martínez-Hernández et al. 1980). Most of the spores and pollen are too old to be identified to modern families and genera, and an artificial system of nomenclature is used (e.g., *Triporopollenites*, pollen with three pores). However, in contrast to the Olmos and other floras where macrofossils of gymnosperms are well represented, there is no winged conifer pollen reported among the thirty pollen types in the Piedras Negras assemblage. This suggests there was a regional mosaic of gymnosperm and angiosperm communities, and moist to moderately dry habitats, over the Cretaceous landscape. These few poorly known floras can all be included within the moister paratropical sites to more seasonally dry forests on edaphic- or slope-conditioned drier sites, along with localized marsh, beach/strand/dune, and aquatic vegetation. A version of the mangrove ecosystem was also likely present (*Acrostichum*, *Spinozonocolpites*/*Nypa*), but as noted, diverse, well-preserved, recently studied fossil floras of Late Cretaceous through early Eocene age are not available for much of Mexico. It is probable that by the end of the early Eocene, local and moderate highlands provided enough slope to allow some vegetation gradient from the lowlands into more upland habitats. The prominent community was likely a version of the lower to upper montane broad-leaved forest.

In the Antilles, there is an Early Cretaceous flora from the Dominican Republic containing the fern *Gleichenites*, the gymnosperms *Brachyphyllum* and *Zamites* (a cycad), and a few angiosperms. The site was originally located in the Isthmian region, and after a brief emergence during which the flora accumulated, it was transported eastward and submerged in the process as indicated by overlying marine deposits of Albian age. The affinities of the flora are widespread but lie primarily with the Cretaceous vegetation of North America. There are no Cretaceous floras known from Central America, and no Paleocene or early Eocene plants known from Mexico, the Antilles, or Central America.

SOUTH AMERICA

At 100 Ma, South America was separating from North America with formation of the Gulf of Mexico, and at 90 Ma it was just moving away along its northeastern coast from Africa. The former common ground between South America and Africa is called West Gondwana, and it is one region favored for the origin of the angiosperms about 135 Ma. South America was connected, or essentially connected, to Antarctica through the shallows of the Drake Passage, and on to Australia (fig. 2.19; Scotese 2004, plates 7–9). This region includes a paleobiotic province called Weddellia (fig. 5.6), and many members shared in the past (e.g., *Casuarina*) and at present reflect the ancient continuity (e.g., *Araucaria*, fig. 2.48; *Nothofagus*,

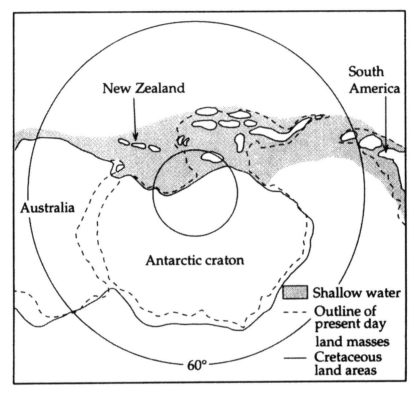

Figure 5.6 Reconstruction of Weddellia during the Cretaceous and Paleogene. Stippled area represents continuous coastal environment surrounded by shallow seas. From Hill and Dettmann 1996 (see also references therein). Used with permission from Yale University Press, New Haven, CT.

southern beech). There were emergent lands in the Guiana and Brazilian shields, the Mato Grosso Plateau, and at scattered sites along the axis of the proto–Andes Mountains. The rest of South America was mostly swampland periodically inundated by marine incursions through the Maracaibo and Paraná inlets. High atmospheric CO_2, warm temperatures, moderate to high rainfall with little seasonality, continental climates (especially in the drier interior northern parts), and broad biotic distributions characterized the continent.

The Cretaceous vegetation of northern South America consisted of gymnosperm-prominent communities. There are macrofossils of the Araucariaceae, Podocarpaceae, cycadophytes, and *Brachyphyllum*, and microfossils similar to *Ephedra* (Mormon tea) representing plants probably growing in edaphically dry habitats in the Early to Middle Cretaceous, for example, in the Barremian of Colombia around 125 Ma. There was an understory of ferns and allied plants, and angiosperms became more widespread in the Cenomanian, around 98 Ma.

In Brazil, sediments of the Rio Acre and Marajó formations document periodic marine incursions into the Amazon Basin, and near Recife there are tsunami deposits and iridium layers from the asteroid impact at Chicxulub. In the Guyanas, Cretaceous-Paleocene sediments include bauxite, coals, and lignites, all of which form under warm-moist climates. The bauxite layers are almost always preceded in the sections by a rise in palm pollen, consistent with moist, warming conditions. There are also layers of evaporates (e.g., gypsum) and oxidized iron indicating local drier sites along the tectonically active and edaphically variable Cretaceous northern coasts of South America.

As mentioned previously, as South America was separating from Africa, geomorphic structures called pull-apart basins developed along the eastern coasts and accumulated plant and animal remains. One of these is the Araripe Basin near Recife, Brazil, where the Middle Cretaceous (Aptian, Albian) Crato Formation contains macrofossils of *Isoëtites* (similar to quillworts of wet-damp habitats), *Equisetites* (horsetail, also mostly of damp habitats), and ferns, along with various gymnosperms such as cycads, a strobilus of an *Araucaria*-like plant, *Brachyphyllum*, various remains attributed to the gymnosperm family Welwitschiaceae (Rydin et al. 2003), some fragmentary angiosperm leaves, and others of Nymphaealean affinities (Mohr et al. 2008). Today, *Welwitschia* is a bizarre-looking extreme xerophyte of the Namib Desert of southwestern Africa. It has two long, strap-shaped leaves growing along the desert surface from a circular, corky center, attached to a tap root that penetrates several meters into soil. Radiocarbon dates on the center give average ages of 500–600 years, and some specimens may be 2000 years old. The ecological relationships between this plant and the Crato specimens are unknown, but the presence of ephedroid pollen (similar to *Ephedra*, another extreme xerophyte) suggests there were locally dry habitats along the margins and in what was then the interior part of the Gondwana continent.

Seeds described as *Musa enseteformis* were included in an interesting early report for the Paleocene of Colombia by E. W. Berry. *Musa* is the genus of bananas, and after its discovery, a banana fruit and later seeds containing *Musa*-like pollen were reported from the Cretaceous of Colombia. Phylogenetically, *Musa* and its relatives are not particularly primitive (basal) plants, and the origin of the cultivated banana is conventionally placed in southeastern Asia. Berry's specimens proved to be recent seeds of the local *Ensete ventricosum*, and the fruit a concretion of nonbiological origin (Manchester and Kress 1993). An extensive middle to late Paleocene flora occurs in the Cerrejón Formation at Guajira, Colombia, at about 60–58 Ma. It is presently under study (e.g., Doria et al. 2008), and already palms (*Nypa*; Gomez-Navarro et al. 2009) and numerous leaf specimens of the Menispermaceae have been described belonging to four species (*Menispermites cerrejonensis*, *M. cordatus*, *M. guajiraensis*, *M. horizontalis*). The family is currently widespread in warm regions of the world. The Araceae was also present (*Montrichardia*; *Pterocardium*; Herrera et al. 2008). The geology and composition suggest a coastal rain forest, as does the gargantuan *Titanoboa*. Microfossils of Paleocene through middle Eocene age from Colombia are described by Jaramillo and Dilcher (2001) and include abundant fern spores and pollen of palms, including *Nypa*, along with pollen similar to *Hibiscus* and *Pelliceria* documenting the presence of an early version of a mangrove ecosystem (still without *Rhizophora*). Most of the Paleocene species are uniquely South American, with only 11.5 percent showing affinities with Africa and 5.2 percent with North America.

A discussion of the paleovegetation of northern South America is convenient for introducing current views on the origin of the lowland neotropical rain forest. Recognition of the formation during its transitory beginning is based in part on upon the composition that includes prominent representation of the angiosperm families Fabaceae (Leguminosae), Moraceae, Lauraceae, and Arecaceae (Palmae; chap. 2). In terms of habitat, the rain forest is found on relatively flat terrane in lowlands with high water tables where the elevation is mostly below 900 m. The climate is warm, with an MAT around 24°C (monthly minimum about 18°C) and moist (minimum of 1800–1500 mm) throughout the year without significant seasonality, and the soils are often sterile, shallow, red, and heavy (clay) laterites. Geographically, these conditions are presently found mostly between 10°N and 10°S. After the rain forest had fully differentiated as an ecosystem, it can further be recognized in the fossil record by a syndrome of morphological features, such as a high percentage of large entire-margined leaves, drip tips, wood with no or poorly developed growth rings, and by the considerable number of macrofossil and microfossil types in the floras, reflecting the great biodiversity of the rain forest community.

When these features were first appearing, however, the time of origin of the lowland neotropical rain forest can only be set within a range of time. Since a prominent angiosperm component is one criterion for its recognition, this puts the maximum age as Late Cretaceous circa 90 Ma, when flowering plants were first becoming widespread over the Earth. The next criterion—the time at which seas had retreated to expose a greater expanse of continental lowlands—was in the early to middle Paleocene around 65–60 Ma. Finally, the most propitious time climatically was during the LPTM/EECL (60 Ma to 55 or 50 Ma), and it was in this interval in the middle to late Paleocene that the above physical, biological, and climatic factors combined to produce a lowland neotropical rain forest recognizable in the modern sense (Morley 2000; Ziegler et al.,2003; Burnham and Johnson 2004; Jaramillo and others 2006; Wing et al., 2009). It is thus one of the oldest of the modern forested ecosystems of the New World.

Argentina offers the most extensive sequence of Cretaceous through early Eocene and younger floras in South America. The Argentinean floras have been studied by Ana and Sergio Archangelsky, Vivian Barreda, Edgardo Romero, and colleagues at the Museo Argentino de Ciencias Naturales "Bernardino Rivadavia" (Buenos Aires) and other museums, universities, and institutes in the country (Archangelsky et al. 2009). There is an Early Cretaceous flora in the Springhill Formation of the Magellanes Basin of southern Argentina and Chile (Baldoni and Archangelsky 1983; Villar de Seoane 2001). It contains familiar gymnosperms (e.g., *Brachyphyllum*) and other plant fossils of widespread distribution. For example, *Otozamites* is a cycadlike plant known in Jurassic and Cretaceous deposits from Antarctica to Oaxaca, Mexico. The microfossils are mostly gymnosperms (saccate or winged conifer pollen 30–50 percent; nonsaccate 15–17 percent), spores of ferns and related groups (30–43 percent); and in the Early Cretaceous Kachaike Formation, there are rare angiosperm leaf fragments. In the Late Cretaceous, ferns and gymnosperms are still abundant, as shown by the Anfiteatro flora of Santa Cruz Province radiometrically dated at 120 Ma (Aptian), and there are fragmentary leaves of the Nymphaeaceae, water lily, type (Romero and Archangelsky 1986). These angiosperm leaf fossils, together with those from the Aptian-Albian Crato Formation of Brazil, are the oldest ones known from South America (Barreda and Archangelsky 2006). By the Middle Cretaceous, in the Cañadon Seco Formation of central Patagonia, there are eleven types of angiosperm pollen. *Nothofagus* pollen is present by the Campanian in Antarctica and Australia; angiosperm pollen reaches 63–79 percent in the Paleocene at some localities, Casuarinaceae is present in Chubut, Argentina, by the early Eocene (Zamaloa et al. 2006), and Proteaceae and *Nothofagus* forests (as opposed to scattered trees) are evident in the middle Eocene. Collectively, these floras show a fern- and gymnosperm-dominated vegetation being supplemented by a diversifying assemblage of flowering plants. The early Eocene Laguna del Hunco flora

of Patagonia (Wilf et al. 2003), with palms, cycads, araucarias, and podo-
carps, indicates elevations in the Southern Andes were still relatively low
and were not yet casting a rain shadow to the east.

The vegetation types in the Cretaceous and Paleocene of South America include megaphylls and notophylls in warm megathermal climates of
northern South America, with microphylls becoming more common in the
cooler microthermal conditions toward the south. Late Cretaceous woods
from Patagonia show distinct growth rings. Later, toward the early Eocene,
conditions warmed, and a more quantitative estimate of climates is afforded
by leaf margin analysis from the Laguna del Hunco flora of Patagonia (Wilf
et al. 2003). This was the time of maximum LPTM/EECL warmth, and
the MAT was around 18°C, increasing to around 24°C through the section.
At present the MAT is 12°C, and the MAP is 300 mm. Minimum winter
temperatures for the Eocene are estimated at more than 10°C, consistent
with the absence of the temperate *Nothofagus* in the flora. Sea surface temperatures in the South Atlantic Ocean are estimated at about 17°C, and
the MAP at about 1200 mm. In vegetation terminology, comparable to that
used for the northern part of the world, after this notophyllous broad-leaved
evergreen forest / paratropical rain forest of the LPTM/EECL interval, the
vegetation at the southern end of the world changed through a microthermal deciduous community of *Nothofagus* forest and gymnosperms (middle
Eocene through the Oligocene), to a colder, seasonally dry shrubland and
herbaceous vegetation (steppe) after about the middle Miocene (Barreda
and Palazzesi 2007). Modernization of the southern flora involved the addition of grasses into various ecosystems; the earliest grasses in southern
South America appear in the early to middle Eocene. In addition to grasses,
dry elements present in the Eocene were *Cassia*, *Celtis*, and some Anacardiaceae and Fabaceae, probably in coastal sands and similar edaphically suitable sites inland. As in northern North America (e.g., in the Eocene Green
River flora), these sites afforded habitats and a source of elements available
for later assemblage into expanding versions, then essentially modern counterparts of steppe (shrubland/chaparral-woodland-savanna). In Antarctica,
the Cretaceous and Paleogene vegetation was a rich assemblage of plants
widespread across to Weddellia and adjacent regions. The dramatic change
in store for Antarctica later in the Tertiary and Quaternary is shown by
only two remaining plants that now constitute the angiosperm flora of the
continent (*Colobanthus quitensis*, Caryophyllaceae, and *Deschampsia antarctica*, Poaceae). Future changes in store for Antarctica as a result of current
global warming is foreshadowed by the dramatic expansion of *Deschampsia*
now covering extensive areas with a layer of grassy green.

In summary, the ecosystems of South America from the Middle Cretaceous through the early Eocene are similar in kinds and distribution, but different in composition, to their counterparts in northern North America, Mexico, the Antilles, and Central America. In the northern, equatorial parts of South America there were tropical forests and paratropical rain forests that toward the early Paleocene had formed a lowland neotropical rain forest in response to increasing and uniform warmth and moisture. It extended north with elements present in the Arctic region and intermingled with tropical to warm-temperate plants moving across the North Atlantic land bridge, and to the south where Australasian plants were moving across Antarctica to Weddellia during the LPCM and EECL. The period around 55 Ma was the most extensive greening of the New World in the last 100 million years. Open and drier sites accommodated elements such as ephedroid and *Welwitschia*-related plants (in northern South America), while wetter habitats across the continent had versions of freshwater herbaceous bog/marsh/swamp and aquatic plant formations (*Equisetum, Isöetes, Azolla, Salvinia*). A kind of mangrove community (without *Rhizophora*), and probably beach/strand/dune vegetation (probably without grasses until about the middle Eocene), grew along the coast.

In the far south in the Cretaceous through about the middle Paleocene, there was a limited polar broad-leaved deciduous forest as one component of what may better be described generally as a lower to upper montane broad-leaved forest. When this forest is considered throughout its southern range, early on there was a mixture of gymnosperms (e.g., *Araucaria, Austrocedrus, Fitzroya*), deciduous angiosperms (e.g., *Nothofagus*), and some limited warm-temperate to subtropical species that in aggregate constitute an austrotropical counterpart to the boreotropical flora. In the late Paleocene and early Eocene, if a south-polar broad-leaved deciduous forest (as opposed to individual elements like *Nothofagus*) was present, it would have been limited to the higher elevations and to southernmost South America and Antarctica. Both in northern North America and in southern South America, the tropical component began to diminish after the early Eocene. In the north, it was eventually replaced by a boreal coniferous forest and tundra. To the south, a cool to cold-temperate deciduous forest (lower to upper montane broad-leaved forest) was present to the west of the Southern Andes, and there were the beginnings of a cold shrubland/woodland steppe to the drier eastern side. These trends in climate and vegetation in the southern fossil floras are reflected in leaf size, which begins to decrease after the middle Eocene.

Recall that twelve modern ecosystems are recognized in this treatment for the New World (chap. 1). Eight were present at the beginning of the Cretaceous, although they were unique or distant versions of the current ones, and seven were present in South America by the end of the early Eocene: The aquatic (Cretaceous) and the lowland neotropical rain forest (Paleocene) are the oldest, and the latter is the oldest forested community to be readily recognizable in terms of its modern (angiosperm prominent) counterpart. The rain forest graded into a lower to upper montane broad-leaved forest in places of moderate altitude and in drier habitats. There were early versions of freshwater herbaceous bog/marsh/swamp, mangrove (without *Rhizophora*), and probably beach/strand/dune ecosystems; and by the middle Eocene, elements of a shrubland/chaparral-woodland-savanna (steppe) were present. Absent were even early versions of desert, tundra, alpine tundra (páramo), coniferous forest (which never developed in South America in the absence of *Abies*, *Picea*, and *Pinus*; *Podocarpus* is typically scattered in temperate and occasionally more lowland forests), and few or no expanses of grassland.

At the beginning of the middle Eocene, significant climatic and physiographic changes took place in the New World that ushered in an additional ecosystem (the coniferous forest in northern North America), continued the modernization of others already present (mangroves with *Rhizophora*, shrubland/chaparral-woodland-savanna), and witnessed the appearance of elements and early versions of desert, grassland, tundra, and alpine tundra (páramo).

: : **REFERENCES** : :

Archangelsky, S., et al. (twelve coauthors). 2009. Early angiosperm diversification: Evidence from southern South America. *Cretaceous Research* 30:1073–82.

Baldoni, A. M., and S. Archangelsky. 1983. Palinología de la Formación Springhill (Cretácico inferior), subsuelo de Argentina y Chile. *Rev. Españ. Micropaleontol.* 15:47–101.

Barreda, V., and S. Archangelsky. 2006. The southernmost record of tropical pollen grains in the mid-Cretaceous of Patagonia, Argentina. *Cretaceous Research* 27:778–87.

Barreda, V., and L. Palazzesi. 2007. Patagonian vegetation turnovers during the Paleogene–early Neogene: Origin of arid-adapted floras. *Bot. Rev.* 73:31–50.

Barron, E. J., S. L. Thompson, and S. H. Schneider. 1981. An ice-free Cretaceous? Results from climate model simulations. *Science* 212:501–8.

Beard, C. 2002. East of Eden at the Paleocene/Eocene boundary. *Science* 295:2028–29. [Summary of Bowen et al. 2002.]

Beerling, D. J., B. H. Lomax, D. L. Royer, G. R. Upchurch Jr., and L. R. Kump. 2002. An atmospheric pCO_2 reconstruction across the Cretaceous-Tertiary boundary from leaf megafossils. *Proc. Natl. Acad. Sci. U.S.A.* 99:7836–40.

Bowen, G. J., et al. (seven coauthors). 2002. Mammalian dispersal at the Paleocene/Eocene boundary. *Science* 295:2062–65.

Boyd, A. 1990. The Thyra Ø flora: Toward an understanding of the climate and vegetation during the early Tertiary in the High Arctic. *Rev. Paleobot. Palynol.* 62:189–203.

———. 1992. *Musopis* n. gen.: A banana-like leaf genus from the early Tertiary of eastern North Greenland. *Am. J. Bot.* 79:1359–67.

———. 1998a. Bennettitales from the Early Cretaceous floras of West Greenland: *Pseudocycas* Nathorst. *Palaeontographica*, Abt. B, 247:123–55.

———. 1998b. Cuticular and impressional angiosperm leaf remains from the Early Cretaceous of West Greenland. *Palaeontographica*, Abt. B, 247:1–53.

Brink, K. S., R. A. Stockey, G. Beard, and W. C. Wehr. 2009. *Cunninghamia hornbyensis* sp. nov.: Permineralized twigs and leaves from the Upper Cretaceous of Hornby Island, British Colombia, Canada. *Rev. Palaeobot. Palynol.* 155:89–98.

Burnham, R. J., and K. R. Johnson. 2004. South American palaeobotany and the origins of neotropical rainforests. In *Plant phylogeny and the origin of major biomes*, ed. R. T. Pennington, Q. C. B. Cronk, and J. A. Richardson. Special issue, *Phil. Trans. R. Soc. Lond. B* 359:1595–1610.

Butler, R. J., P. M. Barrett, P. Kenrick, and M. G. Penn. 2008. Diversity patterns amongst herbivorous dinosaurs and plants during the Cretaceous: Implications for hypotheses of dinosaur/angiosperm co-evolution. *J. Evol. Biol.* 22:446–59.

Carpenter, S. J., J. M. Erikson, and F. D. Holland Jr. 2003. Migration of a Late Cretaceous fish. *Nature* 423:70–74.

Clemens, W. A., and C. W. Allison. 1985. Late Cretaceous terrestrial vertebrate fauna, North Slope, Alaska. Abstract. *Geol. Soc. Am.* 17:548.

Coiffard, C., B. Gomez, and F. Thevenard. 2007. Early Cretaceous angiosperm invasion of western Europe and major environmental changes. *Ann. Bot.* 100:545–53.

Crepet, W. L., and K. C. Nixon. 1994. Flowers of Turonian *Magnoliidae* and their implications. *Plant Syst. Evol.* 8:73–91.

———. 1998. Fossil Clusiaceae from the Late Cretaceous (Turonian) of New Jersey and implications regarding the history of bee pollination. *Am. J. Bot.* 85:1122–33.

Crepet, W. L., K. C. Nixon, E. M. Friis, and J. V. Freudenstein. 1992. Oldest fossil flowers of hamamelidaceous affinity, from the Late Cretaceous of New Jersey. *Proc. Natl. Acad. Sci. U.S.A.* 89:8986–89.

Dillhoff, R. M., E. B. Leopold, and S. R. Manchester. 2005. The McAbee flora of British Columbia and its relation to the early-middle Eocene Okanagan Highlands flora of the Pacific Northwest. *Canadian J. Earth Sci.* 42:151–66.

Doria, G., C. A. Jaramillo, and F. Herrera. 2008. Menispermaceae from the Cerrajón Formation, middle to late Paleocene, Colombia. *Am. J. Bot.* 95:954–73.

Eberle, J. J. 2005. A new "tapir" from Ellesmere Island, Arctic Canada—Implications for northern high latitude palaeobiogeography and tapir palaeobiology. *Palaeogeogr. Palaeocl. Palaeoecol.* 227:311–22.

Estrada-Ruiz, E., L. Calvillo-Canadell, and S. R. S. Cevallos-Ferriz. 2009. Upper Cretaceous aquatic plants from northern Mexico. *Aquat. Bot.* 90:282–88.

Estrada-Ruiz, E., H. I. Martínez-Cabrera, and S. R. S. Cevallos-Ferriz. 2007. Fossil woods from the late Campanian–early Maastrichtian Olmos Formation, Coahuila, Mexico. *Rev. Palaeobot. Palynol.* 145:123–33.

Falcon-Lang, H. J., R. A. MacRae, and A. Z. Csank. 2004. Palaeoecology of Late Cretaceous polar vegetation preserved in the Hansen Point Volcanics, NW Ellesmere Island, Canada. *Palaeogeogr. Palaeocl. Palaeoecol.* 212:45–64.

Ferguson, K. M., T. M. Lehman, and R. T. Gregory. 1991. C- and O-isotopes of pedogeneic soil nodules from two sections spanning the K/T transition in west Texas. Abstract. *Geol. Soc. Am., Abstracts with Programs* 23:302.

Frederiksen, N. O. 1988. *Sporomorph biostratigraphy, floral changes, and paleoclimatology, Eocene and earliest Oligocene of the eastern Gulf Coast.* U.S. Geological Survey Professional Paper 1448. U.S. Department of the Interior, Washington, D.C.

———. 1989. Changes in floral diversities, floral turnover rates, and climates in Campanian and Maastrichtian time, North Slope of Alaska. *Cretaceous Research* 10:249–66.

———. 1991. Rates of floral turnover and diversity change in the fossil record. *Palaeobotanist* 39:127–39.

Frederiksen, N. O., L. E. Edwards, T. A. Ager, and T. P. Sheehan. 2002. Palynology of Eocene strata in the Sagavanirktok and Canning Formations on the North Slope of Alaska. *Palynology* 26:59–93.

Fricke, H. C., R. R. Rogers, R. Backlund, C. N. Dwyer, and S. Echt. 2008. Preservation of primary stable isotope signals in dinosaur remains, and environmental gradients of the Late Cretaceous of Montana and Alberta. *Palaeogeogr. Palaeocl. Palaeoecol.* 266:13–27.

Friis, E. M., K. R. Pedersen, and P. R. Crane. 2001. Fossil evidence of water lilies (Nymphaeales) in the Early Cretaceous. *Nature* 410:357–60.

Gandolfo, M. A., and R. N. Cúneo. 2005. Fossil Nelumbonaceae from the La Colonia Formation (Campanian-Maastrichtian, Upper Cretaceous), Chubut, Patagonia, Argentina. *Rev. Palaeobot. Palynol.* 133:169–78.

Gandolfo, M. A., K. C. Nixon, and W. L. Crepet. 2004. Cretaceous flowers of Nymphaeaceae and implications for complex insect entrapment pollination mechanisms in early angiosperms. *Proc. Natl. Acad. Sci. U.S.A.* 101:8056–60.

Germeraad, J. H., C. A. Hopping, and J. Muller. 1968. Palynology of Tertiary sediments from tropical areas. *Rev. Palaeobot. Palynol.* 6:189–348.

Gomez-Navarro, C., C. Jaramillo, F. Herrera, S. L. Wing, and R. Callejas. 2009. Palms (Arecaceae) from a Paleocene rainforest of northern Colombia. *Am. J. Bot.* 96:1300–1312.

Graham, A. 1962. *Ficus ceratops* Knowlton and its affinities with the living genus *Guarea. J. Paleontol.* 36:521–23.

———, ed. 1972. *Floristics and paleofloristics of Asia and eastern North America.* Elsevier Science Publishers, Amsterdam.

———. 1993. History of the vegetation: Cretaceous (Maastrichtian)–Tertiary. In *Flora of North America,* 57–70.Oxford University Press, Oxford.

Hall, J. W. 1974. Cretaceous Salviniaceae. *Ann. Mo. Bot. Gard.* 61:354–67.

Hay, W. W. 2008. Evolving ideas about the Cretaceous climate and ocean circulation. *Cretaceous Research* 29:725–53.

Hermsen, E. J., M. A. Gandolfo, K. C. Nixon, and W. L. Crepet. 2003. *Divisestylus* gen. nov. (aff. Iteaceae), a fossil Saxifrage from the Late Cretaceous of New Jersey, USA. *Am. J. Bot.* 90:1373–88.

Hernández-Castillo, G. R., and S. R. S. Cevallos-Ferriz. 1999. Reproductive and vegetative organs with affinities to Haloragaceae from the Upper Cretaceous Huepac Chert locality of Sonora, Mexico. *Am. J. Bot.* 86:1717–34.

Hernández-Castillo, G. R., R. A. Stockey, and G. W. Rothwell. 2006. *Anemia quatsinoensis* sp. nov. (Schizaeaceae), a permineralized fern from the Lower Cretaceous of Vancouver Island. *Int. J. Plant Sci.* 167:665–74.

Herrera, F. A., C. A. Jaramillo, D. L. Dilcher, S. L. Wing, and C. Gómez-N. 2008. Fossil Araceae from a Paleocene neotropical rainforest in Colombia. *Am. J. Bot.* 95:1569–83.

Hickey, L. J. 1977. Stratigraphy and paleobotany of the Golden Valley Formation (early Tertiary) of western North Dakota. *Geol. Soc. Am. Mem.* 150:1–183.

Hill, R. S., and M. E. Dettmann. 1996. Origin and diversification of the genus *Nothofagus*. In *The ecology and biogeography of* Nothofagus *forest*, ed. T. T. Veblen, R. S. Hill, and J. Read, 11–24. Yale University Press, New Haven.

Hu, S., D. L. Dilcher, H. Schneider, and D. M. Jarzen. 2006. Eusporangiate ferns from the Dakota Formation, Minnesota, U.S.A. *Int. J. Plant Sci.* 167:579–89.

Jaramillo, C. A., and D. L. Dilcher. 2001. Middle Paleogene palynology of central Colombia, South America: A study of pollen and spores from tropical latitudes. *Palaeontographica*, Abt. B, 258:87–213.

Jaramillo, C. A., M. J. Rueda, and G. Mora. 2006. Cenozoic plant diversity in the neotropics. *Science* 311:1893–96.

Jarzen, D. M. 1980. The occurrence of *Gunnera* pollen in the fossil record. *Biotropica* 12:117–23.

Jarzen, D. M., and M. E. Dettmann. 1989. Taxonomic revision of *Tricolpites reticulatus* Cookson ex Couper, 1953, with notes on the biogeography of *Gunnera* L. *Pollen et Spores* 31:97–112.

Johnson, K. R., and B. Ellis. 2002. A tropical rainforest in Colorado 1.4 million years after the Cretaceous-Tertiary boundary. *Science* 296:2379–83.

Lehman, T. M. 1990. Paleosols and Cretaceous/Tertiary transition in the Big Bend region of Texas. *Geology* 18:362–64.

Leopold, E. S., and H. D. MacGinitie. 1972. Development and affinities of Tertiary floras in the Rocky Mountains. In *Floristics and paleofloristics of Asia and Eastern North America*, ed. A. Graham, 147–200. Elsevier Science Publishers, Amsterdam.

Little, S. A., and R. A. Stockey. 2003. Vegetative growth of *Decodon allenbyensis* (Lythraceae) from the middle Eocene Princeton Chert with anatomical comparisons to *Decodon verticillatus*. *Int. J. Plant Sci.* 164:453–69.

Little, S. A., R. A. Stockey, and G. W. Rothwell. 2006. *Stramineopteris aureopilosus* gen. et sp. nov.: Reevaluating the role of vegetative anatomy in the resolution of leptosporangiate fern phylogeny. *Int. J. Plant Sci.* 167:683–94.

Manchester, S. R., and W. J. Kress. 1993. Fossil bananas (Musaceae): *Ensete oregonense* sp. nov. from the Eocene of western North America and its phytogeographic significance. *Am. J. Bot.* 80:1264–72.

Manchester, S. R., V. Wilde, and M. E. Collinson. 2007. Fossil cashew nuts from the Eocene of Europe: Biogeographic links between Africa and South America. *Int. J. Plant Sci.* 168:1199–1206.

Manchester, S. R., Q.-Y. Xiang, and Q.-P. Xiang. 2007. *Curtisia* (Cornales) from the Eocene of Europe and its phytogeographical significance. *Bot. J. Linn. Soc.* 155:127–34.

Martínez-Hernández, E., L. Almedia-Leñero, M. Reyes-Salas, and Y. Betancourt-Aguilar.

1980. Estudio palinológico para la determination de ambientes en la Cuenca Fuentes–Río Escondido (Cretácico superior), region de Piedras Negras, Coahuila. *Rev. Univ. Nacl. Autón. México. Inst. Geol.* 4:167–85.

Millar, C. I. 1998. Early evolution of pines. In *Ecology and biogeography of* Pinus, ed. D. M. Richardson, 69–91. Cambridge University Press, Cambridge.

Miller, K. G., et al. (nine coauthors). 2005. The Phanerozoic record of global sea-level change. *Science* 310:1293–98.

Mohr, B. A. R., M. E. C. Bernardes de Oliveira, and D. W. Taylor. 2008. *Pluricarpellatia*, a nymphaealean angiosperm from the Lower Cretaceous of northern Gondwana (Crato Formation, Brazil). *Taxon* 57:1147–58.

Morley, R. J. 2000. *Origin and evolution of tropical rain forests.* John Wiley and Sons, New York.

Nixon, K. C., and W. L. Crepet. 1993. Late Cretaceous fossil flowers of Ericalean affinity. *Am. J. Bot.* 80:616–23.

Pagani, M., K. Caldeira, D. Archer, and J. C. Zachos. 2006. An ancient carbon mystery. *Science* 314:1556–57.

Passalia, M. G. 2009. Cretaceous pCO_2 estimation from stomatal frequency analysis of gymnosperm leaves of Patagonia, Argentina. *Palaeogeogr. Palaeocl. Palaeoecol.* 273:17–24.

Perry, J. P. Jr., A. Graham, and D. M. Richardson. 1998. The history of pines in Mexico and Central America. In *Ecology and biogeography of* Pinus, ed. D. M. Richardson, 137–49. Cambridge University Press, Cambridge.

Richards, P. W. 1952. *The tropical rain forest: An ecological study.* Cambridge University Press, Cambridge. [2nd ed., with contributions by R. P. D. Walsh, I. C. Baillie, and P. Greig-Smith, published 1996.]

Riley, M. G., and R. A. Stockey. 2004. *Cardstonia tolmanii* gen. et sp. nov. (Limnocharitaceae) from the Upper Cretaceous of Alberta, Canada. *Int. J. Plant Sci.* 165:897–916.

Romero, E. J., and S. Archangelsky. 1986. Early Cretaceous angiosperm leaves from southern South America. *Science* 234:1580–82.

Rydin, C., B. Mohr, and E. M. Friis. 2003. *Cratonia cotyledon* gen. et sp. nov.: A unique Cretaceous seedling related to *Welwitschia*. *Proc. R. Soc. Lond. B* 270 (suppl. 1): S29–S32.

Scotese, C. R. 2004. Cenozoic and Mesozoic paleogeography: Changing terrestrial biogeographic pathways. In *Frontiers of biogeography: New directions in the geography of nature*, ed. M. V. Lomolino and L. R. Heaney, 9–26. Sinauer Associates, Sunderland, Mass.

Skog, J. E., and D. L. Dilcher. 1992. A new species of *Marsilea* from the Dakota Formation in central Kansas. *Am. J. Bot.* 79:982–88.

Skog, J. E., D. L. Dilcher, and F. W. Potter. 1992. A new species of *Isoetites* from the mid-Cretaceous Dakota Group of Kansas and Nebraska. *Am. Fern J.* 82:151–61.

Smith, M. E., A. R. Carroll, and B. S. Singer. 2008. Synoptic reconstruction of a major ancient lake system: Eocene Green River Formation, western United States. *Geol. Soc. Am. Bull.* 120:54–84.

Stamatakos, J. A., J. M. Trop, and K. D. Ridgway. 2001. Late Cretaceous paleogeography of Wrangellia: Paleomagnetism of the MacColl Ridge Formation, southern Alaska, revisited. *Geology* 29:947–50.

Stockey, R. A., and G. W. Rothwell. 1997. The aquatic angiosperm *Trapago angulata* from the Upper Cretaceous (Maastrichtian) St. Mary River Formation of southern Alberta. *Int. J. Plant Sci.* 158:83–94.

———. 2004. Cretaceous tree ferns of western North America: *Rickwoodopteris hirsuta* gen. et sp. nov. (Cyatheaceae s.l.). *Rev. Palaeobot. Palynol.* 132:103–14.

Stockey, R. A., G. W. Rothwell, and S. A. Little. 2006. Relationships among fossil and living Dipteridaceae: Anatomically preserved *Hausmannia* from the Lower Cretaceous of Vancouver Island. *Int. J. Plant Sci.* 167:649–63.

Taylor, W., M. L. DeVore, and K. B. Pigg. 2006. *Susiea newsalemae* gen. et sp. nov. (Nymphaeaceae): *Euryale*-like seeds from the late Paleocene Almont flora, North Dakota, U.S.A. *Int. J. Plant Sci.* 167:1271–78.

Tiffney, B. H. 1985. The Eocene North Atlantic land bridge: Its importance in Tertiary and modern phytogeography of the Northern Hemisphere. *J. Arn. Arbor.* 66:243–73.

Utescher, T., and V. Mosbrugger. 2007. Eocene vegetation patterns reconstructed from plant diversity—a global perspective. *Palaeogeogr. Palaeocl. Palaeoecol.* 247:243–71.

Vandermark, D., J. A. Tarduno, and D. B. Brinkman. 2007. A fossil champsosaur population from the High Arctic: Implications for Late Cretaceous paleotemperatures. *Palaeogeogr. Palaeocl. Palaeoecol.* 248:49–59.

Vavrek, M. J., R. A. Stockey, and G. W. Rothwell. 2006. *Osmunda vancouverensis* sp. nov. (Osmundaceae), permineralized fertile frond segments from the Lower Cretaceous of British Colombia, Canada. *Int. J. Plant Sci.* 167:631–37.

Villar de Seoane, L. 2001. Cuticular study of Bennettitales from the Springhill Formation, Lower Cretaceous of Patagonia, Argentina. *Cretaceous Research* 22:461–79.

Wang, H., and D. L. Dilcher. 2006. Aquatic angiosperms from the Dakota Formation (Albian, Lower Cretaceous), Hoisington III locality, Kansas, USA. *Int. J. Plant Sci.* 167:385–401.

Wang, H., et al. (nine coauthors). 2009. Rosid radiation and the rapid rise of angiosperm-dominated forests. *Proc. Natl. Acad. Sci. U.S.A.* 106:3853–58.

Weber, R. 1973. *Salvinia coahuilensis* nov. sp. del Cretácico superior de México. *Ameghiniana* 10:173–90.

———. 1975. *Aachenia knoblochi* n. sp., an interesting conifer from the Upper Cretaceous Olmos Formation of northeastern Mexico. *Palaeontographica*, Abt. B, 152:76–83.

———. 1976. *Dorfiella auriculata* f. gen. nov., sp. nov., un genero nuevo de helechos acuaticos del Cretácico superior de México. *Bol. Asoc. Latinoam. Paleobot. Palinol.* 3:1–13.

Westgate, J. W., and C. T. Gee. 1990. Paleoecology of a middle Eocene mangrove biota (vertebrates, plants, and invertebrates) from southwest Texas. *Palaeogeogr. Palaeocl. Palaeoecol.* 78:163–77.

Wheeler, E. A. 1991. Paleocene dicotyledonous trees from Big Bend National Park, Texas: Variability in wood types common in the Late Cretaceous and early Tertiary, and ecological inferences. *Am. J. Bot.* 78:658–71.

Wilf, P. 2000. Late Paleocene–early Eocene climate changes in southwestern Wyoming: Paleobotanical analysis. *Geol. Soc. Am. Bull.* 112:292–307.

Wilf, P., et al. (five coauthors). 2003. High plant diversity in Eocene South America: Evidence from Patagonia. *Science* 300:122–25.

Wing, S. L. 1980. Fossil floras and plant-bearing beds of the central Bighorn Basin. In *Early Cenozoic paleontology and stratigraphy of the Bighorn Basin*, ed. P. D. Gingerich, 119–25. University of Michigan Museum Paleontology, Ann Arbor.

———. 1987. Eocene and Oligocene floras and vegetation of the Rocky Mountains. *Ann. Mo. Bot. Gard.* 74:748–84.

Wing, S. L., J. Alroy, and L. J. Hickey. 1995. Plant and mammal diversity in the Paleocene to early Eocene of the Bighorn Basin. *Palaeogeogr. Palaeocl. Palaeocl.* 115:117–55.

Wing, S. L., et al. (five coauthors). 2009. Late Paleocene fossils from the Cerrejó Formation, Colombia, are the earliest record of neotropical rainforest. *Proc. Natl. Acad. Sci. U.S.A.* 106:18627–32.

Wolfe, J. A. 1972. An interpretation of Alaskan Tertiary flora. In *Floristics and paleofloristics of Asia and eastern North America*, ed. A. Graham, 201–33. Elsevier Science Publishers, Amsterdam.

———. 1975. Some aspects of plant geography of the Northern Hemisphere during the Late Cretaceous and Tertiary. *Ann. Mo. Bot. Gard.* 62:264–79.

———. 1977. *Paleogene floras from the Gulf of Alaska region*. U.S. Geological Survey Professional Paper 997. U.S. Department of the Interior, Washington, D.C.

Wolfe, J. A., and D. L. Dilcher. 2000. Late Paleocene through middle Eocene climates in lowland North America. *GFF (Geologiska Föreningens Förhandlingar)* 122:184–85. [Quarterly English-language journal published by the Geological Society of Sweden.]

Zachos, J. C., G. R. Dickens, and R. E. Zeebe. 2008. An Early Cenozoic perspective on greenhouse warming and carbon-cycle dynamics. *Nature* 451:279–83.

Zamaloa, M. C., et al. (five coauthors). 2006. Casuarinaceae from the Eocene of Patagonia, Argentina. *Int. J. Plant Sci.* 167:1279–89.

Ziegler, A. M., et al. (five coauthors). 2003. Tracing the tropics across land and sea: Permian to present. *Lethaia* 36:227–54.

Additional Readings and Updates

Archangelsky, A., S. Archangelsky, D. G. Poiré, and N. D. Canessa. 2008. Registros palinológicos en la Formación Piedra Clavada (Albiano) en su area tipo, Provincia de Santa Cruz, Argentina. *Rev. Mus. Argent. Cienc. Nat.*, n.s. 10:185–98.

Balthazar, M. V., K. R. Pedersen, P. R. Crane, and E. M. Friis. 2007. *Potomacanthus lobatus* gen. et sp. nov., a new flower of probable Lauraceae from the Early Cretaceous (early to middle Albian) of eastern North America. *Am. J. Bot.* 94:2041–53.

Batten, D. J., and M. E. Collinson. 2001. Revision of species of *Minerisporites, Azolla* and associated plant microfossils from deposits of the upper Palaeocene and Palaeocene/Eocene transition in the Netherlands, Belgium and the USA. *Rev. Palaeobot. Palynol.* 115:1–32.

Beerling, D. J., et al. (six coauthors). 2001. Evidence for the recovery of terrestrial ecosystems ahead of marine primary production following a biotic crisis at the Cretaceous-Tertiary boundary. *J. Geol. Soc.* (London) 158:737–740.

Bercovivi, A., J. Wood, and D. Pearson. 2008. Detailed palaeontologic and taphonomic techniques to reconstruct an earliest Paleocene fossil flora: An example from southwestern North Dakota, USA. *Rev. Palaeobot. Palynol.* 151:136–46.

Borsch, T., and P. S. Soltis. 2008. Symposium: Nymphaeales—the first globally diverse clade? *Taxon* 57:1051–1158.

Boucher, L. D., W. D. Tidwell, B. Handley, and S. R. Manchester. 2004. Cretaceous and Eocene plants of eastern Utah and western Colorado. Botanical Society of America Field Trip 15. Botany 2004 Conference, Snowbird, Salt Lake City Utah, 31 July–5 August. http://www.2004.botanyconference.org/.

Bralower, T. J. 2008. Volcanic cause of catastrophe. *Nature* 454:285–87. ["From the timing, it looks as if an episode of marked oceanic oxygen deficiency during the Cretaceous was the result of undersea volcanism. Studies of such events are relevant to the warming world of today." Re Turgeon and Creaser 2008.]

Collinson, M. E. 2001. Cainozoic ferns and their distribution. *Brittonia* 53:173–235.

———. 2002. The ecology of Cainozoic ferns. *Rev. Palaeobot. Palynol.* 119:51–68.

Corbett, S. L., and S. R. Manchester. 2004. Phytogeography and fossil history of *Ailanthus* (Simaroubaceae). *Int. J. Plant Sci.* 165:671–90. [Early Eocene North America and Asia earliest records.]

Dalton, R. 2007. Time traps. *Nature* 449:20–21. ["The whole world felt the effects of the dinosaur-killing mass extinction 65 million years ago. But a spot in Colorado may have the best record of it."]

De la Parra, F., C. A. Jaramillo, and D. L. Dilcher. 2007. Paleoecological changes of spore producing plants through the Cretaceous-Paleocene boundary in Colombia. *Palynology* 32:258–59.

Del Fueyo, G. M., S. Archangelsky, M. Llorens, and R. Cúeno. 2008. Coniferous ovulate cones from the Lower Cretaceous of Santa Cruz Province, Argentina. *Int. J. Plant Sci.* 169:799–813. [Kachike Formation; *Anthrotaxis ungeri, Kachaikestrobus accuminatus*.]

Del Fueyo, G. M., et al. (nine coauthors). 2007. Biodiversidad de las paleofloras de Patagonia Austral durante el Cretácico Inferior. *Ameghiniana Publicación Especial* 11:101–22.

Denk, T., and I.-C. Oh. 2006. Phylogeny of Schisandraceae based on morphological data: Evidence from modern plants and the fossil record. *Plant Syst. Evol.* 256:113–45.

DeVore, M. L., and K. B. Pigg. 2007. A brief review of the fossil history of the family Rosaceae with a focus on the Eocene Okanogan Highlands of eastern Washington state, USA, and British Columbia, Canada. *Plant Syst. Evol.* 266:45–57.

Dyman, T. S., R. G. Tysdal, W. J. Perry, Jr., D. J. Nichols, and J. D. Obradovich. 2008. Stratigraphy and structural setting of Upper Cretaceous Frontier Formation, western Centennial Mountains, southwestern Montana and southeastern Idaho. *Cretaceous Research* 29:237–48.

Eckert, A. J., and B. D. Hall. 2006. Phylogeny, historical biogeography, and patterns of diversification for *Pinus* (Pinaceae): Phylogenetic tests of fossil-based hypotheses. *Mol. Phylog. Evol.* 40:166 82.

Estrada-Ruiz, E., and S. R. S. Cevallos-Ferriz. 2007. Infructescences from the Cerro del Pueblo Formation (late Campanian), Coahuila, and El Cien Formation (Oligocene-Miocene), Baja California Sur, Mexico. *Int. J. Plant Sci.* 168:507–19.

Fluteau, F., G. Ramstein, J. Besse, R. Guiraud, and J. P. Masse. 2007. Impacts of palaeogeography and sea level changes on mid-Cretaceous climate. *Palaeogeogr. Palaeocl. Palaeoecol.* 247:357–81.

Friis, E. M., et al. (five coauthors). 2007. Phase-contrast x-ray microtomography links Cretaceous seeds with Gnetales and Bennettitales. *Nature* 450:549–52.

Gonzalez, C. C., et al. (five coauthors). 2007. Revision of the Proteaceae macrofossil record from Patagonia, Argentina. *Bot. Rev.* 73:235–66. [Tufolitas Laguna del Hunco Formation, early Eocene, Chubut Province: *Lomatia occidentalis, Orites bivascularis, Roupala patagonica*. Ventana, middle Eocene, Río Negro Province: *Lomatia preferrugina*. Río Nirihuau, late Oligocene–early Miocene, Río Negro Province: *Lomatia patagonica*.]

Graham, A.. 2006. Introduction to *Latin American biogeography—causes and effects*. Proceedings of the 51st Annual Systematics Symposium of the Missouri Botanical Garden (Alan Graham, organizer). *Ann. Mo. Bot. Gard.* 93:173–358.

Harrington, G. J., E. R. Clechenko, and D. C. Kelly. 2005. Palynology and organic-carbon iso-
    tope ratios across a terrestrial Palaeocene/Eocene boundary section in the Williston Basin,
    North Dakota, USA. *Palaeogeogr. Palaeocl. Palaeoecol.* 226:214–32.

Heimhofer, U., P. A. Hochuli, S. Burla, and H. Weissert. 2007. New records of Early Creta-
    ceous angiosperm pollen from Portuguese coastal deposits: Implications for the timing of
    the early angiosperm radiation. *Rev. Palaeobot. Palynol.* 144:39–76.

Iglesias, A. P., et al. (six coauthors). 2007. A Paleocene lowland macroflora from Patagonia
    reveals significantly greater richness than North American analogs. *Geology* 35:947–50.

Jacques, F. M. B. 2009. Fossil history of the Menispermaceae (Ranunculales). *Ann. Paléontol.*
    95:53–69.

Jaramillo, C., et al. (six coauthors). 2007. The palynology of the Cerrejón Formation (upper
    Paleocene) of northern Colombia. *Palynology* 31:153–89.

Karafit, S. J., and R. A. Stockey. 2008. *Paralygodium meckertii* sp. nov. (Schizaeaceae) from
    the Upper Cretaceous (Coniacian) of Vancouver Island, British Columbia, Canada. *Rev.
    Palaeobot. Palynol.* 149:163–73.

Keller, G., S. Abramovich, Z. Berner, and T. Adatte. 2009. Biotic effects of the Chicxulub
    impact, K-T catastrophe and sea level change in Texas. *Palaeogeogr. Palaeocl. Palaeoecol.*
    271:52–68. [Contends that " the Chicxulub impact and the K-T mass extinction are two
    separate and unrelated events, and that the biotic effects of this impact have been vastly
    overestimated."]

Kraus, M. J., and S. Riggins. 2007. Transient drying during the Paleocene-Eocene Thermal
    Maximum (PETM): Analysis of paleosols in the Bighorn Basin, Wyoming. *Palaeogeogr.
    Palaeocl. Palaeoecol.* 245:444–61.

Labandeira, C. C., K. R. Johnson, and P. Wilf. 2002. Impact of the terminal Cretaceous event
    on plant-insect associations. *Proc. Natl. Acad. Sci. U.S.A.* 99:2061–66.

Little, S. A., R. A. Stockey, and G. W. Rothwell. 2006. *Solenostelopteris skogiae* sp. nov. from
    the Lower Cretaceous of Vancouver Island. *J. Plant Res.* 119:525–32.

Lomax, B. H., D. J. Beerling, G. R. Upchurch Jr., and B. L. Otto-Bliesner. 2000. Terrestrial
    ecosystem responses to global environmental change across the Cretaceous-Tertiary
    boundary. *Geophys. Res. Lett.* 27:2149–52.

Manchester, S. R. 2001. Leaves and fruits of *Aesculus* (Sapindales) from the Paleocene of
    North America. *Int. J. Plant Sci.* 162:985–98. [Describes leaves of *A. hickeyi* sp. nov. from
    the Paleocene Fort Union Formation of North Dakota and Wyoming.]

———. 2002. Leaves and fruits of *Davidia* (Cornales) from the Paleocene of North America.
    Syst. Bot. 27:368–82.

Manchester, S. R., and L. J. Hickey. 2007. Reproductive and vegetative organs of *Browniea* gen.
    n. (Nyssaceae) from the Paleocene of North America. *Int. J. Plant Sci.* 168:229–49.

Manchester, S. R., Q.-Y. Xiang, T. M. Kodrul, and M. A. Akhmetiev. 2009. Leaves of *Cornus*
    (Cornaceae) from the Paleocene of North America and Asia confirmed by trichome
    characters. *Int. J. Plant Sci.* 170:132–42. [Recognizes *C. Swingii* sp. nov. from the Paleocene
    of Wyoming, Montana, and North Dakota, and *C. krassilovii* sp. nov. from the Paleocene
    Tsagayan flora of Russia.]

McClain, A. M., and S. R. Manchester. 2001. *Dipteronia* (Sapindaceae) from the Tertiary of
    North America and implications for the phytogeographic history of the Aceroideae. *Am. J.
    Bot.* 88:1316–25.

Miller, I. M., M. T. Brandon, and L. J. Hickey. 2006. Using leaf margin analysis to estimate the mid-Cretaceous (Albian) paleolatitude of the Baja BC block. *Earth Planet. Sci. Lett.* 245:95–114.

Miller, K. G., et al. (six coauthors). 2004. Upper Cretaceous sequences and sea-level history, New Jersey Coastal Plain. *Geol. Soc. Am. Bull.* 116:368–93.

Mourier, T., et al. (seventeen coauthors). 1988. The Upper Cretaceous–lower Tertiary marine to continental transition in the Bagua Basin, northern Peru, paleontology, biostratigraphy, radiometry, correlations. *Newsl. Stratigr.* 19:143–77.

Nichols, D. J., and K. R. Johnson. 2008. *Plants and the K-T boundary*. Cambridge Paleobiology Series. Cambridge University Press, Cambridge.

Peppe, D. J., J. M. Erickson, and L. J. Hickey. 2007. Fossil leaf species from the Fox Hills Formation (Upper Cretaceous: North Dakota, USA) and their paleogeographic significance. *J. Paleontol.* 81:550–67.

Pigg, K. B., and M. L. DeVore. 2005. *Paleoactaea* gen. nov. (Ranunculaceae) fruits from the Paleogene of North Dakota and the London Clay. *Am. J. Bot.* 92:1650–59.

Pigg, K. B., R. M. Dillhoff, M. L. DeVore, and W. C. Wehr. 2007. New diversity among the Trochodendraceae from the early/middle Eocene Okanogan Highlands of British Columbia, Canada, and northeastern Washington State, United States. *Int. J. Plant Sci.* 168:521–32.

Poinar, G., and R. Poinar. 2008. *What bugged the dinosaurs? Insects, disease, and death in the Cretaceous*. Princeton University Press, Princeton.

Pons, D., P.-Y Berthou, J. Melo Filgueira, and J. Alcantara Sampaio. 1994. Palynologie de unites lithostrtigraphiques "Fundão," "Crato" et "Ipubi" (Aptien Supéieur à Albien Inférieur-Moyen, Bassin d'Arapire, NE du Brésil): Enseignements paléoécologiques, stratigraphiques et climatologiques. *Géologie de l'Afrique et de l'Atlantique Sud: Actes Colloques Angers*, 1994, 383–401.

Prámparo, M. B., M. E. Quattrocchio, M. A. Gandolfo, M. C. Zamaloa, and E. J. Romero. 2007. Historia evolutiva de las angiospermas (Cretácico-Paleogeno) en Argentina a traves de los registrol paleoflorísticos. *Ameghiniana Publicaçión Fspecial* 11.157–72.

Prasad, V., C. A. E. Strömberg, H. Alimohammadian, and A. Sahni. 2005. Dinosaur coprolites and the early evolution of grasses and grazers. *Science* 310:1177–80.

Scott, A. C., B. H. Lomax, M. E. Collinson, G. R. Upchurch, and D. J. Beerling. 2000. Fire across the K-T boundary: Initial results from the Sugarite Coal, New Mexico, USA. *Palaeogeogr. Palaeocl. Palaeoecol.* 164:381–95.

Smedmark, J. E. E., and A. A. Anderberg. 2007. Boreotropical migration explains hybridization between geographically distant lineages in the pantropical clade Sideroxyleae (Sapotaceae). *Am. J. Bot.* 94:1491–1505.

Smith, J. F., A. C. Stevens, E. J. Tepe, and C. Davidson. 2008. Placing the origin of two species-rich genera in the Late Cretaceous with later species divergence in the Tertiary: A phylogenetic, biogeographic and molecular dating analysis of *Piper* and *Peperomia* (Piperaceae). *Plant Syst. Evol.* 275:9–30.

Stockey, R. A. 2006. The fossil record of basal monocots. *Aliso* 22:91–106.

Stockey, R. A., T. C. Lantz, and G. W. Rothwell. 2006. *Speirseopteris orbiculata* gen. et sp. nov. (Thelypteridaceae), a derived fossil filicalean from the Paleocene of western North America. *Int. J. Plant Sci.* 167:729–36. [Paskapoo Formation, 57 Ma, central Alberta, Canada.]

Stockey, R. A., G. W. Rothwell, and A. B. Falder. 2001. Diversity among taxodioid conifers: *Metasequoia foxii* sp. nov. from the Paleocene of central Alberta, Canada. *Int. J. Plant Sci.* 162:221–34.

Stockey, R. A., G. W. Rothwell, and K. R. Johnson. 2007. *Cobbania corrugata* gen. et sp. nov. (Araceae): A floating aquatic monocot from the Upper Cretaceous of western North America. *Am. J. Bot.* 94:609–24.

Stockey, R. A., and N. J. P. Wiebe. 2008. Lower Cretaceous conifers from Apple Bay, Vancouver Island: *Picea*-like leaves, *Midoriphyllum piceoides* gen et sp. nov. (Pinaceae). *Botany* (NRC Canada) 86:649–57.

Turgeon, S. C., and R. A. Creaser. 2008. Cretaceous oceanic anoxic event 2 triggered by a massive magmatic episode. *Nature* 454:323–26.

Ufnar, D. F., G. A. Ludvigson, L. González, and D. R. Gröcke. 2008. Precipitation rates and atmospheric heat transport during the Cenomanian greenhouse warming in North America: Estimates from a stable isotope mass-balance model. *Palaeogeogr. Palaeocl. Palaeoecol.* 266:28–38.

Villar de Seoane, L., and S. Archangelsky. 2008. Taxonomy and biostratigraphy of Cretaceous megaspores from Patagonia, Argentina. *Cretaceous Research* 29:354–72.

Westerhold, T., et al. (seven coauthors). 2008. Astronomical calibration of the Paleocene time. *Palaeogeogr. Palaeocl. Palaeoecol.* 257:377–403.

Wheeler, E. A., and T. M. Lehman. 2005. Upper Cretaceous–Paleocene conifer woods from Big Bend National Park, Texas. *Palaeogeogr. Palaeocl. Palaeoecol.* 226:233–58. [Aguja, Javelina (Upper Cretaceous) and Black Peaks (Paleocene) formations.]

Wilf, P., K. C. Beard, K. S. Davies-Vollum, and J. W. Norejko. 1998. Portrait of a late Paleocene (early Clarkforkian) terrestrial ecosystem: Big Multi Quarry and associated strata, Washakie Basin, southwestern Wyoming. *Palaios* 13:514–32.

Wilf, P., C. C. Labandeira, K. R. Johnson, P. D. Coley, and A. D. Cutter. 2001. Insect herbivory, plant defense, and early Cenozoic climate change. *Proc. Natl. Acad. Sci. U.S.A.* 98:6221–26.

Wilf, P., et al. (six coauthors). 2000. Timing the radiations of leaf beetles: Hispines on gingers from latest Cretaceous to Recent. *Science* 289:291–94.

Williams, C. J., B. A. LePage, A. H. Johnson, and D. R. Vann. 2009. Structure, biomass, and productivity of a late Paleocene Arctic forest. *Proc. Acad. Nat. Sci. Phila.* 158:107–27.

Wing, S., et al. (five coauthors). 2005. Transient floral change and rapid global warming at the Paleocene-Eocene boundary. *Science* 310:993–96. [Notes biotic changes correlated with a 5C°–10°C warming around 55.8 Ma.]

Yang, Y. 2007. The nomenclature of fossil Ephedraceae. *Taxon* 56:1271–73. [Macrofossils of *Ephedrites* and *Liaoxia*.]

6

Transition

Middle Eocene through the Early Miocene

Momentous events occurred during the interval of the middle Eocene through the early Miocene (50–16.3 Ma), including a precipitous drop in temperature beginning around 52 Ma and continuing through the end of the Eocene (35–34 Ma). This period was followed by a less eventful transition time during the Oligocene and early Miocene, 35–34 Ma through 16.3 Ma (fig. 3.4). The cumulative effect of the changes was to significantly alter the old Paleogene world and set the stage for the next inevitable series of developments in the Earth's biota later in the Neogene. Atmospheric CO_2 concentration diminished to the point that climates cooled, seasonality increased, and temperate conditions spread. This was due in part to a slowing of plate movement, from an annual average of about 12–15 cm in the Jurassic and Early Cretaceous to about 2–3 cm at present. With a reduced output of CO2, less heat remained trapped in the Earth's atmosphere, and temperatures fell.

Superimposed on this trend was a period of significant uplift of major mountain systems, which increased the weathering and erosion of silicate rocks, further reducing CO_2 concentration, and providing new temperate habitats in the highlands. The Rocky Mountains and the Sierra Madre were at least half their modern elevations by the end of the Eocene, and early uplift in the Sierra Nevada began at the Eocene-Oligocene boundary at 33 Ma. The principal rise of the Andes Mountains would be later in the Miocene. During this interval of the middle Eocene through the early Miocene, South America separated from Antarctica through the Drake Passage (32–28 Ma), and it remained an island continent for some 29 million years until reconnected with North America around 3.5 Ma.

With cooler temperatures, there was also less evaporation of water from the ocean surface, resulting in less atmospheric moisture and greater seasonality. The continued drainage of epicontinental seas further promoted continentality. Paleosols from the Badlands of South Dakota show a decrease in annual precipitation from 1000 mm in the late Eocene, to 500–900 mm in the early Oligocene, to 450–500 mm later in the Oligocene, to 250–450 mm at the Oligocene-Miocene boundary (Retallack 1983, 1990). Sea levels were falling because reduced amounts of crustal material were being generated at the midocean ridges, and because early glacial ice was forming in the Eocene on Antarctica, Greenland, and in the Arctic. More widespread glaciations began around 35 Ma (fig. 3.4), especially on Antarctica, accounting for the notable drop in sea level around 30 Ma (fig. 3.1). Furthermore, the inexorable process of evolution continued to introduce new forms with varied ecological characteristics. Thus, the hallmarks of this interval were cooler and more seasonal climates, increased topographic heterogeneity, and continued phylogenetic and ecologic diversification of plant lineages and communities

In addition to these trends of a global scope, there were others that affected more specifically the biotas of northern North America. Spreading along the Mid-Atlantic Ridge continued into the North Atlantic until by around 45 Ma the connection between North America and Europe was becoming significantly disrupted. This was at the same time that climates were cooling. The greatest impact of the cooling was in the polar regions because greater amounts of solar energy are absorbed during the longer distance traveled to the Earth's surface at the poles than at the equator (I, 28–29). Also, a given amount of energy is distributed over the increasingly slanted surface toward the poles than at the equator, which is more perpendicular to the sun. Palynofloras in Greenland across the Eocene-Oligocene boundary (33.5 Ma) indicate winter temperatures declined from more than 5°C to 0°C–2°C, and MAP fluctuated around 120 cm (Eldrett et al. 2009). The combined effect of land disruption and cooling climates was to dramatically decrease the exchange of life-forms, especially tropical organisms and those with limited capacity for dispersal, between North America and Europe across the North Atlantic land bridge after the Eocene, and to increase the migrations of temperate organisms and those with greater capacity for dispersal, over the Bering land bridge between North America and Asia in the later Tertiary and Quaternary. The result was a realignment in the affinities of the North American fauna, recognized in the biogeographic literature as the "Grande Coupure." There was a 60 percent change in the land mammal fauna due to increased migration of rodents and large mammals from Asia. The shift was from small mammals and arboreal forms of the forest, including many primates, to large herding ungulates of more open habitats—the beginning of the North American land mammal megafauna.

Another "coupure," but less "grande," was the contraction of tropical vegetation toward the middle latitudes, into the lowlands, and along the coasts, with a corresponding expansion of temperate and deciduous vegetation from polar and highland habitats. There was also a noticeable modernization of the flora at the generic level. A large number of plant fossils, particularly fossil pollen, of early Eocene and older ages are difficult to relate to extant genera, hence the use of an artificial system of nomenclature. In the middle Eocene, however, most of the fossil specimens are similar morphologically to extant genera and to some species. When the purpose of a paleobotanical (plant macrofossil) or paleopalynological (plant microfossil) study is geological—zonation of formations, stratigraphic correlation—an artificial system is still often used for middle Eocene and younger plant fossils. If the intent is biological, however—lineage history, vegetation and environmental reconstruction, biogeography—the long-term, immensely time-consuming efforts at building a comprehensive, vouchered reference collection and establishing the biological affinities of the specimens is both possible and necessary if the full potential of the methods are to be realized (chap. 4).

Modern lineages make for modern communities, and in the middle Eocene as older vegetation types were disappearing, new and familiar ecosystems appeared, versions of others became more recognizable, and individual elements evolved that would coalesce later in the Tertiary. The middle Eocene was a pivotal time when development of the vegetation accelerated from essentially ancient to essentially modern at the level of genera, plant formations, and ecosystems.

In terms of global trends and new paradigms, the Oligocene and early Miocene represent a 20-million-year period of adjustment between the conditions of the Paleocene and early Eocene and the dramatically new environments that began to develop in the middle Miocene (15 Ma)—a time

of relative pause. It is not generally commented on but this seemingly uneventful transitory period—its relative quiescence suggested by the Oligocene interval on the paleotemperature curve (fig. 3.4)—was important. If
the trend in extinctions and redistributions from older times had continued
unabated into the ice ages, without the benefit of a transition interval, we
would likely be living under a very different set of global ecosystems.

NORTH AMERICA (NORTH OF MEXICO)

To the far north, these adjustments are evident in fossil floras from Alaska
(Wolfe 1972, 1977, 1992). Data presented in table 6.1 show a decrease in
MAT between the late Paleocene to the early Oligocene from about 12.3°C
to 9.0°C, and a decrease in growing season precipitation from around
1450 mm to 600 mm. The boreotropical flora developed a less tropical and
more boreal aspect, with *Alnus* (alder) and *Ulmus* (elm), for example, and
there were an increasing number of deciduous, temperate-habitat plants
shared with eastern Asia. In the early to middle Eocene McAbee flora of
British Colombia, the oldest remains of *Fagus* are recorded (Dillhoff et al.
2005; see also Mindell et al. 2009). Many older fossils referred to *Ulmus*
are now treated as *Chaetoptelea*, and the oldest elm fossils in North America
are from the late part of the early Eocene (Denk and Dillhoff 2005). By the
beginning of the Oligocene, the vegetation of the far north was essentially a
deciduous forest—*Acer* (maple), *Betula* (birch), *Carya* (hickory, pecan), *Cas-*

Table 6.1. Alaskan Paleogene floras and inferred climates

Assemblage	Age	MAT	CMMT	WMMY	MART	MGSP
Redoubt	E Oligo	9.0	−4.0	22.0	26.0	60?
Rex Creek	L Eoc	12.3	0.3	24.3	24.0	40?
Katalla	L Eoc	16.2	6.6	25.6	19.2	120
Carbon Mt.	M-L Eoc	8.7	−0.8	18.2	19.0	?
Aniakchak	M Eoc	13.7	3.3	24.1	20.8	>145
Charlotte Rdg.	M Eoc	17.2	8.6	25.8	17.2	>145
Kulthieth	E Eoc	19.4	12.2	26.2	13.6	>145
Chickaloon	L Paleo	12.3	1.1	23.5	22.4	>145

Source: Adapted from Wolfe 1994.
Note: MAT = mean annual temperature (°C); CMMT = cold-month mean temperature (°C);
WMMT = warm-month mean temperature (°C); MART = mean annual range of temperature
(°C); MGSP = mean growing season precipitation (cm); E = early; M = middle; L = late.

tanea (chestnut), *Fagus* (beech), *Juglans* (walnut), *Liquidambar* (sweetgum), *Platanus* (sycamore), *Prunus* (cherry), *Quercus* (oak), and *Ulmus*—brought about largely by the disappearance of tropical forms. The deciduous forest differed from its modern counterpart now in eastern North America by the presence of a few more tropical holdovers, particularly in coastal areas, and especially by a number of present-day Asian angiosperms (*Pachysandra, Platycarya, Pterocarya*) and deciduous gymnosperms (*Ginkgo, Glyptostrobus, Metasequoia*). Modernization of the New World deciduous forest would occur toward the end of the Miocene and in the Pliocene, when most of the mesothermal to megathermal and present-day Asian genera disappeared.

To the south is the middle Eocene Princeton Chert flora of southern British Columbia, with representative Lauraceae, *Abies* (fir), and *Pinus* (pine), circa 50–45 Ma (Erwin and Stockey 1991; Crane and Stockey 1987; Pigg and Stockey 1996; Smith et al. 2006; Little et al. 2009), along with the Republic flora of northeastern Washington, with *Thuja* (arborvitae), *Abies, Picea* (spruce), *Tsuga* (hemlock), and *Betula*, circa 50–49 Ma (Wolfe and Wehr 1987; Pigg et al. 2001). The Republic flora is important for tracing the history of New World ecosystems because it is the first assemblage where fir, spruce, pine, arborvitae, hemlock, and birch are recorded together in one region. These elements were coalescing in the highlands of the northern Rocky Mountains during the middle Eocene to form a recognizable coniferous forest ecosystem. From this site it would later spread northward into the lowlands to form the boreal forest, and north and south into the uplands to form the western montane coniferous forest with continued rise of the Rocky Mountains and other western cordilleras.

The trends in climate, orogeny, and vegetation from this interval are known in sufficient detail that the kind of paleocommunity can often be anticipated given its age, location, and the physical environment of the region, for example, at a western site of low to moderate elevation, close to the coast, in the absence of significant highlands to the west, near the early to middle Eocene boundary. These parameters are met by the Clarno flora in the John Day Basin of north-central Oregon (Manchester 1994; Wheeler and Manchester 2002). The age (44 Ma), its location within 100 km of the paleo–Oregon coastline, and the absence of highlands in the Coast Ranges and the Cascade Mountains would suggest a paratropical forest beginning the transition to a mixed deciduous angiosperm-gymnosperm forest. In fact, among the approximately 145 genera and 173 species in this large and important flora, 52 percent have entire-margined leaves, 43 percent are lianas, and broad-leaved evergreens are abundant. There are *Sabal* (palms), *Ensete* (Musaceae, banana family), other primarily tropical to subtropical families such as the Annonaceae, Araliaceae, Burseraceae, and Lauraceae, and one crocodilian. In addition, however, there are also fossils of *Pinus*, *Cornus* (dogwood), *Nyssa* (tupelo, or sour gum), *Juglans*, *Magnolia*, *Platanus* (sycamore), *Prunus* (cherry), *Celtis* (hackberry), and *Vitis* (grape). Twenty-four percent of the genera are found in the early to middle Eocene of Europe, but strongest affinities are with Asia (*Ginkgo*, *Actinidia*, *Alangium*, *Ampelocissus*, *Eucommia*, *Iodes*, *Schizandra*), and there are Asian species among other genera. The MAT is estimated at 16°C, at the lower limit of a paratropical forest, and fossil woods show some seasonality in precipitation. As expected, this was a vegetation retaining many attributes of the earlier Eocene tropical communities, but showing the effects of seasonality and cooling temperatures that would next convert the regional biota into a temperate deciduous forest. In the slightly younger (38.8 Ma) John Day Formation, the MAT is estimated at about 14°C (Manchester 2000). Present was *Quercus*, among the oldest records for the genus, and *Pinus* (Cretaceous/Paleocene), that along with *Juniperus* in the late Eocene floras of Montana would differentiate elements available to form dry shrubland/chaparral-woodland-savanna later in the Tertiary with increased seasonality and rain shadow effects from the rising mountains. The modern sagebrush (*Artemisia*) desert of the region would not appear until the Pliocene and Quaternary.

Moving inland from the coast, upward in elevation, and into late Eocene time (to about 40 Ma), the Copper Basin flora of northeastern Nevada, as expected, is a microthermal mixed deciduous hardwood-gymnosperm assemblage of *Abies*, *Picea*, *Pinus*, *Larix*, and *Tsuga*, with a warmer, more mesothermal flora lower on the slopes: *Pseudotsuga* (Douglas fir), *Mahonia*, *Rhododendron*, and *Sassafras* (Axelrod 1966). The MAT is estimated at around 11°C.

The middle Eocene Green River flora of northern Colorado-Utah on the east-facing slopes of the Rocky Mountains circa 45 Ma (Smith et al. 2008) provides insight on vegetation developing under a rain shadow. The mountains in the vicinity were approximately 1.3 km higher than the basin, and the flora grew around a series of lakes covering an area of 65,000 km². There were seventy-seven depositional cycles that generally conform to the 100,000-year precessional cycle of the Milankovitch variations, and the sedimentology shows that in twenty-five of these episodes the lakes dried up to form playas and salt flats. The Green River Formation is known for its exquisitely preserved fish fauna, and the flora includes plants of drier habitats like *Ephedra*, *Pinus*, *Bursera*, *Canavalia* (*C. maritima* is a common pantropical shore plant), *Cardiospermum*, *Celtis*, *Quercus*, *Sapindus*, and *Caesalpinia*. This assemblage suggests an open oak-piñon pine woodland, or a type of savanna with few grasses. *Typha* (cattail) and *Populus* were present

(Manchester et al. 2006), adding to the inventory of angiosperms in fresh-water marsh and riverine vegetation. MacGinitie 1969 calls the vegetation Orizaban in analogy with the seasonally dry woodland-savanna on Mount Orizaba in Mexico; Wolfe 1985 calls it a semideciduous tropical dry forest similar to the short-lived middle Eocene seasonally dry tropical forest of the southeastern United States (Bullock et al. 1995; Graham and Dilcher 1995). The Copper Basin and the Green River floras show the mosaic of vegetation types of increasingly modern aspect at the formation level that were developing in the American west during the middle to late Eocene. Further details are contributed by other large and important fossil floras in the region that are often associated with extensive faunas. The Floris-sant flora to the southeast in central Colorado is dated to the late Eocene or early Oligocene, around 35 Ma (Gregory-Wodzicki 2001; Leopold and Clay-Poole 2001; Manchester 2001; Meyer 2001; Meyer and Smith 2008). It includes a rich mesic forest growing along streams and lakes, and a drier evergreen oak-pine woodland on the slopes. The MAP is estimated at 508–635 mm and winter dry, and the MAT at 18°C (MacGinitie 1953) to a seem-ingly low 12.5°C (CLAMP) depending on the paleoelevation assigned to the site (I, 204–10). The Beaverhead Basins floras (Becker 1961, 1969) are to the northwest in southeastern Montana and are also late Eocene to early Oligocene in age. They show a similar aquatic, lake and riverside (*Typha*), and floodplain vegetation (*Glyptostrobus, Taxodium, Salix*) in the lowlands, a mesic deciduous forest on the slopes (*Ginkgo, Metasequoia, Sequoia, Acer, Quercus; Fraxinus,* ash; *Ulmus; Zelkova,* an Asian tree related to elm), and a western montane coniferous forest at the highest elevations (*Abies, Picea*). Also present were a substantial number of drier-habitat plants such as *Ju-niperus* (juniper), *Arctostaphylos, Berberis* (barberry), *Cercocarpus, Mahonia,* and *Potentilla.* The driest of the floras in the Beaverhead Basins is the Ruby flora, which is also the youngest (and so further along the trend toward dryness and seasonality), and it adds *Ephedra* (Mormon tea, an extreme xerophyte) and *Prosopis* (mesquite). These assemblages show that by the end of the Eocene, versions of a shrubland/chaparral-woodland-savanna ecosystem and elements of a desert vegetation occupied the coarse soils, rain shadow habitats, and steeper slopes of the drier intermontane basins of the Rocky Mountain region, and they would coalesce and spread in the later Tertiary and Quaternary.

To the far northeast, at about 81°N, fossil floras (Axelrod 1984) and fau-nas (McKenna, 1983) provide an estimate of 13°C MAT in the middle to late Eocene, while in the southeastern United States at about 35°N MAT was 24°C, for a gradient of 0.28°C/1° latitude. The present gradient is

0.5°C/1° latitude, and the steeper trend intensified near the beginning of the middle Eocene. Between these two sites is the Brandon Lignite of west-central Vermont, which has been well-known since the 1840s as a source of fuel for the Brandon Iron and Car Wheel Company. It is important because there are so few floras of Tertiary age in the northeastern United States. The exact age is difficult to determine because it is underlain by Precambrian quartzite, overlain by Pleistocene drift, and the sediments are not suitable for radiometric dating. The current estimate is early Miocene (Tiffney and Traverse 1994). It contains an exceptional assemblage of fruits and seeds (e.g., Tiffney 1977; Tiffney and Traverse 1994) and pollen (Traverse 1955). Among the fossils are *Carya*, *Castanea* (chestnut), *Ilex* (holly), *Nyssa*, *Quercus*, *Ulmus*, and *Vitis* (grape). In addition, there is a present-day Asian component of *Glyptostrobus*, *Alangium*, *Engelhardia*, *Pterocarpus*, and *Sargentodoxa* (a deciduous vine of southeastern Asia). Other plants represented are *Cyrilla*, *Gordonia*, *Persea*, and *Symplocos*, currently of the southeastern United States and absent in areas with significant frost. The MAT is estimated at around 17°C (at present 7.6°C) or intermediate between those to the far north and in the southeastern United States. It was a mixed deciduous forest and had broad geographic affinities with Europe and especially with eastern Asia.

The Eocene vegetation of the Jackson, Clairborne, and Wilcox Group in southeastern United States is known from early studies by Berry (1916), and is currently being revised by David Dilcher and colleagues (e.g., Dilcher 1973; Dilcher and Lott 2005). The previously reported *Taxodium* has been found to be *Podocarpus*; *Aralia* is *Dendropanax*; and the Proteaceae, thought to be represented by four species, is not present. The vegetation in the uplands was a warm-temperate forest trending toward temperate to cool-temperate forest later in the Oligocene. The lowland vegetation was deposited in oxbow lakes and other basins varying distances from the coast, and it was a moist to seasonally dry community later becoming warm-temperate, with pine forest developing on edaphically dry coastal sands. The middle Eocene Claiborne Group (Laredo Formation) of south Texas has shark, skates, rays, *Tarpon*, and Crocodylidae, along with the present-day Old World mangrove *Nypa* (Westgate and Gee 1990). At this time *Nypa* (fig. 6.1) also occurred along the Gulf Coast to Florida (Jarzen and Dilcher 2006), and southward to northern South America.

An important development in the modernization of New World ecosystems in middle Eocene and slightly later times was the first appearance of *Rhizophora* in mangrove communities occupying coastal brackish-water habitats (Ellison 2008; Graham 1995, 2006). Prior to that time, *Nypa* (*Spi-*

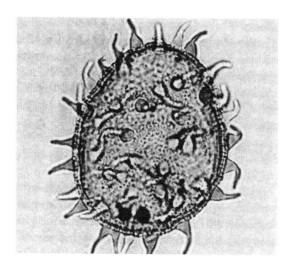

Figure 6.1 *Spinozonocolpites prominatus (Nypa)* from the Eocene Laredo Formation (Claiborne Group; Casa Blanca flora and fauna), Webb County, southeastern Texas. From Westgate and Gee 1990. Used with permission from Elsevier Science Publishers, Amsterdam.

nozonocolpites) and a pollen type of unknown affinities called *Brevitricolpites variabilis* in northern South America are found in lignite sediments. But in the middle to late Eocene, these species disappeared and *Rhizophora* became the prominent member of mangrove vegetation. This transition converted pre–middle Eocene brackish-water coastal communities to essentially modern counterparts of mangrove vegetation. Thus, the mangrove community joins the coniferous forest of western montane northern North America as another modern ecosystem appearing in the middle Eocene. Also present was the lowland neotropical rain forest from Paleocene times, along with communities of freshwater aquatic plants, present in the Cretaceous and diverse, for example, in the middle Eocene Princeton chert (Cevallos-Ferriz et al. 1991). There was a limited but expanding deciduous forest differentiating from the old polar broad-leaved deciduous forest and coalescing with elements elsewhere. Likely present in near-modern aspect was a freshwater herbaceous bog/marsh/swamp and, difficult to detect because of poor taphonomic (preservation) conditions, a beach/strand/dune plant formation. Versions of the shrubland/chaparral-woodland-savanna were present in the Green River flora, and briefly in the southeastern United States as a dry tropical forest during the middle Eocene. This makes eight of the twelve New World ecosystems, or recognizable variations, present by the end of the early Miocene, along with elements of four others (desert, grassland, alpine tundra, and tundra). The latter two coalesced as conditions became colder and drier later in the Tertiary.

MEXICO, THE ANTILLES, CENTRAL AMERICA

Widespread seasonally drying climates, expressed primarily as less winter rainfall, are evident from sediments in northern Baja California (Peterson and Abbott 1979). The paleosols show a sequence from quartz-kaolinite and cation-depleted soils in the Paleocene and early Eocene, indicating a humid tropical climate (MAP 1250–1900 mm), to caliche, vermiculite, and smectite clays in the late Eocene, characteristic of drier and seasonal climates (MAP 630 mm). The present MAP is 250 mm. These trends in the later Eocene were toward drier and more seasonal climates. Climatically, conditions were not arid, but local arid habitats existed because of the augmenting effects of topography and soil. Drying climates intensified, and dry to arid vegetation continued to develop and expand, especially in middle Miocene through Pliocene times, and during each of the dry glacial intervals of the Quaternary.

There are few extensive and recently studied fossil floras available from the middle Eocene through the early Miocene in northern Latin America, especially ones integrated into multifaceted investigations most effective for reconstructing and tracing the history of ecosystems. The Eocene floras, in particular, often contain marginally preserved fossils that are difficult to identify accurately. Two sites have provided information on Oligocene vegetation of northern Latin America. One is the Los Ahuehuetes flora from the Pié de Vaca Formation in the state of Puebla, Mexico. It is probably late Oligocene to possibly early Miocene in age. The plant macrofossils are currently under study by Susana Magallón-Puebla and Sergio Cevallos-Ferriz, and the microfossils by Enrique Martínez-Hernández and colleagues at UNAM (e.g., Magallón-Puebla and Cevallos-Ferriz 1994; Martínez-Hernández and Ramírez-Arriaga 1999). The flora contains at least one present-day temperate Asian plant familiar from fossil floras to the north. The genus *Eucommia* is native to western and central China, and its presence in southern Mexico documents the extension of present-day Asian elements into northern Latin America during the Tertiary. In addition to this and other mesic species in the flora, drier vegetation is indicated by *Haplorhus*, *Cercocarpus*, *Prosopis* (mesquite), and *Sophora*. The flora was deposited near the confluence of the southern Sierra Madre Oriental and the just-rising eastern Transvolcanic Belt in a region of at least moderate topographic diversity. In the same area today, there is mesic vegetation, as well as drier communities resulting from edaphic conditions, slope, exposure, and rain shadow habitats such as the *Nolina-Hechtia-Agave* near desert on the border of Puebla and Veracruz. The fossil flora documents the continuing development of the drying and seasonal environments seen earlier, for example, in the middle Eocene Green River flora of Colorado and Utah. It further provides a picture of the mosaic of vegetation at this latitude in the transitional times of the Oligocene (Pälike et al. 2006) that included a shrubland/chaparral-woodland-savanna formation.

To the south in the state of Chiapas, just north of Tuxtla Gutiérrez and San Cristóbal, there are microfossil-bearing lignites in the La Quinta Formation of Oligo-Miocene age around the village of Simojovel (figs. 6.2 and 6.3a–d. This was the fateful cemetery site where we were held at bay by the residents until our intentions were made clear. The sediments contain abundant fossil spores and pollen. Langenheim, Hackner, and Bartlett (1967) studied material from the site, providing several biological identifications, and since then further studies have allowed reconstruction of the vegetation and paleoenvironments in greater detail (Graham 1999; Graham and Palacios Chávez 1996). The lignites as sediments, and their content of pollen, including the brackish-water mangroves *Pelliceria* (figs. 6.4 and 6.5) and *Rhizophora* (fig. 6.6; see also fig. 2.24), place the Oligo-Miocene shoreline about 90 km inland from its present position, reflecting high sea levels resulting from the absence of extensive continental glaciers. Also present were spores of the floating fern *Ceratopteris* (figs. 6.3a, 2.25) and the aquatic angiosperm *Pachira* that give a modern aspect to this community by the Oligocene. Pollen of some warm-temperate (e.g., *Alfaroa-Oreomunnea*; fig. 6.3c) and tropical plants was present (e.g., the tree fern *Sphaeropteris-Trichipteris*, fig. 6.3b; *Crudia*, fig. 6.3d), but dominants of the present-day lowland neotropical rain forest in Mexico were not found, so the moist forests likely represented the wet phase of a lower to upper montane broad-leaved forest ecosystem. Most tropical trees comprised by the canopy today derived from South America, and as noted in chapter 2, the connection across the Isthmian region would not be completed until about 3.5 Ma. There was no extensive dry vegetation or grassland represented in the La Quinta microfossil assemblage.

Also absent or poorly represented was pollen of *Abies* and *Picea*, suggesting no high altitudes in the vicinity in Oligo-Miocene time. The present elevation in the Front Ranges of Chiapas is about 2000 m, and the highest peak is San Cristóbol de La Casas at 3004 m. Folded strata are early Miocene through Pliocene-Pleistocene age, so the Oligocene and Oligo-Miocene elevations are estimated at about 1000–1200 m. Present climatic conditions at Simojovel are a MAT of 24°C and a MAP of 2500 mm. It was warmer and wetter in the Oligocene (fig. 3.4) so the Oligocene temperature is estimated at 25°C–26°C and the precipitation at 2700–3000 mm. Given the

Figure 6.2 Collection locality, Oligo-Miocene La Quinta Formation, Simojovel, Chiapas, Mexico. From Graham 1999, fig. 2. Used with permission from the Botanical Society of America, St. Louis.

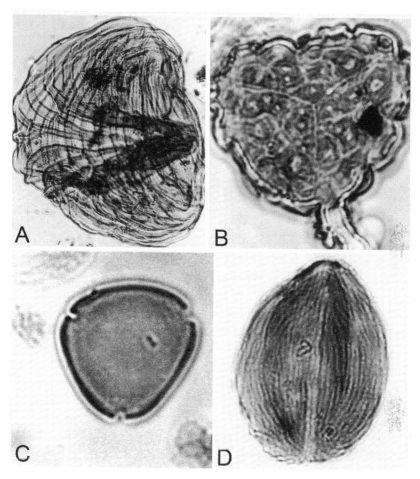

Figure 6.3 Plant microfossils from the Oligo-Miocene La Quinta Formation, Simojovel, Mexico. (a) *Ceratopteris*. (b) *Sphaeropteris-Trichipteris*. (c) *Alfaroa-Oreomunnea*. (d) *Crudia*. From Graham 1999, figs. 2, 8, 9, 7, 18. Used with permission from the Botanical Society of America, St. Louis.

paleovegetation, orogenic history, and coastal position of the locality, it is likely that the environments of southern Mexico in Oligocene and early Miocene times were near tropical and trending toward greater seasonality and significantly increasing habitat diversity.

Only a few fossil floras of middle Eocene through early Miocene age have been studied from the Caribbean region. These are the middle Eocene Saramaguacán assemblage from Cuba (Areces-Mallea 1988; Graham et al. 2000), the Guys Hill flora from Jamaica (Graham 1993), and the late

Figure 6.4 Pelliceria rhizophorae from the Golfo Dulce, Costa Rica. From Allen 1956. Used with permission from the University Press of Florida, Gainesville.

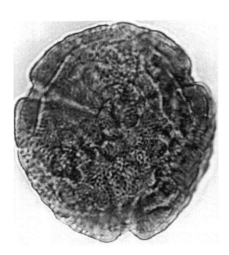

Figure 6.5 Pelliceria pollen from the Oligocene San Sebastian Formation, Puerto Rico. From Graham and Jarzen 1969. Used with permission from the Missouri Botanical Garden Press, St. Louis.

Figure 6.6 Rhizophora pollen from the Oligocene San Sebastian Formation, Puerto Rico. From Graham and Jarzen 1969. Used with permission from the Missouri Botanical Garden Press, St. Louis.

Eocene Gatuncillo flora from the canal region of Panama (Graham 1985). They mostly serve to document the expected coastal mangrove to mid-elevation tropical communities growing there during the Eocene. In the middle Oligocene, there is a large flora of macrofossils (91 taxa; Hollick 1928) and microfossils (165 types; Graham and Jarzen 1969) known from the San Sebastian Formation at the western end of the central cordillera of Puerto Rico. At the time of deposition, the locality was near sea level, and the deposits contain the mangroves *Pelliceria* and *Rhizophora*. *Pelliceria* is presently found along the western coast of Central America and northern Colombia. Its disappearance from the northern Caribbean region by the Miocene, along with that of *Nypa* in the Eocene, represented a step in the modernization of the mangrove ecosystem during the Tertiary. There was also a lower to upper montane broad-leaved forest on the adjacent slopes with *Podocarpus*, *Acacia*, and *Bursera*.

As a further note, recall from chapter 1 that the lower to upper montane broad-leaved forest is designated to include midelevational forests in regions of topographic and climatic diversity and at its lower limits grades either into wetter and more tropical vegetation or into drier vegetation, and at its upper limits into cooler deciduous or treeless communities (alpine tundra or páramo). The many gradations within this formation, and between it and other formations, cannot be consistently identified in fossil floras, so the more inclusive lower to upper montane broad-leaved forest formation is recognized. In the Appalachian part of northern North America, the deciduous forest is a prominent component of this formation that in the east does not grade upward into a treeless alpine tundra because of the moderate elevations, or downward into a tropical forest or drier shrubland or near-desert vegetation. Hence, it is designated there as a separate, well-defined formation. In western northern North America where there are alpine and desert habitats, the deciduous forest is less extensive and well-defined because it is often intermingled with gymnosperms to form a mixture of deciduous angiosperm and evergreen coniferous forest, or it occurs as gallery (streamside) vegetation through coniferous and drier-land communities. As noted, in Latin America, however, there are numerous instances where warm-temperate deciduous forests grade into tropical forest or drier woodland, or into ceja (stunted forest at its uppermost limits) or alpine tundra / páramo. The designation lower to upper montane broad-leaved forest is especially useful in the modern vegetation of Latin America, and in fossil floras where more subtle distinctions are not warranted. In the Antilles, on the slopes of Puerto Rico during San Sebastian times, the wet lower phase of the formation grew there during the transition times of the Oligocene, and it is represented at present in the Luquillo Mountains (chap. 2). It was probably more tropical on the limited emergent lands of the Antilles in the Eocene, and somewhat more seasonally temperate, especially at the higher altitudes, in Miocene and later Tertiary times and in the glacial intervals of the Quaternary. These conditions and trends were muted, however, by the insular nature and maritime climates of the islands.

In most instances, paleontological information does not define the tectonic and orogenic history of land fragments but must accommodate itself to geological reconstructions and the findings of geophysics. In the case of Puerto Rico, however, the fossil plant record does contribute to refining the submergent-emergent history of the island. Subsidence took place between the middle Eocene and the middle Oligocene, and the geology suggests that most of Puerto Rico may have been under water during this time. However, the presence of lignites with mangrove fossils, along with the extensive terrestrial San Sebastian flora, reveals an emergent coast and documents densely vegetated uplands in the middle Oligocene.

SOUTH AMERICA

There are few extensive, concentrated, well-preserved macro- or microfossil floras of middle Eocene to early Miocene age in South America that have been studied either sufficiently or recently to reveal the histories of ecosystems. Also, in some instances, the primary nomenclature for fossils from this interval is based on an artificial system with biological affinities appended in the discussions. (See II, tables 7.1, 7.2, 7.6, and appendices 2.1, 2.2 for a list of plant macro- and microfossils for South America, along with the associated literature.) Collectively, fossil evidence indicates that swamps, lagoons, and riverine habitats occupied much of the Amazon Basin, and that flow of the major rivers continued to the west into the Pacific until about the middle Miocene. In the Eocene, *Nypa* and other palms and *Pelliceria* were represented, along with *Retibrevitricolpites* of unknown affinities, the abundance of which in lowland coastal deposits (e.g., lignites) suggests it occupied the mangrove habitat. Thus, a version of mangrove vegetation was present until the middle to late Eocene when *Rhizophora* established an essentially modern counterpart. There was a mudflat association at scattered sites mostly inland to the mangroves that included the present-day emergent aquatic *Crenea*. This plant grows today along the northern coast of South America in saline soils subject to periodic drying. It is known in the fossil record as *Verrutricolporites rotundiporis*, and it extends back to the late Eocene in the Maracaibo Basin region associated with

Rhizophora (Germeraad et al. 1968). A late Eocene assemblage of permineralized woods and other fossils from 39.4 Ma, the El Bosque Petrificado Piedra Chamana in the Huambos Formation of Peru, is under study (Woodcock et al. 2009).

The presence of a lowland neotropical rain forest is suggested by logs up to 4 m in length in the late Eocene Itaquaquecetuba Formation of the São Paulo Basin (Fittipaldi et al. 1989). Outcrops of sandstone, and sandy coastal areas, may have supported communities of savanna or grassland by the end of the early Miocene, but following the sequence elsewhere, their development and expansion was favored by enhanced cooling, drying, and greater seasonality beginning in about the middle Miocene.

At the base of the moderately elevated Andes Mountains in the early Miocene, there were likely wet and super-wet areas resulting from the configuration of the east-facing slopes of the mountains vis-à-vis winds angling southwest across the Amazon lowlands (Killeen et al. 2007; fig. 6.7). For most of the length of the Northern and Central Andes, the winds intercept the mountains at an angle. Recall that at the "elbow" of the Andes in Bolivia, however, the interception is more at a right angle, and precipitation is especially high on the slopes and on the lands below. On these slopes, there was likely a tropical rain forest or the wet phase of a lower to upper montane broad-leaved forest, with drier vegetation in the Andean valleys, as at present, where soils, slope, exposure, and rain shadows augment the reduced moisture. To the west of the mountains, there were other dry areas, in part, as the result of a limited rain shadow, and in part to the cooling Humboldt Current that was beginning to form because of closure of the Drake Passage, thermal isolation of Antarctica from warmer equatorial waters, southern drift of the continent, early continental glaciations, and cold meltwaters starting to flow into the ocean. The dryness along the western coast appearing about or just prior to the middle Miocene, circa 15 Ma, is evident from sediments in the central Andean fore-arc basins of Chile (Sáez et al. 1999). Slightly older sediments from Peru of early to middle Miocene age are alluvial deposits and do not show appreciable dryness. Early Miocene water temperature along coastal Chile around 24 Ma was 5°C higher than at the present mean annual minimum of 20°C at 45°S (Nielsen and Glodny 2009). Thus, the Atacama Desert was just beginning to form in the early Miocene. There is no tundra or páramo evident at this time.

There is a quantum leap from the meager information available at scattered sites in northern South America to the paleobotanical "hot spot" of Argentina. Assemblages ranging in age from the Cretaceous through the Oligocene are known from Santa Cruz Province (Hunicken 1955), and

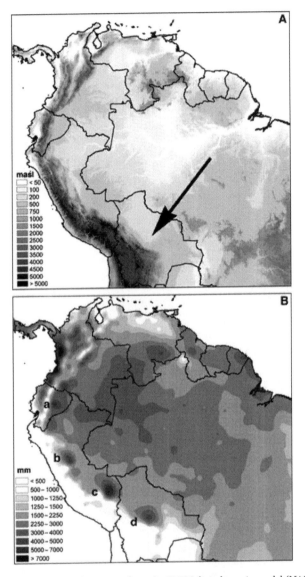

Figure 6.7 Northwestern South America from the SRTM digital terrain model (NASA/JPL). Topography shown in map A; arrow indicates the elbow of the Andes. Map B shows area of high annual precipitation (>5000 mm) at Chapare, Bolivia. From Killeen and others 2007. Used with permission from Wiley-Blackwell Publishing, Oxford.

Viviana Barreda and colleagues (e.g., Barreda et al. 2007) are presently revising the paleofloras. Those from the Río Guillermo Formation are now considered early Oligocene in age and will be important in outlining biotic and environmental history during this transitory interval. The temperature history for the middle Eocene is difficult to trace in detail for southern South America, but palynofloras in the Salta Basin suggest climates were cooling (Quattrocchio and Volkheimer 1990). This is consistent with leaf-margin evidence from Chile (Romero 1986), where in the early Eocene Lota and Coronel floras entire-leaf margins are 70 percent (paratropical flora), while by middle Eocene Río Turbio times they are 40 percent, and in the late Eocene Ñirihuau flora of northwestern Patagonia they are 27 percent (seasonally cool forest). In place of forested vegetation east of the substantially elevated Southern Andes, and with grasses widespread by the end of this interval, the plant formation was an expanding shrubland/chaparral-woodland-savanna (i.e., steppe).

Thus, the ecosystems of South America at the end of the early Miocene included the lowland neotropical rain forest recognizable as an ecosystem since the Paleocene, circa 58 Ma, reaching its maximum extent at the LPTM/EECL, and retreating from its northern and southern margins since the early Eocene. By this time, it primarily occupied its present range in the Amazon Basin and along the east-central coast of Brazil (the Atlantic forest), with elements extending into other moist lowland habitats. There was also mangrove (with *Rhizophora* since the middle to early Eocene), lower to upper montane broad-leaved forest (with a deciduous component of *Nothofagus* in the south and west of the Southern Andes Mountains), shrubland/chaparral-woodland-savanna (steppe east of the Southern Andes; monte east of the Central Andes), beach/strand/dune, freshwater herbaceous bog/marsh/swamp, and aquatic ecosystems of essentially modern aspect. Grasses had been present since the early Eocene and versions of local grassland were likely present (e.g., early pampas), but they expanded and modernized with further drying and seasonality beginning in middle Miocene times and later. There was also an early version of the Atacama Desert, and it further developed with continued uplift of the Central Andes Mountains. At this time, individual elements to early versions of tundra and páramo were present at the highest southern latitudes and at the summits of the highest mountains.

The available record for South America is diffuse across the north, extensive in the south, and meager in between. It provides a broad but rapidly improving impression of South American vegetation during the transition times of the middle Eocene, Oligocene, and early Miocene. It thus indicates that trends in biotas, climates, and geologic history, already in motion at the end of the early Eocene, continued in a muted mode through the early Miocene. Then, beginning about 16.3 Ma, the pace of change picked up considerably in the middle Miocene.

: : REFERENCES : :

Allen, P. H. 1956. *The rain forests of Golfo Dulce.* University of Florida Press, Gainesville.

Areces-Mallea, A. E. 1988. Palinomorfos de la Costa del Golfo de Northamérica en el Eoceno medio de Cuba. *Rev. Tecnol.* 18:15–25.

Axelrod, D. I. 1966. *The Eocene Copper Basin flora of northeastern Nevada.* University of California Publications in Geological Sciences 59. University of California Press, Berkeley.

———. 1984. An interpretation of Cretaceous and Tertiary biota in polar regions. *Palaeogeogr. Palaeocl. Palaeoecol.* 45:105–47.

Barreda, V., et al. (nineteen coauthors). 2007. Diversificación y cambios de las angiospermas durante el Neogeno en Argentina. *Ameghiniana Publicación Especial* 11:173–91.

Becker, H. F. 1961. *Oligocene Plants from the upper Ruby River basin, Southwestern Montana.* Memoir 82. Geological Society of America, Boulder.

———. 1969. Fossil plants of the Tertiary Beaverhead Basins in southwestern Montana. *Palaeontographica,* Abt. B, 127:1–142.

Berry, E. W. 1916. *The lower Eocene floras of southeastern North America.* U.S. Geological Survey Professional Paper 91. U.S. Department of the Interior, Washington, D.C.

Bullock, S. H., H. A. Mooney, and E. Medina, eds. 1995. *Seasonally dry tropical forests.* Cambridge University Press, Cambridge.

Cevallos-Ferriz, S. R. S., R. A. Stockey, and K. B. Pigg. 1991. The Princeton Chert: Evidence for in situ aquatic plants. *Rev. Palaeobot. Palynol.* 70:173–85.

Crane, P. R., and R. A. Stockey. 1987. *Betula* leaves and reproductive structures from the middle Eocene of British Colombia, Canada. *Can. J. Bot.* 65:2490–2500.

Denk, T., and R. M. Dillhoff. 2005. *Ulmus* leaves and fruits from the early-middle Eocene of northwestern North America: Systematics and implications for character evolution within Ulmaceae. *Can. J. Bot.* 83:1663–81.

Dilcher, D. L. 1973. A paleoclimatic interpretation of the Eocene floras of southeastern North America. In *Vegetation and vegetational history of northern Latin America,* ed. A. Graham, 39–59. Elsevier Science Publishers, Amsterdam.

Dilcher, D. L., and T. A. Lott. 2005. A middle Eocene fossil plant assemblage (Powers Clay Pit) from western Tennessee. *Bull. Flor. Mus. Nat. Hist.* 45:1–43.

Dillhoff, R. M., E. B. Leopold, and S. R. Manchester. 2005. The McAbee flora of British Colombia and its relation to the early-middle Eocene Okanagan Highlands flora of the Pacific Northwest. *Can. J. Earth Sci.* 42:151–66.

Eldrett, J. S., D. R. Greenwood, I. C. Harding, and M. Huber. 2009. Increased seasonality through the Eocene to Oligocene transition in northern high latitudes. *Nature* 459:969–973. [See also in the same issue Harding and Eldrett, "Making the Paper," 887.]

Ellison, J. C. 2008. Long-term retrospection on mangrove development using sediment cores and pollen analysis: A review. *Aquat. Bot.* 89:93–104.

Erwin, D. M., and R. A. Stockey. 1991. Silicified monocotyledons from the middle Eocene Princeton chert (Allenby Formation) of British Columbia, Canada. *Rev. Palaeobot. Palynol.* 70:147–62.

Fittipaldi, F. C., M. Guímarães Simões, A. M. Giulietti, and J. Rubens Pirani. 1989. Fossil plants from the Itaquaquecetuba Formation (Cenozoic of the São Paulo Basin) and their possible paleoclimatic significance. Bol. *IG-USP (Instituto Geociências–Universidade São Paulo) Publ. Esp.* 7:183–203.

Germeraad, J. H., C. A. Hopping, and J. Muller. 1968. Palynology of Tertiary sediments from tropical areas. *Rev. Palaeobot. Palynol.* 6:189–348.

Graham, A. 1985. Studies in neotropical paleobotany. Part 4, The Eocene communities of Panama. *Ann. Mo. Bot. Gard.* 72:504–34.

———. 1993. Contribution toward a Tertiary palynostratigraphy for Jamaica: The status of Tertiary paleobotanical studies in northern Latin America and preliminary analysis of the Guys Hill Member (Chapelton Formation, middle Eocene) of Jamaica. In *Biostratigraphy of Jamaica*, ed. R. M. Wright and E. Robinson, Geological Society of America Memoir 182, 443–61. Geological Society of America, Boulder.

———. 1995. Diversification of Gulf/Caribbean mangrove communities through Cenozoic time. *Biotropica* 27:20–27.

———. 1999. Studies in neotropical paleobotany. Part 8, An Oligo-Miocene palynoflora from Simojovel (Chiapas, Mexico). *Am. J. Bot.* 86:17–31.

———. 2006. Paleobotanical evidence and molecular data in reconstructing the historical phytogeography of Rhizophoraceae. In *Latin American biogeography—causes and effects*. Proceedings of the 51st Annual Systematics Symposium of the Missouri Botanical Garden (Alan Graham, organizer). *Ann. Mo. Bot. Gard.* 93:325–34.

Graham, A., D. Cozadd, A. Areces-Mallea, and N. O. Frederiksen. 2000. Studies in neotropical paleobotany. Part 14, A palynoflora from the middle Eocene Saramaguacán Formation of Cuba. *Am. J. Bot.* 87:1526–39.

Graham, A., and D. L. Dilcher. 1995. The Cenozoic record of tropical dry forest in northern Latin America and the southern United States. In *Seasonally dry tropical forests*, ed. S. H. Bullock, H. A. Mooney, and E. Medina, 124–45. Cambridge University Press, Cambridge.

Graham, A., and D. M. Jarzen. 1969. Studies in neotropical paleobotany. Part 1, The Oligocene communities of Puerto Rico. *Ann. Mo. Bot. Gard.* 56:308–57.

Graham, A., and R. Palacios Chávez. 1996. Additions and preliminary study of an Oligo-Miocene palynoflora from Chiapas, Mexico. *Rheedea* 6:1–12.

Gregory-Wodzicki, K. M. 2001. Paleoclimatic implications of tree-ring growth characteristics of 34.1 Ma *Sequoioxylon pearsallii* from Florissant, Colorado. *Proc. Denver Mus. Nat. Sci.*, ser. 4:163–86.

Hollick. A. 1928. Paleobotany of Porto Rico. *New York Academy of Sciences Scientific Survey of Porto Rico and the Virgin Islands* 7:177–393.

Hunicken, M. 1955. Depositos Neocretacicos y Tertiarios del extremo SSW de Santa Cruz (Cuenca Carbonifera Río Turbio). *Museo Argentino de Ciencias Naturales "Bernardino Rivadavia"* 4:1–164.

Jarzen, D. M., and D. L. Dilcher. 2006. Middle Eocene terrestrial palynomorphs from the Dolime Minerals and Gulf Hammock Quarries, Florida, U.S.A. *Palynology* 30:89–110.

Killeen, T. J., M. Douglas, T. Consiglio, P. M. Jørgensen, and J. Mejia. 2007. Dry spots and wet spots in the Andean hotspot. Special issue, *J. Biogeogr.* 34:1357–73.

Langenheim, J., B. L. Hackner, and A. S. Bartlett. 1967. Mangrove pollen at the depositional site of Oligo-Miocene amber from Chiapas, Mexico. *Bot. Mus. Leafl. Harvard Univ.* 21:289–324.

Leopold, E. B., and S. T. Clay-Poole. 2001. Florissant leaf and pollen floras of Colorado compared: Climatic implications. *Proc. Denver Mus. Nat. Sci.*, ser. 4:17–69.

Little, S. A., R. A. Stockey, and B. Penner. 2009. Anatomy and development of fruits of Lauraceae from the middle Eocene Princeton Chert. *Am. J. Bot.* 96:637–51.

MacGinitie, H. D. 1953. *Fossil plants of the Florissant Beds, Colorado.* Contributions to Paleontology 599. Carnegie Institution of Washington, Washington, D.C.

———. 1969. *The Eocene Green River flora of northwestern Colorado and northeastern Utah.* University of California Publications in Geological Sciences 83. University of California Press, Berkeley.

Magallón-Puebla, S., and S. R. S. Cevallos-Ferriz. 1994. *Eucommia constans* n. sp. fruits from upper Cenozoic strata of Puebla, Mexico: Morphological and anatomical comparison with *Eucommia ulmoides* Oliver. *Int. J. Plant Sci.* 155:80–95.

Manchester, S. R. 1994. *Fruits and seeds of the middle Eocene Nut Beds flora, Clarno Formation, Oregon.* Palaeontographica Americana 58. Paleontological Research Institute, Ithaca.

———. 2000. Late Eocene fossil plants of the John Day Formation, Wheeler County, Oregon. *Oregon Geology* 62:51–63.

———. 2001. Update on the megafossil flora of Florissant, Colorado. *Proc. Denver Mus. Nat. Sci.*, ser. 4:137–61.

Manchester, S. R., W. S. Judd, and B. Handley. 2006. Foliage and fruits of early poplars (Salicaceae: *Populus*) from the Eocene of Utah, Colorado, and Wyoming. *Int. J. Plant Sci.* 167:897–908.

Martínez-Hernández, E., and E. Ramírez-Arriaga. 1999. Palinoestratigrafía de la región de Tepexi de Rodríguez, Puebla, México—implicaciones cronoestratigráficas. *Rev. Mex. Cienc. Geol.* 16:187–207.

McKenna, M. C. 1983. Holarctic landmass rearrangement, cosmic events, and Cenozoic terrestrial organisms. *Ann. Mo. Bot. Gard.* 70:459–89.

Meyer, H. W. 2001. A review of the paleoelevation estimates for the Florissant flora, Colorado. *Proc. Denver Mus. Nat. Sci.*, ser. 4:205–18.

Meyer, H. W., and D. M. Smith, eds. 2008. *Paleontology of the late Eocene Florissant Formation, Colorado.* GSA Special Paper 435. Geological Society of America, Boulder, CO.

Mindell, R. A., R. A. Stockey, and G. Beard. 2009. Permineralized *Fagus* nuts from the Eocene of Vancouver Island, Canada. *Int. J. Plant Sci.* 170:551–60.

Nielsen, S. N., and J. Glodny. 2009. Early Miocene subtropical water temperatures in the southeast Pacific. *Palaeogeogr. Palaeocl. Palaeoecol.* 280:480–88.

Pälike, H., et al. (eight coauthors). 2006. The heartbeat of the Oligocene climate system. *Science* 314:1894–98.

Peterson, G. L., and P. L. Abbott. 1979. Mid-Eocene climate change, southwestern California and northwestern Baja California. *Palaeogeogr. Palaeocl. Palaeoecol.* 26:73–87.

Pigg, K. B., and R. A. Stockey. 1996. The significance of the Princeton Chert permineralized flora to the middle Eocene upland biota of the Okanogan Highlands. *Wash. Geol.* 24:32–36.

Pigg, K. B., W. C. Wehr, and S. M. Ickert-Bond. 2001. *Trochodendron* and *Nordenskioldia* (Trochodendraceae) from the middle Eocene of Washington state, U.S.A. *Int. J. Plant Sci.* 162:1187–98.

Quattrocchio, M., and W. Volkheimer. 1990. Paleogene paleoenvironmental trends as reflected by palynological assemblage types, Salta Basin, NW Argentina. *N. Jb. Geol. Paläont. Abh.* 181:377–96.

Retallack, G. J. 1983. *Late Eocene and Oligocene paleosols from Badlands National Park, South Dakota.* GSA Special Paper 193. Geological Society of America, Boulder.

———. 1990. *Soils of the past: An introduction of paleopedology.* Unwin-Hyman, Boston.

Romero, E. J. 1986. Paleogene phytogeography and climatology of South America. *Ann. Mo. Bot. Gard.* 73:449–61.

Sáez, A., L. Cabrera, A. Jensen, and G. Chong. 1999. Late Neogene lacustrine record and palaeogeography in the Quillagua-Llamara Basin, Central Andean fore-arc (northern Chile). *Palaeogeogr. Palaeocl. Palaeoecol.* 151:5–37.

Smith, M. E., A. R. Carroll, and B. S. Singer. 2008. Synoptic reconstruction of a major ancient lake system: Eocene Green River Formation, western United States. *Geol. Soc. Am. Bull.* 120:54–84.

Smith, S. Y., R. A. Stockey, H. Nishida, and G. W. Rothwell. 2006. *Trawetsia princetonensis* gen. et sp. nov. (Blechnaceae): A permineralized fern from the middle Eocene Princeton Chert. *Int. J. Plant Sci.* 167:711–19.

Tiffney, B. H. 1977. Fruits and seeds of the Brandon Lignite: Magnoliaceae. *Bot. J. Linn. Soc.* 75:299–323.

Tiffney, B. H., and A. Traverse. 1994. The Brandon Lignite (Vermont) is of Cenozoic, not Cretaceous, age! *Northeastern Geology* 16:215–20.

Traverse, A. 1955. *Pollen analysis of the Brandon Lignite of Vermont.* Report of Investigations 5151. U.S. Dept. Interior, Bureau of Mines, Washington, D.C.

Westgate, J. W., and C. T. Gee. 1990. Paleoecology of a middle Eocene mangrove biota (vertebrates, plants, and invertebrates) from southwest Texas. *Palaeogeogr. Palaeocl. Palaeoecol.* 78:163–77.

Wheeler, E. A., and S. R. Manchester. 2002. Woods of the Eocene Nut Beds Flora, Clarno Formation, Oregon, USA. *IAWA Journal*, suppl. 3, 1–188.

Wolfe, J. A. 1972. An interpretation of Alaskan Tertiary floras. In *Floristics and paleofloristics of Asia and eastern North America*, ed. A. Graham, 201–33. Elsevier Science Publishers, Amsterdam.

———. 1977. *Paleogene floras from the Gulf of Alaska region.* U.S. Geological Survey Professional Paper 997. U.S. Department of the Interior, Washington, D.C.

———. 1985. Distribution of major vegetational types during the Tertiary. In *The carbon cycle and atmospheric CO$_2$: Natural variations Archean to present*, ed. E. T. Sundquist and W. S. Broecker, AGU Monograph 32, 357–76. American Geophysical Union, Washington, D.C.

———. 1992. Climatic, floristic, and vegetational changes near the Eocene/Oligocene boundary in North America. In *Eocene-Oligocene climatic and biotic evolution*, ed. D. R. Prothero and W. A. Berggren, 421–36. Princeton University Press, Princeton.

———. 1994. Tertiary climatic changes at middle latitudes of western North America. *Palaeogeogr. Palaeocl. Palaeoecol.*108:195–205.

Wolfe, J. A., and W. Wehr. 1987. Middle Eocene dicotyledonous plants from Republic, northeastern Washington. *U.S. Geol. Surv. Bull.* 1597:1–25.

Woodcock, D., et al. (five coauthors). 2009. Geologic and taphonomic context of El Bosque Petrificado Piedra Chamana (Cajamarca, Peru). Abstract. *Geol. Soc. Am. Bull.* 121:1172–78, doi: 10.1130/826359.1. Published online 24 April 2009.

Additional Readings and Updates

Bestland, E. A. 2000. Weathering flux and CO_2 consumption determined from palaeosol sequences across the Eocene-Oligocene transition. *Palaeogeogr. Palaeocl. Palaeoecol.* 156:301–26.

Boucher, L. D., S. R. Manchester, and W. S. Judd. 2003. An extinct genus of Salicaceae based on twigs with attached flowers, fruits, and foliage from the Eocene Green River Formation of Utah and Colorado, USA. *Am. J. Bot.* 90:1389–99. [*Pseudosalix handleyi.*]

Bowen, G. J. 2007. When the world turned cold. *Nature* 445:607–8.

Chen, I., and S. R. Manchester. 2007. Seed morphology of modern and fossil *Ampelocissus* (Vitaceae) and implications for phytogeography. *Am. J. Bot.* 94:1534–53.

Elliott, L. L., R. A. Mindell, and R. A. Stockey. 2006. *Beardia vancouverensis* gen. et sp. nov. (Juglandaceae): Permineralized fruits from the Eocene of British Colombia. *Am. J. Bot.* 93:557–65.

Erwin, D. M., and H. E. Schorn. 2005. Revision of the conifers from the Eocene Thunder Mountain flora, Idaho, U.S.A. *Rev. Palaeobotany Palynology* 137:125–45.

Estrada-Ruiz, E., and S. R. S. Cevallos-Ferriz. 2007. Infrutescenses from the Cerro del Pueblo Formation (late Campanian), Coahuila, and El Cien Formation (Oligocene-Miocene), Baja California Sur, Mexico. *Int. J. Plant Sci.* 168:507–19.

Florindo, F., A. Nelson, and A. M. Haywood, eds. 2008. *Antarctic cryosphere and Southern Ocean climate evolution (Cenozoic-Holocene)*. Special issue, *Palaeogeogr. Palaeocl. Palaeoecol.* 260:1–298. [Papers from the European Geosciences Union General Assembly 2009.]

Harding, I. C., and L. S. Chant. 2000. Self-sedimented diatom mats as agents of exceptional fossil preservation in the Oligocene Florissant lake beds, Colorado, United States. *Geology* 28:195–98.

Hedges, S. B. 2006. Paleogeography of the Antilles and origin of West Indian terrestrial vertebrates. In *Latin American biogeography—causes and effects*. Proceedings of the 51st Annual Systematics Symposium of the Missouri Botanical Garden (Alan Graham, organizer). *Ann. Mo. Bot. Gard.* 93:231–44.

Hernandez-Castillo, G. R., R. A. Stockey, and G. Beard. 2005. Taxodiaceous pollen cones from the early Tertiary of British Colombia, Canada. *Int. J. Plant Sci.* 166:339–46. [*Homalcoia littoralis*, Appian Way locality.]

Iturralade-Vinent, M. A., and R. D. E. MacPhee. 1999. Paleogeography of the Caribbean region: Implications for Cenozoic biogeography. *Bull. Am. Mus. Nat. Hist.* 238:1–95.

Katinas, L., J. V. Crisci, M. C. Tellería, V. Barreda, and L. Palazzesi. 2007. Early history of Asteraceae in Patagonia: Evidence from fossil pollen grains. *New Zealand J. Bot.* 45:605–10.

Kim, S., D. E. Soltis, P. S. Soltis, and Y. Suh. 2004. DNA sequences from Miocene fossils: An ndhF sequence of *Magnolia latahensis* (Magnoliaceae) and an *rbcL* sequence of *Persea pseudocarolinensis* (Lauraceae). *Am. J. Bot.* 91:615–20. [Clarkia fossil beds, Idaho, 17–20 Ma.]

Little, S. A., and R. A. Stockey. 2003. Vegetative growth of *Decodon allenbyensis* (Lythraceae) from the Miocene Princeton Chert with anatomical comparisons to *Decodon verticillatus*. *Int. J. Plant Sci.* 164:453–69.

Little, S. A., R. A. Stockey, and R. C. Keating. 2004. *Duabanga*-like leaves from the middle Eocene Princeton Chert and comparative leaf histology of Lythraceae sensu lato. *Am. J. Bot.* 91:1126–39.

Liu, Z., et al. (eight coauthors). 2009. Global cooling during the Eocene-Oligocene climate transition. *Science* 323:1187–90.

Lorente, M. A. 1986. *Palynology and palynofacies of the upper Tertiary in Venezuela*. Dissertationes Botanicae 99. J. Cramer, Berlin.

Manchester, S. R., and E. J. Hermsen. 2000. Flowers, fruits, seeds, and pollen of *Landeenia* gen. nov., an extinct sapindalean genus from the Eocene of Wyoming. *Am. J. Bot.* 87:1909–14. [Recognizes *L. aralioides* (MacGinitie) comb. nov. for the middle Eocene Bridger Formation, southwestern Wyoming.]

Manchester, S. R., and W. C. McIntosh. 2007. Late Eocene silicified fruits and seeds from the John Day Formation near Post, Oregon. *PaleoBios* 27:7–17.

Meyer, H. W. 2003. *The fossils of Florissant*. Smithsonian Books, Washington, D.C., and London.

Mindell, R. A., R. A. Stockey, and G. Beard. 2006. Anatomically preserved staminate inflorescences of *Gynoplatananthus oysterbayensis* gen. et sp. nov. (Platanaceae) and associated pistillate fructifications from the Eocene of Vancouver Island, British Columbia. *Int. J. Plant Sci.* 167:591–600. [Appian Way locality.]

———. 2007. *Cascadiacarpa spinosa* gen. et sp. nov. (Fagaceae): Castaneoid fruits from the Eocene of Vancouver Island, Canada. *Am. J. Bot.* 94:351–61. [Appian Way locality.]

Mindell, R. A., R. A. Stockey, G. Beard, and R. S. Currah. 2007. *Margaretbarromyces dictyosporus* gen sp. nov.: A permineralized corticolous ascomycete from the Eocene of Vancouver Island, British Columbia. *Mycol. Res.* 111:680–84. [Appian Way locality.]

Mindell, R. A., R. A. Stockey, G. W. Rothwell, and G. Beard. 2006. *Gleichenia appianensis* sp. nov. (Gleicheniaceae): A permineralized rhizome and associated vegetative remains from the Eocene of Vancouver Island, British Columbia. *Int. J. Plant Sci.* 167:639–47. [Appian Way locality.]

Morrill, C., E. E. Small, and L. C. Sloan. 2001. Modeling orbitan forcing of lake level change: Lake Gosiute (Eocene), North America. *Global Planet. Change* 29:57–76. [Green River Formation, southwest Wyoming: "[L]ake evaporation ~25% higher when perihelion occurs at the summer solstice."]

O'Brien, N. R., H. W. Meyer, K. Reilly, A. M. Ross, and S. Maguire. 2002. Microbial taphonomic processes in the fossilization of insects and plants in the late Eocene Florissant Formation, Colorado. *Rocky Mountain Geology* 37:1–11.

Otto, A., J. D. White, and B. R. T. Simoneit. 2002. Natural product terpenoids in Eocene and Miocene conifer fossils. *Science* 297:1543–45. [The results show that "fossil conifers can contain polar terpenoids, which are valuable markers for (paleo)chemosystematics and phylogeny."]

Rankin, B. D., R. A. Stockey, and G. Beard. 2008. Fruits of Icacinaceae from the Eocene Appian Way locality of Vancouver Island, British Columbia. *Int. J. Plant Sci.* 169:305–14.

Schorn, H. E. 1998. *Holodiscus lisii* (Rosaceae): A new species of Ocean Spray from the late Eocene Florissant Formation, Colorado, USA. *PaleoBios* 18:21–24.

Smith, S. Y., and R. A. Stockey. 2003. Aroid seeds from the middle Eocene Princeton Chert (*Keratosperma allenbyense*, Araceae): Comparisons with extant Lasioideae. *Int. J. Plant Sci.* 164:239–50.

———. 2007. Establishing a fossil record for the perianthless Piperales: *Saururus tuckerae* sp. nov. (Saururaceae) from the middle Eocene Princeton Chert. *Am. J. Bot.* 94:1642–57.

Taggart, R. E., and A. T. Cross. 1990. Plant successions and interruptions in Miocene volcanic deposits, Pacific Northwest. In *Volcanism and fossil biotas*, ed. M. G. Lockley and A. Rice, GSA Special Paper 244, 57–68. Geological Society of America, Boulder.

Trusswell, E. M., and M. K. Macphail. 2009. Polar forests on the edge of extinction: What does the fossil spore and pollen evidence from East Antarctica say? *Austral. Syst. Bot.* 22:57–106.

Wheeler, E. A., and S. R. Manchester. 2007. Review of the wood anatomy of extant Ulmaceae as context for new reports of late Eocene *Ulmus* woods. *Bull. Geosci., Czech Geol. Surv.* 82:329–42.

Wible, J. R., G. W. Rougier, M. J. Novacek, and R. J. Asher. 2007. Cretaceous eutherians and Laurasian origin for placental mammals near the K/T boundary. *Nature* 447:1003–6.

Yang, H., and L. J. Hickey, eds. 2007. Metasequoia: *Back from the Brink? An update*. Proceedings of the Second Symposium on *Metasequoia* and Associated Plants, 6–10 August 2006, New Haven. *Bull. Peabody Mus. Nat. Hist.* (Yale Univ.) 48:179–426.

7

Modernizing

Middle Miocene through the Pliocene

The concept of "tipping points" is useful because it aptly conveys the way events frequently happen in complex systems. The phrase entered the lexicon of popular culture with Malcolm Gladwell's 2000 book, *The Tipping Point: How Little Things Can Make a Big Difference*, and it is now used so extensively that some pundits have suggested trying to identify tipping point's tipping point. The underlying premise is that when systems are pertubated, even those with some resilience to change, and even when there is a lag time between cause and effect, there nonetheless comes a point of no return when moderately paced processes or events suddenly accelerate. The buildup may be slow, affording lengthy time for planning, debate, diversion, denial, and delay, but if the process and its effects are real, the eventual result will be the emergence of a new steady state. The concept is now widely used in discussions of environmental degradation and global warming (e.g., Foley 2005; Kemp 2005). In this regard, the geological record preserves clear information about previous changes in climate, their frequency, causes, and their effects on the biota. One such point was reached in the middle Miocene, when temperatures began an accelerated decline toward a new climate state.

The paleotemperature curve for 16.3 to 2.6 Ma shows that the relatively slow pace of changes in the Oligocene and early Miocene picked up in the middle Miocene around 15 Ma (fig. 3.4) and began altering the Earth's biota at an ever-increasing rate. Some physical changes were gradual, such as the erosion of the Appalachian Mountains down to about modern elevations by the end of the Pliocene. Others were faster, such as uplift of the Rocky Mountains that began a rapid rise between 12 and 4 Ma. The Sierra Nevada attained two-thirds of their current height since 10 Ma. At 3 Ma they stood at around 2100 m, rising about 950 m since that time. The California Coast Ranges, Cascade Mountains, and Alaska Ranges all reached significant altitudes within the past 6 million years, creating new temperate zones, and increasing habitat diversity, the weathering of silicate rocks, continentality, and the extent of land affected by rain shadow. Another subtle but important change was greater albedo, or reflection of heat, as the dense, evergreen, heat-absorbing tropical vegetation, widespread in the Paleocene and Eocene, was replaced over large areas by heat-reflecting open or deciduous vegetation and exposed soil surfaces in the Oligocene and early Miocene, and especially after the middle Miocene.

The most rapid change, however, was in climate. At the beginning of the middle Miocene, 17–15 Ma, temperatures increased to what is known as the Middle Miocene Climatic Optimum, or MMCO (Riishuus et al. 2006). The MMCO was followed by a steep decline around 15 Ma that can be characterized as a tipping point in global climates. The cause was a continuing drop in atmospheric CO_2 concentration, with less compensating input because of slowing plate movement and no sudden release of methane as had been the case in the early Eocene. Polar glaciations were just underway in the Eocene, but they were extensive on Antarctica by the Oligocene, and likely present on Greenland as shown by seasonal sea ice at the Eocene-Oligocene transition 38 Ma (Eldrett et al. 2009; "An Insight into Amber," 2008). Ice masses had appeared elsewhere in the Arctic by the middle to late Miocene, expanded in the Mio-Pliocene, and were in full glacial mode by the late Pliocene. With glaciation, oceans further cool from meltwater, creating reduced evaporation and precipitation, and resulting in greater seasonality. The effects on New World vegetation will be discussed below, but the trend was global, including, for example, New Zealand (Lee et al. 2001), as shown by widespread increases in seasonally dry forests, grasslands, and in populations of grazing animals. In Turkey, phytolith assemblages indicate cooling, drying, and the spread of savanna/grassland in the eastern Mediterranean region through the Miocene (Strömberg et al. 2007). The shift can also be widely detected from expanding dry paleosols, a change in the isotopic carbon content of mammalian tooth enamel, and by increasing dust in ocean sediments from the middle Miocene onward (Rea et al. 1985). The beginning of the middle Miocene represents another pivotal point in ecosystem history because it was a time favorable to the development, coalescence and radiation of drier, seasonal communities and lineages.

Although the focus here is on the New World, it is widely known that climate systems are teleconnected (Markgraf 2001), meaning that events in one part of the world usually have global impact. This is now widely recognized through the highly publicized effects of El Niños (Nash 2002). Another example is the glacial expansion beginning in the late Miocene that caused a marine regression of about 40–50 m. This isolated the Mediterranean Sea, resulting in an accumulation of salt (gypsum) in the basin that reduced salinity in the ocean by 6 percent. Reduced salinity in turn allowed Northern Hemisphere seawater to freeze at higher temperatures, created more heat-reflecting snow and ice, and served as a positive feedback to global cooling and seasonality. Much of the life of the Mediterranean Sea died during this event, giving rise to the Messinian Salinity Crisis, named after the interval of the Miocene when it occurred, around 4.8 Ma (Rouchy et al. 2001; Rouchy et al. 2007). Although it happened in the Mediterranean Sea, the effects were transmitted throughout the teleconnected climate and biotic systems of the Earth. A repeated lesson from the past is that the effects of regional environmental alterations, whether an El Niño, a salinity crisis, or a large-scale change or removal of vegetation, will be experienced well beyond the area of origin.

Cooling after the middle Miocene was not a straight-line trend. There was the MMCO between 17 and 15 Ma. At 4–3 Ma in the middle Pliocene, atmospheric CO_2 values are estimated at 360–440 ppmv, and temperatures increased to around 3.5°C warmer than at present (Salzmann et al. 2008). The causes were a brief increase in atmospheric CO_2 concentration in the middle Miocene and greater oceanic transport of heat from the lower latitudes by the Gulf Stream with closure of the Isthmus of Panama in the middle Pliocene. In the middle Pliocene, sea level was about 35 m higher than at present along the Atlantic coast; winter temperatures in the Arctic were an estimated 20°C–22°C warmer; summer temperatures were 6°C–8°C warmer (Zubakov and Borzenkova 1990); and MAP increased. After this warm interval, temperatures resumed the trend toward cooling, led by a decline in atmospheric CO_2 (Lunt et al. 2008); and at 3 Ma near the end of the Pliocene, ice volume in the Arctic had reached two-thirds that of the Pleistocene (Crowley and North 1991). Haug and colleagues (2005) place the beginning of significant Northern Hemisphere glaciations at 2.7 Ma, based on alkenone unsaturation ratios and diatom oxygen isotope ratios from cores in the western subarctic Pacific Ocean. Kleiven and others (2002) note an intensification of glaciations in the North Atlantic region between 3.5 and 2.4 Ma. The North American Laurentide ice sheet was present by 2.6 Ma; there was permafrost in the boreal region by 2.4 Ma; tillites (glacial rock debris) are found in Puget Sound by around 2.1 Ma, and in Yellowstone National Park by 2 Ma; and ice briefly extended down the Mississippi River Valley as far as Iowa around 2.1 Ma. After a warm period around 1.7 Ma, cold conditions returned, continuing the sequence of eighteen to twenty Quaternary glaciations. These glaciations were under the influence of Milankovitch cycles lasting roughly100,000 years, interrupted by brief 10,000-year-long interglacials like that of the present, and modulated by changes in ocean circulation and fluctuations in CO_2. We are currently about 11,000 years into this latest interglacial, and past patterns suggest that were it not for increasing warmth created by use of fossil fuels and burning of forests, the Earth would be entering about the twenty-first glacial cycle of the past 2.6 million years. The concern is that by delaying the event, its eventual effects may be more lasting, drastic, and chaotic than the dramatic-enough changes of the last interglacial-glacial transition as demonstrated by the isotope and fossil record. Some believe present climates may be "approaching a bifurcation point, that is, a point at which the system will undergo a transition to a new stable climate state of permanent midlatitude northern hemisphere glaciation" (Crowley and Hyde 2008).

NORTH AMERICA (NORTH OF MEXICO)

In the far northwest, sequences of insects, ostracodes, soils, and fossil floras all reveal the fluctuating but inexorable trend toward colder, winter-drier, and increasingly seasonal conditions (Axelrod et al. 1991). The Alaskan Seldovia Point flora is early to middle Miocene in age, 16–14 Ma, and it preserves a deciduous forest *Larix*, with some evergreen boreal forest trees such as *Abies*, *Picea*, and *Tsuga* (see fig. 7.1). These species began coalescing inland in the mid to late Miocene around 10–6 Ma into a recognizable version of the boreal coniferous forest association. In the middle Miocene Ballast Brook Formation on Banks Island, Northwest Territories, Canada (74°N), large trees of *Glyptostrobus*, *Picea*, and *Pinus* were present with

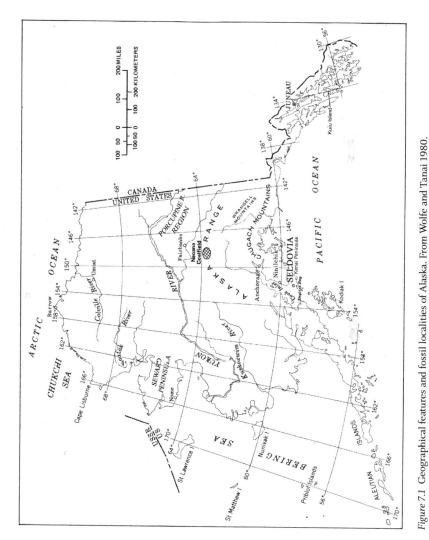

Figure 7.1 Geographical features and fossil localities of Alaska. From Wolfe and Tanai 1980.

maximum heights to 21 m (Williams et al. 2008). From the mid to late Miocene onward, the boreal forests moved episodically across the high northern
latitudes of the world. They were also spreading south into montane areas
of the west. *Alnus* (dominant) and some thermophilous (warmth-requiring)
angiosperms persisted in the Kenai lowlands at nearby coastal Homer in
southern Alaska: *Carya, Corylus, Ilex, Juglans, Myrica, Ostrya/Corylus, Ptero-*
carya (Asian), *Quercus, Ulmus/Zelkova* (Reinink-Smith and Leopold 2005).
Later in Alaska, the cold trend was augmented by blockage of warm air
from the south and from off the Pacific Ocean by rise of the Alaska Range.
Before around 5–4 Ma, southern Alaska was relatively flat, but at about
this time the mountains began a period of significant uplift that contributed to cooling and drying of the adjacent and interior lands. Arctic mammals preserved circa 5.5–4.5 Ma in a beaver (*Arctomeles, Dipoides*) pond
on Ellesmere Island (78°33′N) show a boreal forest with larch, alnus, and
birch, together with sixteen species of beetles, rabbit (*Hypolagus*), shrew
(*Arctisorex*), bear (*Urus*), wolverine (*Plesiogulo*), an Asiatic horse (*Plesiohip-*
parion), and a deerlet (*Moschus*) that lives today in Siberia (Tedford and
Harrington 2003). Winter temperatures are estimated to have been about
15°C higher than today, and summer temperatures 10°C higher. A study of
tree ring width and isotopic composition in larch give a comparable MAT
of −5.5°C ± 1.9°C, or about 14°C warmer than at present (Ballantyne et al.
2006). After 5–4 Ma temperatures cooled rapidly.

Other events in the modernization of the northern floras during this
time include the disappearance of most Asian exotics (e.g., *Glyptostrobus,*
Pterocarya), and many thermophilous deciduous angiosperm trees (e.g.,
Carya, Fagus, Liquidambar, Nyssa, Tilia, Ulmus) from the high north in the
Mio-Pliocene. The sequence in the Seldovia Point flora is from a deciduous forest, to mixed conifer and deciduous forest, to coniferous forest. The
estimated trend in MAT, based on modern analog and CLAMP methods, is
from 9°C to 3°C. The present MAT at Fort Yukon is −6.4°C. In the Porcu-
pine River region (fig. 7.1), fossil woods of the middle Miocene are Taxodiaceae and *Pinus* with growth rings more than 1 cm wide, and in the late
Pliocene they are *Picea* with growth rings less than 1 mm wide. Elsewhere,
on the Seward Peninsula, the Pliocene Lava Camp flora, 75 km south of the
Arctic Circle, is a boreal forest with cold-temperate deciduous trees (*Alnus,*
Betula, Salix), along with herbaceous elements that appear around 7–6 Ma
and are characteristic of tundra and near-tundra communities (the sedges
Carex and *Cyperus, Epilobium, Oenothera, Symphoricarpos, Vaccinium*). These
began to spread by 3–2 Ma as a recognizable local shrub tundra of dwarf
Alnus, Betula, Salix, and herbaceous tundra at the coldest and highest sites

in the far north. Inland around Fairbanks, the vegetation was boreal forest at around 2 Ma (Péwé et al. 2009).

Thus, the broad and overlapping trends in ecosystem history in northwestern North America include the appearance of an early montane coniferous forest in the highlands of the northern Rocky Mountains at about 45 Ma; the early presence of elements and then versions of the boreal coniferous forest in the lowlands to the north between 10 and 6 Ma; and its subsequent spread into other montane regions as shown in the Beaverhead Basins floras in nearby Montana, and somewhat less strong representation in the Florissant flora more distant in central Colorado, both of late Eocene to early Oligocene age (35 Ma). The appearance of tundra in the lowlands is difficult to date specifically because with the early formation of local glaciers in the Eocene and afterward there were narrow fringing zones of soil rubble, lichens, and various herbaceous plants that constituted treeless vegetation in a cold climate (i.e., tundra). However, there were more kinds and widely distributed tundra elements, and local tundra/near-tundra vegetation, in the coldest northern and high-altitude sites around 7–6 Ma; and a recognizable and relatively widespread tundra ecosystem between 3 and 2 Ma. In the highlands, altitudes suitable for tundra conditions were available at scattered sites along the cordilleras after about the late Eocene. The climatic component was added after about 15 Ma, and this is taken as the approximate beginning of the alpine tundra ecosystem (earliest in the highlands to the north, later to the south).

In the northeast, vegetation changes were influenced by similar trends in climate, but without the simultaneous extensive orogenic activity of the west. For example, the Holmatindur Tuff floras of eastern Iceland (10.3–9.5 Ma) show a sequence from temperate forest (*Taxodium, Carpinus-Ostrya, Carya, Fagus, Nyssa*) to a cool to cold-temperate forest (*Picea*, decrease in *Fagus*), to a cold-temperate forest with intermingled subarctic woodland (increase in *Picea, Alnus,* and *Betula*). The same trends are evident toward the southeast in the middle Miocene Martha's Vineyard flora of Massachusetts and the late Miocene Brandywine flora of southern Maryland (I, 246–50). However, without the high elevations, comparable rain shadow effects, and the intensive landscape modifications from volcanism that were present in the west, no dry grasslands, shrublands, or near-desert vegetation developed in the east. There was also greater moisture brought into the midcontinent region from low pressure systems in the Gulf of Mexico. The Appalachian montane coniferous forest probably appeared first in the highlands to the north in the late Eocene, and was confined to the highest altitudes toward the south. It extended downslope and further southward with cooling in the middle Miocene and later times, then fluctuated with the climate changes of the Quaternary. However, the number of Tertiary floras available to document these trends in northeastern North America is not great.

Those that are available, such as the early Miocene Brandon Lignite flora of Vermont, and others to the south, like the Eocene Wilcox flora in Kentucky and Tennessee, reveal deciduous forest elements in the highlands during and after about the middle Eocene. A widespread and extensive deciduous forest with Asian exotics developed in the Oligocene, Miocene, and Pliocene. By the Oligocene, and especially in the Miocene, the deciduous forest expanded from its Cretaceous-Paleocene polar broad-leaved deciduous forest and boreotropical progenitors and coalesced with other members elsewhere to become the most widespread ecosystem across the middle latitudes of the Northern Hemisphere. It extended from eastern North America north of Mexico, across the Plains in scattered populations, into western North America, over Beringia into eastern Asia, and across to Europe. The Peace Creek flora in central Florida, after 2.8 Ma, near the beginning of extensive Northern Hemisphere glaciations, includes some of the last occurrences of present-day Asian taxa in North America—from Japan, *Pterocarya* and *Sciadopitys* (Japanese umbrella pine), and from southeast Asia possibly *Ginkgo* and *Dacrydium* (Hansen et al. 2001).

Tertiary vertebrate fossils from deciduous forest habitats in eastern North America are few, but the late Miocene to early Pliocene Gray Fossil Site, Washington County, Tennessee, 7–4.5 Ma, is yielding new and interesting discoveries (DeSantis and Wallace 2008; Shunk et al. 2006; Liu, in press). They document faunistic affinities between eastern North American and eastern Asia that parallel those long recognized in the plant fossil record (Graham 1966, 1972; Wen 1999; chap. 9 below). The finds include the lesser panda (*Pristinailurus bristoli*), peccary (*Tayassuidae*), rhino (*Teleoceras*), camel (cf. *Megatylopus*), Eurasian badger (*Arctomeles dimologontus*), as well as the greatest concentration of fossil tapirs (*Tapirus polkensis*) on record (Wallace and Wang 2004). These lacustrine sediments will undoubtedly continue to yield material important for better understanding the Neogene environments of the southeastern United States and the disjunct biotic affinities with eastern Asia.

In the midcontinent region from the Eocene through the Miocene, there was a shift from dense forest to savanna to grassland (Strömberg 2002, 2004, 2005). There were open areas with grasses and shrubs, a version of a grassland/savanna, between a seemingly early 27 Ma and 17 Ma based on phytoliths (Strömberg 2002), and certainly by 18 Ma based on grazing mammals (Janis et al. 2002). Grasslands were prominent in the plains by at least 13 Ma (Leopold and Denton 1987; Fox 2000), reflected in part in growth increments in *Gomphotherium* tusks—the gomphotheriums were ancestors to mammoths and elephants. In response to the climatic changes of the Miocene, the Equinae (horse subfamily) underwent rapid speciation by dispersal, rather than vicariance, from one species (*Parahippus leonensis*) to seventy species (Maguire and Stigall 2008). The grassland/savanna became progressively more grassy with the gradual loss of trees and shrubs in later Miocene and Pliocene times circa 8 Ma. Both the floras and faunas tell of increasing cold conditions toward the mid to late Pliocene. For example, around 10 Ma, a volcanic eruption in southwestern Idaho deposited 3 m of ash in western Nebraska, preserving the Ash Bed Fauna. It includes camel, five species of horses, zebra, and other browsing and grazing mammals. At Big Springs, Nebraska, around 2 Ma, the fauna includes lemming, shrew, and mastodon. During this interval, tooth enamel of the ungulates shows a shift from primarily C3 to C4 grasses (Cerling et al. 1997) involving the complex response to increasing seasonality and decreasing temperatures and humidity. Rodent communities near the Pliocene-Pleistocene transition (2.06–1.95 Ma) in southwestern Kansas change from one with a strong subtropical southern component to one of temperate aspect that also coincides with expansion of C4 grasses (Martin et al. 2008). The data also suggest a more precise date for the Blancan/Rancholabrean NALMA at this time.

The dramatic changes in northern North American Tertiary biotas, first from tropical to temperate, then, in the midcontinent region to savanna and grassland, and in the west from temperate to lower altitude shrubland to desert, are familiar to those who have visited localities with fossil plants or have seen dioramas depicting the ancient life. For example, a collector in southeastern Oregon, sitting today in a desert of sagebrush, will be retrieving from rocks in which the sagebrush is rooted fossils of ash, basswood, beech, birch, dogwood, elm, hickory, *Magnolia*, maple, *Sassafras*, sweetgum, sycamore, and walnut, together with several present-day Asian plants such as *Ginkgo*, *Glyptostrobus*, *Metasequoia*, *Eucommia*, *Platycarya*, and *Pterocarya*. The trend from deciduous forest to shrubland/chaparral-woodland-savanna to desert is consistently evident in mid- to low-elevation mid-Tertiary floras between the Rocky Mountains and the rising coastal mountains. To the east of the Rocky Mountains, the trend was toward grassland, with deciduous forest confined mostly to lake and river margins. The altitudinal cline in vegetation revealed by the fossil assemblages is from grassy woodland-savanna at low elevations, upward into a piñon pine and

juniper woodland, an oak-sagebrush shrubland, and finally to a western coniferous forest at the highest elevations. Asian exotics mostly disappeared
from the Rocky Mountains, High Plains, and the midcontinent region by
about 8 Ma, persisting somewhat later in the moister habitats to the west
and in the southeastern United States.

West of the Rocky Mountains, the change from Oligocene and Miocene
deciduous forest to Pliocene shrubland/chaparral-woodland-savanna and
near desert is recorded in the numerous floras and faunas around the Colorado and Columbia plateaus. Recall that in volcanically active regions, lava
will dam streams, creating lakes. Volcanic ash will rapidly fill these lakes
with fine-grained sediments, providing ideal conditions for preservation of
the surrounding biota. One such assemblage is the middle Miocene Succor Creek flora of southeastern Oregon of 15–14 Ma (Graham 1963, 1965;
Fields 1992).

The Succor Creek flora (fig. 2.13) is one of the largest Tertiary plant assemblages known for North America, and it has been studied for both macrofossils and microfossils, yielding a total of about 160 species. The biodiversity was enhanced by the spatial heterogeneity and the varied paleosols
that developed on the surface of the Columbia River Basalts. The minimum
winter temperature is estimated at 1.7°C–4.4°C, the maximum summer
temperatures at 24.4°C–27°C, and the MAP at 1270–1520 mm based on
paleobotanical evidence. Paleosol analysis yield regional MAT estimates of
17°C and MAP of 1400–1500 mm (Takeuchi et al. 2007; to 17.7°C, Retallack 2004). The leaves are 25–30 percent entire-margined (generally temperate), with warm-temperate evergreen species in the lowlands (*Cedrela,
Mahonia, Oreopanax, Quercus*), and cool-temperate deciduous species like
Acer on the moist slopes (see fig. 2.14), along with *Alnus, Betula, Carpinus, Carya, Castanea, Cornus, Diospyros* (persimmon), *Fagus, Fraxinus, Hy-
drangea, Juglans, Liquidambar, Magnolia, Nyssa, Ostrya, Platanus, Populus*
(aspen), *Prunus*, other *Quercus, Salix, Sassafras, Tilia*, and *Ulmus*, including
a contingent of present-day Asian plants like *Ginkgo, Glyptostrobus, Metasequoia, Ailanthus, Pachysandra, Pterocarya*, and *Zelkova*. There were a few dry
elements (*Ephedra*), and other genera characteristic of habitats from damp
(*Taxodium*), to wet (*Equisetum, Carex, Typha*), to aquatic (*Nymphaea, Pota-
mogeton*). At the highest elevations there was a western coniferous forest of
Abies, Picea, Thuja, and *Tsuga*.

Subtle shifts in climate, altitudes, and vegetation in the region are evident even within short spans of time. In the adjacent Trout Creek Flora
(13.8 Ma), present-day Mexican (*Cedrela, Oreopanax*) and Asian genera
(*Ginkgo, Glyptostrobus, Pterocarya*) were not present, and cooler conditions

are indicated by an increase in conifers typical of higher elevations in the western cordillera. Drying at the lower elevations in the latest Miocene is revealed by the Teewinot Lake flora at Jackson Hole, Wyoming (8 Ma). It is shrubland to near desert with some scattered arid elements such as *Ephedra* and *Sarcobatus* (greasewood). Pollen from the Idaho Group on the Snake River Plain extends the drying record to 3–2 Ma near the Pliocene-Quaternary boundary. By this time, deciduous hardwoods are rare. There is an impoverished conifer forest, as well as components of a Great Basin desert-shrubland (*Artemisia*, *Sarcobatus*), with up to 60 percent grasses. Typical MATs in the region, still 3°C–4°C warmer than the present, were declining. Grasslands to the west of the Rocky Mountains, such as the Palouse Prairie of eastern Washington, Oregon, and adjacent Idaho, appear around 3 Ma as a result of the uplift of the Cascade Ranges to a height sufficient to cast a rain shadow. These grasslands developed about 10 million years later than those appearing around 13 Ma on the drier plains to the east of the older Rocky Mountains.

Thus, at the beginning of the middle Miocene (16.3 Ma), the ecosystems of northern North America are shrubland/chaparral-woodland-savanna, with desert elements; grassland, by about 13 Ma in the Plains; beach/strand/dune, mostly unrecorded because of habitat and preservation conditions; freshwater herbaceous bog/marsh/swamp; and aquatic communities. There was a coniferous forest ecosystem, both boreal and montane, and probably a Gulf Coast pine forest association in sandy habitats (as opposed to pine as an element on sandy soils in deciduous forest vegetation). This association would increase after Eocene cooling diminished tropical vegetation and would develop further during the early Miocene, middle Pliocene, and interglacial warm periods. There was lowland tundra (transitional and locally variable at 15 Ma, recognizable and more extensive by 7–6 Ma), and alpine tundra (earliest in the highlands toward the north beginning around 15 Ma).

Later, near the end of the Pliocene (2.6 Ma), the northern North America ecosystems are desert, shrubland/chaparral-woodland-savanna, grassland (with Palouse Prairie added around 3 Ma west of the Rocky Mountains), beach/strand/dune, freshwater herbaceous bog/marsh/swamp, aquatic, coniferous forest, tundra, and alpine tundra. Mangrove with *Rhizophora* is not widely evident as an ecosystem in the Tertiary of northern North America (see Jarzen and Dilcher 2006, middle Eocene Florida), and its presence today in southern peninsula Florida is probably a result of the most recent of repeated introductions during the interglacial warm periods. Time, as well as climate, may also be a factor because *Rhizophora* likely came from the Old World—oldest records in Latin America are in the middle to late Eocene—and progressively moved northward. It eventually reached the northern part of its range, where it came and went in the southeastern United States with swings in Neogene and Quaternary temperatures.

MEXICO, THE ANTILLES, CENTRAL AMERICA

In northern Latin America, there are several floras available for reconstructing middle Miocene through Pliocene vegetation and terrestrial environments for particular times and places. However, these are not sufficiently distributed stratigraphically to reveal detailed ecosystem evolution through time, or concentrated geographically to reveal their composition or distribution at any given time over a broad area. That history must be extrapolated from several widely separated fossil floras and cautiously interpreted from multiple lines of evidence and within the context of global to regional climatic trends and local orogenic history. A case in point is the Mint Canyon fossil flora, now located along the coast of southern California. It is part of a terrane to the west of the San Andreas Fault that has been transported about 300 km from the south. Although it therefore provides an indication of late Miocene vegetation and environments in the Sonoran region of present-day coastal Sonora and Sinaloa, there are no other diverse and well-preserved Neogene floras in the region to document long-term trends. The Mint Canyon flora is a sclerophyllous (thick-leaved) shrubland/ chaparral and woodland of *Acacia*, *Bursera*, *Caesalpinia*, *Cardiospermum*, *Cassia*, *Celtis*, *Fouquieria*, *Mahonia*, *Persea*, *Robinia*, and *Sapindus*, with a gallery forest of *Juglans*, *Platanus*, *Populus*, and *Salix*. The MAP is estimated at 500 mm (currently 228 mm in Guaymas, Sonora). The appearance of the shrubland/chaparral-woodland-savanna as a recognizable ecosystem in Mexico was probably in about the middle Miocene and derived from elements preadapted to arid conditions by rain shadow and edaphic factors like those found in the middle Eocene Green River flora of Colorado-Utah and in the Oligo-Miocene Pié de Vaca flora of Puebla. It likely reached its full development between the middle Miocene and Pliocene, fluctuated with the cold/dry and warm/moist intervals of the Quaternary, and it is currently expanding with desertification brought on by human activity.

To the southeast, middle Pliocene vegetation in coastal southern Veracruz and the adjacent highlands is preserved in the Paraje Solo Formation. This is a large flora of over 100 taxa identified from about 150 types of spores and pollen (Graham 1976). It conforms to most expectations about paleovegetation for this time and place, but it also includes two

Figure 7.2 Pollen of *Picea* (spruce) from the middle Pliocene Paraje Solo Formation near Coatzacoalcos, Veracruz, Mexico.

surprises. By the Pliocene, the eastern end of the Transvolcanic Belt had reached moderately high elevations, so there was an altitudinal cline from mangrove-fringed coastlands with *Acrostichum, Hibiscus, Laguncularia,* and *Rhizophora* (over 96 percent in some samples) upward to montane coniferous forest with *Abies, Picea* (fig. 7.2), and probably some high-altitude pines. In between, the ecosystems included beach/strand/dune with *Acacia, Casearia, Coccoloba, Terminalia,* and Poaceae (grasses); freshwater herbaceous bog/marsh/swamp with Arecaceae (palms), Cyperaceae (sedges), and *Typha*; aquatic *Ceratopteris, Ludwigia, Pachira, Utricularia*; and lower to upper montane broad-leaved forest; along with a prominent deciduous forest association of northern temperate elements like *Alnus, Juglans, Liquidambar, Populus, Quercus,* and *Ulmus.* As expected at this moist coastal site, there was no evidence of local desert, shrubland/chaparral-woodland-savanna, or grassland ecosystems, although they were well-established elsewhere by this time. Likewise, there was no evidence of páramo, although elements may have been present at the highest elevations and undistinguishable in the fossil record from non-páramo taxa or not represented in the lowland basin of deposition; and there was no tundra, which has never existed in lowlands of the New World outside the polar regions or along the margins of glaciers.

One of the unexpected results was the absence of pollen from any of the nine to eleven genera of trees that now form the dominants of the lowland neotropical rain forest in Veracruz. In the present arrangement, the rain forest grows immediately behind the mangroves and borders the coastal depositional basin. Communities farther inland and higher in elevation, such as the deciduous forest of *Quercus* and *Liquidambar*, and even the high-altitude coniferous forest of *Abies* and *Picea*, were represented, so the rain forest as presently constituted really was absent, or present only as a highly modified community. In its place there was the low wet phase of the lower to upper montane broad-leaved forest. Another surprise was the presence of *Picea* (spruce). It no longer grows in the vicinity, but it is found in the highlands of the Sierra Madre Occidental far to the north.

The most parsimonious explanation for these patterns—the absence (or presence of a modified version) of rain forest, the good representation of midaltitude deciduous forest and high-altitude montane coniferous forest, and the occurrence of spruce far south of its present range—is cooler temperatures. However, this creates a problem regarding context information provided by the global paleotemperature trend, and the results from oxygen isotope and paleoecological studies of western Atlantic marine waters, which show that the mid-Pliocene between 4 and 3 Ma was a period of warmth within an overall trend toward cooler conditions. It has been mentioned that when such anomalies arise, especially in the modern era when paleontological information is incorporated into such a wide variety of taxonomic and environmental studies, it is incumbent on investigators to (1) acknowledge an awareness of the apparent inconsistency, (2) suggest a plausible explanation, or (3) defer conclusions pending additional data. In this case, there is an explanation for cool conditions along the coast of Veracruz during an otherwise warm period.

The Isthmian region of Panama began closing with the formation of a submerged sill 9–7 Ma, and closure was essentially completed by about 3.5 Ma, when the land bridge was established. One effect was to intensify northward flow of the Agulhas / Gulf Stream. When such changes in ocean circulation occur, one consequence is upwelling of cold bottom waters along coastal areas (e.g., Prange and Schulz 2004). The correspondence of paleobotanical evidence for cool conditions, at the same time the isthmus was closing, suggests upwelling as a plausible explanation. Other likely factors were higher sea levels that reduced the available real estate between the coast and the better-drained uplands, and the barely completed Isthmian land bridge over which were just migrating many South American taxa that now comprise most of the tropical canopy trees of southeastern Mexico (Wendt, 1993).

In the Antilles, a source of fossils is Dominican amber from the early to middle Miocene, 20–15 Ma (Iturralde-Vinent 2001). The specimens are beautifully preserved, and a new methodology is providing remarkable photographs with striking detail. Previously, x-ray CT-scans gave photographic sections that could be assembled into 3-D images. A new instrument called a synchrotron—for example, at the ESRF (European Synchrontron Radiation Facility, Grenoble, France)—uses subatomic particles (electrons) to generate more powerful x-rays that in combination with new 3-D printers produce images of exquisitely fine detail without having to section the specimens ("An Insight into Amber," 2008; "Seeing the Light," 2008). In addition to insects, the instrument has been used to examine fossil plants in amber from France dating to 100 Ma, as well as fossils from elsewhere. The different organisms represented in amber are important for establishing lineage histories (bacteria, fungi, bees, arachnids, bryophytes, ferns, flowers of *Acacia*, *Hymenaea*, and others), and the number identified from the Dominican amber is rapidly increasing and soon will be sufficient to reconstruct ecosystems in combination with ancillary and context information. The new method is only one example of advances in technology yielding an ever more precise record of life in the past. One of the many interesting revelations of the Dominican amber is that one infrared spectrophotometry pattern, rather than being like the local *Hymenaea courbaril*, is most similar to the African *H. verrucosa* (Hueber and Langenheim 1986). There are a number of other examples of Antillean plants with African affinities, and their presence will be considered further in chapter 9.

There is another flora that reveals Antillean vegetation and environment at a specific time and place, but without others in the vicinity, it is not possible to trace the plant communities through Neogene time. The Artibonite flora of Haiti at 7–5 Ma (Bowin 1975) is located just north of Mirebalais and consists of plant macrofossils studied in the early twentieth century (Berry 1923), along with microfossils studied more recently (Graham 1990). There is no pollen of *Pelliceria* or *Rhizophora*, so the organic-rich sediments were deposited in an inland to upland bog or swamp located beyond the influence of marine waters (fig. 7.3). The vegetation was a lower to upper montane broad-leaved forest including some plants that now grow in the cloud forest, which has a lower limit of about 1400 m. The present high elevations average around 1800 m (the highest is around 2900 m), so there has been some uplift since Mio-Pliocene time. Pollen of *Pinus* reaches 16 percent in one sample, indicating a montane coniferous forest association; spores of ferns make up 36 percent of the assemblage, suggesting a freshwater herbaceous bog/marsh/swamp formation; and there is pollen of the aquatic *Hygrophila*. The MAT is estimated at 23°C–26°C. Considering that the midelevation communities represented in the San Sebastian

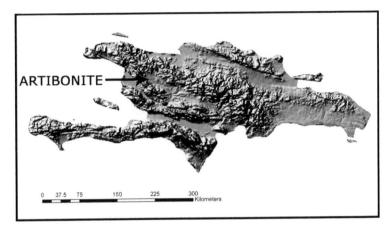

Figure 7.3 Physiography of Hispaniola (Haiti and the Dominican Republic, Greater Antilles), showing location of the Artibonite flora. Map created by Trisha Distler, Missouri Botanical Garden, St. Louis, based on the U.S. Geological Survey's Shuttle Radiography Topography Mission (SRTM) 90m DEM digital elevation database.

Formation of adjacent Puerto Rico were more tropical, comparable to those growing presently under a MAT of about 28°C, the general trend seems roughly in accord with cooling in the Neogene, muted by the insular environment of the Antilles.

In Central America, fossil floras are also mostly scattered geographically and stratigraphically. The Padre Miguel and the Herrería floras of eastern Guatemala are similar in age, with the Padre Miguel probably being slightly older (late Miocene to Mio-Pliocene). The Padre Miguel is an upland flora with pollen of *Picea* and other cool-temperate plants from the north (*Juglans, Ulmus*). The Herrería flora is from the lowlands, as indicated by the marine dinoflagellate *Operculodinium*, as well as by *Rhizophora* pollen that reaches nearly 100 percent in one sample. The MAT for the more highly elevated Padre Miguel assemblage is estimated at 2°C–3°C cooler than at present, while the MAT of the early Pliocene coastal Herrería flora is estimated to be 2°C–3°C warmer (Graham 1998). These are approximations, but they are consistent with the cooling trend, altitudinal difference, and the absence of appreciable upwelling that characterized the later middle Pliocene Paraje Solo flora of southeastern Mexico.

An exception to the scattered nature of northern Latin America fossil floras is a series from the late Eocene, early to middle Miocene, late Miocene, and Quaternary in the canal region of Panama. Samples from the Tertiary sediments were collected from roadside and canalside exposures, and from cores drilled to monitor subsurface changes in this strategically important and tectonically active region (fig. 7.4). They were all deposited near the low latitude of 9°N when the landscape consisted of peninsulas and islands, so it is expected that climatic changes and the vegetation responses would be subtle. The late Eocene Gatuncillo, along with the early to middle Miocene Culebra, Cucaracha, and La Boca floras (the more recent dating follows Retallack and Kirby 2007), reveal mangrove along the coasts, lowland neotropical rain forest, the low and wet phases of a lower to upper montane broad-leaved forest on the slopes, the likely presence of beach/strand/dune, freshwater herbaceous bog/marsh/swamp, and aquatic ecosystems, all growing over a landscape without appreciable highlands.

Figure 7.4 Drilling apparatus and cores through the Tertiary formations of the canal region of Panama.

As expected in such a setting, there is no evidence of desert, shrubland/chaparral-woodland-savanna, or grassland communities.

By the late Miocene, represented by the Gatún Formation, the land fragments were more consolidated, and they had been elevated somewhat by subduction, compression forces generated by microshifts between the North and South American continents, lowering of sea levels, and the accumulation of lava and ash generated as the Caribbean plate slid slowly to the east. Two subtle changes in the vegetation resulted from the appearance of this slightly higher land. One was the initial presence of a few northern deciduous elements, such as the ecologically broad *Alnus* and *Quercus*. The great distance involved in migrating from the north was also a likely factor in determining their relatively late arrival. The other change was the first indication of a slightly drier Pacific (southern) side, compared to a wetter Atlantic (northern) side of Panama. This is recorded in the plant fossil record by an increase in grass pollen from rare or absent to 7.5 percent. Among the causes was deflection of the trade winds by the early Cordillera Central. By the end of the Pliocene, all twelve ecosystems had long been established somewhere in the New World, and those absent from southern Central America are explained by its near-equatorial position, low elevations, and the insular configuration of the landscape. In summary, northern Latin America at the end of the Pliocene included at least elements of ten ecosystems: desert, shrubland/chaparral-woodland-savanna, grassland, mangrove, beach/strand/dune, aquatic, lowland neotropical rain forest, lower to upper montane broad-leaved forest, coniferous forest, and páramo.

Like their counterparts to the north, deserts began forming in the late Miocene and Pliocene from older preadapted elements, and further coalesced in the warm interglacials. The collective evidence suggests that geologically, deserts are among the more recent of the modern New World ecosystems.

The representatives of shrubland/chaparral-woodland-savanna grow under less severe conditions of cold/heat and dryness than do desert taxa, and therefore they appeared earlier. Elements of this vegetation were already present by the Eocene to the north, and probably also in the Sierra Madre region, although fossils floras and associated faunas documenting their presence are few. During the Oligocene, they were coalescing in basins and leeward habitats, and they began to spread with mid-Miocene drying and increasing rain shadows, augmented by slope, exposure, and soil. In northern Latin America, this middle to late Miocene dry shrubby community, or tropical dry forest, was modernized by continued drying in the Pliocene and Quaternary.

To some extent the development of grasslands, along with dry woody communities, was influenced by a factor other than those noted above in chapter 2. The funnel-shape configuration of North America southward to the Isthmus of Tehuantepec is a physiographic factor that affects moisture availability. This means that continentality is greatest in this region and decreases south of the Tehuantepec lowlands, where the interior is more under the influence of maritime climates modified by local and discontinuous highlands of moderate elevation. Grasslands as a recognizable and extensive ecosystem probably developed from a savanna-like version of the community present by the middle Miocene to one of increasingly modern aspect during the Pliocene and Quaternary.

The mangrove community was modernized with the introduction of *Rhizophora*, probably from the Old World, in the middle Eocene.

With the introduction of grasses, probably in the early Eocene, the beach/strand, dune community was modernized.

Aquatic communities, with an angiosperm component, have been present since the Cretaceous.

The rain forest of the neotropical lowlands originated in the equatorial latitudes and dates from the Paleocene.

Versions of the lower to upper montane broad-leaved forest were present by the Late Cretaceous, but representatives were restricted to areas where topographic diversity provided an altitudinal cline in habitats and climate, primarily moisture. The introduction of a northern deciduous component correlates with the cool events of the middle Miocene and later times (see chap. 9 below).

Elements of the coniferous forest like *Abies* and *Picea* were present from the middle Eocene in the highlands of the northern Rocky Mountains. They probably appeared somewhat later to the south in the still-rising Sierra Madre of northern Mexico, possibly in the late Eocene, although their exact pathway and time of arrival cannot be determined from the floras available. The present widespread pine-oak woods in the midaltitudes of Mexico and northern Central America, as an association within the coniferous forest, or as part of the broad-leaved deciduous forest, may also date from this time because the oldest record of both *Pinus* (Cretaceous/Paleocene) and *Quercus* (middle Eocene) are to the north. Pine-oak was present as an association in the late Eocene / early Oligocene Florissant flora of Colorado. An estimate of the earliest time both genera intermingled as an association in Mexico is in the Oligocene because by the Oligo-Miocene they are identified individually or together in fossil floras from Puebla, Chiapas, and in drill-core sediments offshore from Guerrero (II, table 4.1).

In northern Latin America, the páramo community probably developed in the Pliocene in the highest elevations of the Transvolcanic Belt, and together with desert, it is among the latest of ecosystems to appear in modern form in northern Latin America. This is consistent with its record in South America, and with systematic studies that indicate modern páramo species are often recently derived from lower altitude progenitors, rather than a product of ancient in situ evolution.

SOUTH AMERICA

Changes in the biota of South America during the middle Miocene through the Pliocene are anchored in three major events that altered the landscape and supplemented trends in global climates. The first involved closure of the marine barrier between South America and North America (Coates 1997; Jackson et al. 1996). The event can be dated to around 3.5 Ma, recognizing that the few and discontinuous upland temperate habitats developed later than the more extensive and continuous tropical lowland habitats. Also, the Darién region at the border between eastern Panama and northwestern Colombia closed last, and even today it is still virtually at sea level. Consequently, continuity through the Darién was likely disrupted or reduced through tectonic/orogenic activity, and from sea level fluctuations in the Quaternary. The biological consequences of reuniting two continents, and the intermingling of biotas that had been separated from the Late Cretaceous through the late Pliocene, will be discussed in chapter 9, but suffice it to say they were considerable as plants and animals moved in both directions across the new land bridge.

The second event was uplift of the Andes Mountains to modern elevations. Among the many results were that drying along the west coast of the Central Andes intensified and arid habitats expanded to form the Atacama Desert of Peru and northern Chile. Other coastal areas to the north, beyond the cold Humboldt Current, became wetter, as warmer moisture-laden westerly winds rose off the Pacific Ocean in the region of the Colombian Chocó (fig. 2.45). An increasingly formidable barrier was established to the east-west migration of low- to midelevation organisms (a gradually developing vicariant event), and to the north-south movement of high-altitude and páramo organisms (a geographically isolating event), while the Andean valleys provided, and continue to provide, migrational Cis- and trans-Andean pathways for organisms adapted to dry habitats.

One of the most dramatic physical changes wrought by the uplift of the Northern Andes was a reversal in the flow of the Orinoco and the Amazon rivers from into the Pacific Ocean to into the Atlantic Ocean beginning in the middle Miocene around 15 Ma. Later in the Miocene, the uplift of the eastern cordillera of the Northern Andes Mountains blocked the flow of the Maracaibo and other rivers to form Lake Maracaibo.

By the Pliocene, alterations in landscape provided habitats for all the ecosystems except coniferous forest. These included páramo, as revealed by macrofossils and microfossils of the sedges *Eleocharis* and *Scirpus, Gunnera, Juncus, Ranunculus, Rubus, Xyris* (Wijninga 1995, 1996a–d; Wijninga and Kuhry 1990). The timing of uplift in the Northern Andes is known from several approaches, including fossils and fission-track zircon dates from a *Podocarpus* mire that give an estimated elevation of 2000 m around 4 Ma, and 2500 m around 3 Ma or near modern heights (Wijninga 1996d). The vertical arrangement of the vegetation in the Northern Andes approaching Plio-Pleistocene times is shown in figure 7.5.

The third event was a change in climate in Antarctica and southern South America, as shown in the McCurdo Valley region of the Transant-

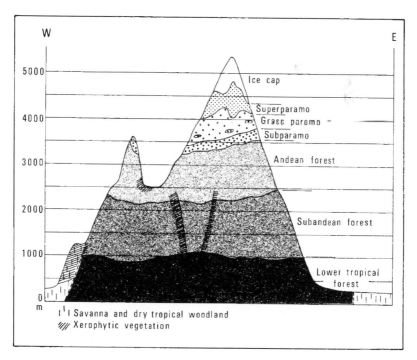

Figure 7.5 Altitudinal zonation of vegetation in the Cordillera Oriental of Colombia. From Wijninga and Kuhry 1990. Used with permission from Elsevier Science Publishers, Amsterdam.

arctic Mountains by a combination of diatoms, palynomorphs, mosses, ostracodes, and insects dated at between 14.07 and 13.85 Ma. They reveal a
cooling of at least 8°C and record "the last vestige of a tundra community
that inhabited the mountains before stepped cooling that first brought a
full polar climate to Antarctica" (Lewis et al. 2008, 10676). Tillites embedded in basalts of Miocene to Quaternary age in Patagonia document that
glaciation in the lowlands was underway there in the middle Pliocene by at
least 3.5 Ma and mark the time when tundra elements, and probably early
versions of tundra vegetation, were present in the lowlands of far southern
South America. Farther north, the combination of high elevations and cooling global climates resulted in glacial conditions in the Andean uplands in
the later Pliocene and Pleistocene. These conditions are widely established
to have fluctuated within the past 2 million years as a result of various
climate-forcing mechanisms discussed below (the Younger Dryas, a mid-
Holocene dry period, the Little Ice Age, Heinrich events). In terms of individual lineages, the interactions of climate, geology, and time are proving to
be of increasingly well-documented evolutionary importance. For example,
Ribas and colleagues (2007) studied the parrot genus *Pionus* using cyto-
chrome *b* (*cyt b*) and NADH dehydrogenase 2 (*ND2*) mitochondrial genes.
Their results showed that diversity in the montane lineages is "directly attributable to events of Earth history," that "the three lineages were transported passively to high elevations by mountain building, and that subsequent diversification within the Andes was driven primarily by Pleistocene
climatic oscillations and their large-scale effects on habitat change" (Ribas
et al. 2007, 2399; see also Ohlemüller et al. 2008).

In terms of ecosystem history, with a few notable exceptions, middle
Miocene through Pliocene fossil floras are widely scattered, and many have
not been studied recently using modern methodologies or interpreted within
a global context of available information. Others, such as the Solimões Basin of Brazil and the Amazonas Basin of Colombia, give a better indication
of conditions at specific places and processes influencing ecosystem development (Hoorn 1993, 1994a–b, 1996). In these fossil assemblages, *Rhizophora* is important because it identifies the time of marine incursions into
the basins. A reconstruction of the middle to late Miocene setting is shown
in figure 7.6. It was a dynamic landscape as revealed both by frequent
changes in the vegetation and by alternating sequences of freshwater and
brackish-water fish faunas (Monsch 1998). Fluctuations continued until
the late Miocene, when inundations ceased except in the borderlands. This
event identifies the time when uplift of the Northern Andes Mountains had
tilted the lowlands to the point that widespread tectonic-induced marine

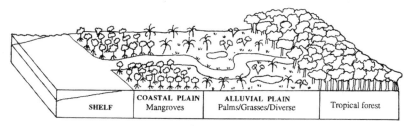

| SHELF | **COASTAL PLAIN**
Mangroves | **ALLUVIAL PLAIN**
Palms/Grasses/Diverse | Tropical forest |

Figure 7.6 Reconstruction of Miocene landscape and communities of northwest Amazonia. From Hoorn 1993. Used with permission from Elsevier Science Publishers, Amsterdam.

incursions deep into the interior were no longer possible. Later, eustatic-induced regressions of ocean levels along the Bragança Peninsula of Brazil caused seaward movement of mangroves 1130–1510 and from 1560 into the 1800s, correlated with the Little Ice Age (Cohen, Behling, and Lara 2005; Cohen, Filho, et al. 2005).

Some of the same information used to reconstruct the origin and spread of New World ecosystems is also applicable to tracking the history of individual lineages and the physical environment. As noted previously, all such applications of paleobotanical data benefit from incorporating faunal studies to generate biotic histories, and from integrating geological and climatic data to provide insight into ecosystem evolution. To illustrate the point, paleodrainage patterns in the Amazon Basin are revealed by sediments from several formations from the Falcón Basin, east of the Maracaibo Basin, to the Venezuela Basin between Caracas and Trinidad and Tobago (Díaz de Gamero 1996). In the middle Eocene, deposits of the Misoa Delta identify a large river flowing north into the Maracaibo Basin and draining the Cordillera Central and the Guiana Highlands. With continued uplift of the eastern cordillera of the Northern Andes, delta formation shifted southward, as indicated by deposits in the late Eocene to Oligocene Carbonera Formation in the Llanos region of Colombia and Venezuela. Then in the mid-early Miocene, deposition shifted east into the northwestern part of the Falcón Basin. With further uplift in the late part of the middle Miocene, the river outlet moved still farther east to its present position off the coast of northeastern Venezuela. These sediments reveal the meandering history of the Orinoco River from the late Eocene through the middle Miocene.

Biologists are interested in inundations and shifting drainage patterns because such shifts are among the several ways populations can be partitioned (geographically isolated) and reunited. Other processes at work include orogeny, tectonics, arch formation, and disappearance of arches through erosion or sedimentation, as we shall see in chapter 9. Together with the myriad effects of climate change—temperature, precipitation, glaciation, water-table fluctuations—they create a dynamic system favorable to the generation of novel genotypes, molecularly distinct entities, and new phenotypes (morphologically different forms). These novelties are acted upon by evolutionary processes such as mutation, hybridization, competition, and extinction to produce new species and account for the biodiversity of ecosystems. It is ironic that not so many years ago, the lowland neotropical rain forest, the most diverse of all ecosystems, was characterized as environmentally and biologically stable and unchanging. One of the significant contributions of geological and biological investigations, including vegetational history studies, has been to document that tropical ecosystems have undergone considerable alterations in range, habitat, and environment throughout Cretaceous and Cenozoic time.

In the Central Andes, a plant fossil of unspecified age, but worthy of mention, was obtained by W. F. Parks of St. Louis, Missouri, from a curio shop in Cuzco, Peru. He sent it to F. H. Knowlton (1919), who described it as *Zea antiqua*, an ear of corn (fig. 7.7), and it was later cited in papers dealing with the origin of maize. Among features that might have raised concern, besides having been purchased in a curio shop, were that it was made of baked clay, it still bore the marks of the tool used to bore a small hole into the hollow center, and it contained pellets that rattled when the object was shaken. It recalls Beringer's predisposition to accept discarded

Figure 7.7 Supposed fossil ear of corn reported from Cuzco, Peru. From Brown 1934. Used with permission from the Washington Academy of Sciences, Washington, DC.

models of creation, and it joins the banana from Colombia as examples of fossil food plants that never fulfilled their promise (Brown 1934).

A site that has provided useful information on the Neogene history of ecosystems in the Central Andes is near the settlement of Pislepampa about 20 km northeast of Cochabamba, Bolivia, at an elevation of about 3600 m. Fieldwork in Bolivia usually begins with an international fight into La Paz, the highest capital city in the world at 4050 m, and our group was no exception. A vehicle was waiting for us at the airport, and we went directly from the pressurized cabin of the plane to Cochabamba, and the next morning to the collecting site. At these heights oxygen, is unsettlingly scarce; it was also cloudy, dark, very cold and windy, and it is almost always misty or raining. Fortunately, the samples from Pislepampa, studied in collaboration with Kathryn Gregory-Wodzicki of the Lamont-Doherty Earth Observatory, contained a diverse assemblage of plant microfossils 6–7 million years old that revealed interesting details about the uplift history of the Central Andes (Graham et al. 2001).

The modern vegetation of Bolivia includes the cloud forest formation, or the bosque montano húmedo, which grows at an elevation between about 1400 and 3200 m. At its upper limit, it grades into the puna; and at its lower limit, it intermingles with tropical plants that extend upward along moist river valleys. In the Pislepampa microfossil flora, three pollen types were of special interest, and they provide an example of the way environmental/vegetation reconstructions begin based on paleobotanical data. Pollen similar to *Oreopanax* (fig. 7.8a) and *Prumnopitys/Podocarpus* (fig. 7.8b) was recovered, and both species grow today in the cloud forest (fig. 7.9). In addition, pollen similar to *Cavanillesia* (Bombacaceae) was present, and this genus is presently found in the tropical lowlands and extends into the uplands along river valleys. Finding fossils of plants from 7–6 Ma growing together today where the ecological parameters can be measured and, thus, applied to the paleocommunity, is a kind of Holy Grail for tropical vegetation history studies. The fossil flora revealed a paleocommunity of cloud forest near its lower limits of about 1400 m. The locality is now at 3600 m, so this information, if used alone, would suggest an uplift of the Central Andes of around 2200 m since about 7 Ma. When all other paleobotanical, geological, ancillary, and context information from the region is taken into account (Gregory-Wodzicki 2000; Gregory-Wodzicki et al. 1998), the consensus is that from one-third to one-half of the present altitude of the Central Andes has been attained since 10 Ma. Between 5 and 2 Ma, the eastern cordillera rose at the annual rate of 0.5–3 mm and continues to rise with a crustal shortening of 10–25 mm. This is geological evidence that

Figure 7.8 Fossil pollen of (a) *Oreopanax* (Araliaceae) and (b) *Prumnopitys (Podocarpus)* from the Mio-Pliocene Pislepampa flora, Bolivia. *Oreopanax* and *Prumnopitys* growing together (fig. 7.9) in the modern cloud forest of Santa Cruz, Bolivia.

speaks clearly of the recent origin of páramo habitats and, thus, of páramo vegetation. As noted earlier, it is further consistent with taxonomic studies of modern organisms that indicate páramo species are mostly derived recently from progenitors growing at lower elevations, rather than from a long period of in situ evolution at high altitudes.

Other aspects of vegetation history can be inferred from isotopic and faunal evidence. MacFadden and others (1994) studied the ^{13}C in tooth enamel from fossil mammals ranging from 25 million to 7500 years old that included sites between 3200 and 4000 m elevation in the Bolivian Andes. The results reveal a change in food resources from shrubs and nongrassy herbs in the early to middle Miocene, to grassy páramo in the late Miocene and Pliocene. This is evidence from yet another line of inquiry suggesting the geologically recent origin of páramo. The consistency of information emerging from so many different ancillary sources is not only fascinating, but it supports the general accuracy of individual approaches, and it establishes a context for interpreting results from other studies. For example, Hughes and Eastwood used DNA sequence data to identify a monophyletic group of 81 species of *Lupinus* (lupine, bluebonnet) endemic to the Andes Mountains. The age of the clade is dated to 1.76–1.18 Ma, giving a diversification rate of 2.49–3.72 species per million years and "providing the most spectacular example of plant species diversification documented to date" (Hughes and Eastwood 2006, 10334). The conclusions are supported by the geologic and varied biologic evidence documenting the recent origin of the High Andes.

In northwestern Argentina, climates were responding to the rise of the Southern Andes to a height of around 5500 m toward the end of the Neogene, and the effects are preserved in sediments of the Santa María Basin (Kleinert and Strecker 2001). They reveal cooling and seasonally dry climates with trees and shrubs circa 7 Ma, then herbs and grasses between 5 and 3 Ma, and finally vegetation of cold and arid environments at 2.6 Ma (latest Pliocene). Dry-habitat caviomorph rodents are also prominent in the late Pliocene San Andrés Formation of east-central Argentina (Verzi and Quintana 2005), and sediments from the Lauca Basin of northern Chile show increasing desiccation between 6.4 and 3.7 Ma (Gaupp et al. 1999).

The expectations of ecosystem change in southern Argentina (e.g., in Chubut and Santa Cruz Provinces, Patagonia) are for cooling, drying, and more seasonal climates between the early and late Miocene; a change from warm and humid shrubby vegetation to cold and drier grassy communities; and a corresponding increase in dry-land faunas, browsers, and grazers. In the early Miocene Chenque Formation, there is pollen resembling *Acacia* and *Anadenanthera* that today are common in warm humid climates (Barreda and Caccavari 1992; Palamarczuk and Barreda 1998). In the early Miocene Santa Cruz Formation, fossil vertebrate faunas are present. From the base of the section toward the top, however, they show a decrease in diversity; euhypsodont species increase; and brachyodont, glyptodontids, and toxodontids become more diverse. Megathroideid body size is reduced, and protheroteid diversity decreases. These faunal shifts are interpreted as responses to an environment that was becoming seasonally dry (Tauber 1997a, b; Barreda and Bellosi 2003). Elsewhere in the vicinity, increasing pollen of the herbaceous grasses and shrubby Amaranthaceae-Chenopodiaceae and Asteraceae also indicate drying and more seasonal climates, as does pollen of *Ephedra* reflecting edaphically drier coastal dunes, probably with strong winds and high evapotranspiration, and inland sandy habitats. The same sense of dramatic change in climate and vegetation in Neogene time, described for the Succor Creek flora in the Basin and Range Province in Oregon, is also apparent at the localities in Patagonia. The collector works in the cold, dry, windy steppe, and from strata in which this vegetation is rooted are found the remains of *Acacia*, *Anadenanthera*, and diverse and large Megathroideids typical of warm-temperate environments. Multiple lines of inquiry then reveal that reduced levels of CO_2, thermal isolation of Antarctica, mountain uplift, and changes in ocean and atmospheric circulation contributed to the development of cooler and seasonal climates between the middle Miocene and the Pliocene. Upon reflection, it is of considerable satisfaction to realize we now know a lot about how, when, and why these changes happened.

In summary, the ecosystems of South America at the end of the Pliocene were desert, shrubland/chaparral-woodland-savanna, including the extensive tropical savannas of Venezuela and steppe in Patagonia; grassland, including the pampas of Argentina; mangrove, beach/strand/dune; freshwater herbaceous bog/marsh/swamp; aquatic; lowland neotropical rain forest; lower to upper montane broad-leaved forest, with cloud forest, elfin forest, and ceja associations at progressively higher elevations, and some northern deciduous forest elements now intermingled (e.g., *Alnus*, *Quercus*, *Juglans*), páramo, and tundra. Similar communities were present in the Northern Hemisphere at the time and collectively constitute a living envelope of modern aspect at the ecosystem level.

The assemblage and sequence of appearance of New World ecosystems was conditioned by the nature and patterns of climate change (e.g., Milankovitch variations) interacting with the mostly independent consequences of landscape development (atmospheric CO_2 concentration, ocean transport of heat, barriers, and bridges), and all were acting on the reservoir of new biotypes generated through evolution. The current result of these processes operating over the past 100 million years is a biological world that for convenience has been classified into the twelve ecosystems recognized in this text. A broad outline of their history is summarized in table 7.1. At present, that history is a big screen picture without high definition; that is, it is a little blurry. But the number of fossil floras and faunas being investigated through innovative techniques is increasing, and the pattern is becoming more clear.

The geologic record documents that over intervals measured in hundreds of thousands to millions of years in the Late Cretaceous and Neogene, environments changed at a comparatively gradual pace that was periodically interrupted by catastrophic events that altered conditions more suddenly.

Table 7.1. Elements (E), early versions (EV), and essentially modern versions (EM) of New World ecosystems and representative taxa in the Late Cretaceous and Tertiary periods

System	Interval			
	Cretaceous	Paleocene–early Eocene	Middle Eocene–early Miocene	Middle Miocene–Pliocene
D	E: *Ephedra*, Welwitschiaceae	E	E	E-EM (Mio-Plio.)
S	E: *Ephedra*, Bennettiales	E: Bennettiales extinct; add Poaceae, possibly *Pinus*, *Quercus* sp.	EV (depending on habitat; e.g., rain shadow): add *Juniperus*, *Acacia*, *Arctostaphylos*, *Berberis*, *Bursera*, *Caesalpinia*, *Cardiospermum*, *Celtis*, *Cercocarpus*, *Mahonia*, *Potentilla*, *Prosopis*, *Sapindus*, *Sophora*	EM midcontinent, woodland, grassy savanna mid-Miocene (13 Ma), then more grassy; west of Rocky Mountains, *Artemisia* (sagebrush), *Sarcobatus* (grease-wood) woodland to near desert
G		E: Poaceae	EV	EM 13 Ma midcontinent; 3 Ma west of Rocky Mountains
M	E: *Acrostichum*; *Nypa*	EV: add *Brevitricolpites* (e)	EM: *Brevitricolporites* disappears; add *Rhizophora*, *Hibiscus*, *Pelliceria*	EM
B	E (preservation limited)	EV, Poaceae	EM: *Sabal*, *Canavalia*, Poaceae	EM

(*continued*)

Table 7.1. (continued)

System	Interval			
	Cretaceous	Paleocene-early Eocene	Middle Eocene-early Miocene	Middle Miocene-Pliocene
F	E: *Equisetum*, ferns	EV add grasses, sedges, *Typha, Sparganium, Decodon*	EM	EM
A	EV: *Azolla, Isoetes, Marselia, Salvinia, Ceratophyllum, Cabombaceae, Dorftella* (e), Haloragidaceae, Nelumbonaceae, Nymphaeaceae, *Trapago*	EM: add *Porosia* (e), Poaceae, *Limnobiophyllum* (e: Lemnaceae)	EM add *Pachria, Ceratopteris*	EM
LR	EV: north, notophyllous broad-leaved evergreen forest, mixed with polar deciduous forest elements to form boreotropical forest; toward south, paratropical rain forest, tropical forest	EM (re structure, ecology, habitat, composition), esp. equatorial America	EM	EM
LM	EV: polar broad-leaved deciduous forest (main part of LM in north; *Nothofagus* in south)	EM (LM, more toward south, Latin America, beyond north deciduous forest; depending on topographic	EM	EM (most Asian components, thermophilous taxa from north gone by Pliocene, 8 Ma)

CF	E (Pinus (K, Pa), Araucaria (presently Southern Hemisphere), Brachyphyllum (e)	EV	diversity; north deciduous forest includes Asian components; *Acer, Alnus, Betula, Carya, Corylus, Cornus, Juglans Liquidambar, Quercus*; Asian: *Ginkgo. Metasequoia, Eucommia* (to Mexico), *Platycarya, Pterocarya, Zelkova*	EM
AP		E	Western montane **EM**: *Abies, Picea, Pinus, Thuja, Tsuga, Betula* (45 Ma)	Boreal EM (10–6 Ma)
T		E	Boreal **EV** (10 Ma)	**EV** 15 Ma to EM Pliocene North, **EV** 7–6 Ma (*Carex, Cyperus, Epilobium, Oenothera, Symphoiocarpus, Vaccinium*; **EM** 3–2 Ma; south, **EM** Mio-Pliocene, 6–5 Ma

Sequence Summary

LR	EM
LM	EM
A	EM

(continued)

Table 7.1. (continued)

System	Interval			
	Cretaceous	Paleocene-early Eocene	Middle Eocene-early Miocene	Middle Miocene-Pliocene
M		EV	**EM**	
CF			EM (montane)	**EM** (boreal)
F		EV	**EM**	
B	?	EV	**EM**	
S			EV	**EM**
G			EV	**EM**
D			E	**EM**
AT				**EM**
T				**EM**

Note: D = desert; S = shrubland-chaparral/woodland-savanna; G = grassland; M = mangrove; B = beach/strand/dune; F = freshwater herbaceous bog/marsh/swamp; A = aquatic; LR = lowland neotropical rain forest; LM = lower to upper montane broad-leaved forest; CF = coniferous forest; AP = alpine tundra (páramo); T = tundra; e = extinct group; bold = estimated time range when the ecosystem first appears in recognizable form.

The fossil record shows that during this time there was extinction of entire ecosystems, extensive modification of others, and that by the end of the Tertiary period the eight ancient ecosystems of the Cretaceous had been modified into or replaced by twelve essentially modern environments and biotic communities that presently extend across the landscapes of the New World.

The record further tells us that changes were beginning to happen at a faster pace in the late Neogene, raising expectations about events to follow in the near future. One of the hallmarks of vegetation history studies for the Quaternary is that the high resolution allows detecting in comparatively great detail the effects of rapid and significant as well as comparatively minor changes in climate on the Earth's vegetation cover. It may be recalled that in the mid-1980s extensive press coverage showed midwestern farmers sending loads of hay to the southeast where crops had been devastated by an El Niño–induced drought. The change corresponded to a temperature increase of about 0.5°C. This compares to changes of 6°C or more just in the recent Holocene. The Neogene was another pivotal time in ecosystem history because it set the stage for the latest modification in the range and composition of the Earth's biota, particularly at the level of plant associations, in response to the considerable environmental fluctuations of the Quaternary period. The Quaternary, in turn, provides near-term analogs for the effects of ongoing and anticipated environmental modification of the Earth's ecosystems on which we depend and of which we are a part.

: : **REFERENCES** : :

Axelrod, D. I., M. T. Kalin Arroyo, and P. H. Raven. 1991. Historical development of temperate vegetation in the Americas. *Rev. Chil. Hist. Nat.* 64:413–46.

Ballantyne, A. P., N. Rybczynski, P. A. Baker, C. R. Harington, and D. White. 2006. Pliocene Arctic temperature constraints from the growth rings and isotopic composition of fossil larch. *Palaeogeogr. Palaeocl. Palaeoecol.* 242:188–200.

Barreda, V. D., and E. Bellosi. 2003. Ecosistémas terrestres del Mioceno temprano de la Patagonia central, Argentina: Primeros avances. *Rev. Mus. Argent. Cienc. Nat.*, n.s. 5:125–34.

Barreda, V. D., and M. Caccavari. 1992. Mimosoideae (Leguminosae) occurrences in the early Miocene of Patagonia (Argentina). *Palaeogeogr. Palaeocl. Palaeoecol.* 94:243–52.

Berry, E. W. 1923. Tertiary fossil plants from the Republic of Haiti. *Proc. U.S. Natl. Mus.* 62:1–10.

Bowin, C. 1975. The geology of Hispaniola. In *The ocean basins and margins*, ed. A. E. M. Narin and F. G. Stehli, 501–52. Plenum, New York.

Brown, R. W. 1934. The supposed fossil ear of maize from Cuzco, Peru. *J. Wash. Acad. Sci.* 24:293–96.

Cerling, T. E., et al. (six coauthors). 1997. Global vegetation change through the Miocene/Pliocene boundary. *Nature* 389:153–58.

Coates, A. G., ed. 1997. *Central America: A natural and cultural history.* Yale University Press, New Haven.

Cohen, M. C. L., H. Behling, and R. J. Lara. 2005. Amazonian mangrove dynamics during the last millennium: The relative sea-level and the Little Ice Age. *Rev. Palaeobot. Palynol.* 136:93–108.

Cohen, M. C. L., P. Filho, R. J. Lara, H. Behling, and R. Angulo. 2005. A model of Holocene mangrove development and relative sea-level changes on the Bragança Peninsula (northern Brazil). *Wetlands Ecology and Management* 13:433–43.

Crowley, T. J., and W. T. Hyde. 2008. Transient nature of late Pleistocene climate variability. *Nature* 456:226–30.

Crowley, T. J., and G. R. North. 1991. *Paleoclimatology.* Oxford Monographs in Geology and Geophysics, no. 16. Oxford University Press, Oxford.

DeSantis, L. R. G., and S. C. Wallace. 2008. Neogene forests from the Appalachians of Tennessee, USA: Geochemical evidence from fossil mammal teeth. *Palaeogeogr. Palaeocl. Palaeoecol.* 266:59–68.

Díaz de Gamero, M. L. 1996. The changing course of the Orinoco River during the Neogene: A review. *Palaeogeogr. Palaeocl. Palaeoecol.* 123:385–402.

Eldrett, J. S., D. R. Greenwood, I. C. Harding, and M. Huber. 2009. Increased seasonality through the Eocene to Oligocene transition in northern high latitudes. *Nature* 459:969–73.

Fields, P. F. 1992. Floras of Idaho's past: Geographic and paleobotanic overview of the middle Miocene Succor Creek flora as an example. *J. Idaho Acad. Sci.* 28:44–56.

Foley, J. A. 2005. Tipping points in the tundra. *Science* 310:627–28.

Fox, D. L. 2000. Growth increments in *Gomphotherium* tusks and implications for late Miocene climate change in North America. *Palaeogeogr. Palaeocl. Palaeoecol.* 156:327–48.

Gaupp, R., A. Kött, and G. Wörner. 1999. Palaeoclimatic implications of Mio-Pliocene sedimentation in the high-altitude intra-arc Lauca Basin of northern Chile. *Palaeogeogr. Palaeocl. Palaeoecol.* 151:79–100.

Gladwell, M. 2000. *The tipping point: How little things can make a big difference.* Little, Brown, London.

Graham, A. 1963. Systematic revision of the Sucker Creek and Trout Creek Miocene floras of southeastern Oregon. *Am. J. Bot.* 50:921–36.

———. 1965. *The Sucker Creek and Trout Creek Miocene floras of southeastern Oregon.* Kent State University Bulletin, Research Series 9. Kent State University Press, Kent.

———. 1966. Plantae rariores camschatcenses: A translation of the dissertation of Jonas P. Halenius, 1750. *Brittonia* 18:131–39.

———, ed. 1972. *Floristics and paleofloristics of Asia and eastern North America.* Elsevier Science Publishers, Amsterdam.

———. 1976. Studies in neotropical paleobotany. Part 2, The Miocene communities of Veracruz, Mexico. *Ann. Mo. Bot. Gard.* 63:787–842.

———. 1990. Late Tertiary microfossil flora from the Republic of Haiti. *Am. J. Bot.* 77:911–26.

———. 1998. Studies in neotropical paleobotany. Part 11, Late Tertiary vegetation and environments of southeastern Guatemala: Palynofloras from the Mio-Pliocene Padre Miguel Group and the Pliocene Herrería Formation. *Am. J. Bot.* 85:1409–25.

Graham, A., K. M. Gregory-Wodzicki, and K. L. Wright. 2001. Studies in neotropical paleobotany. Part 15, A Mio-Pliocene palynoflora from the Eastern Cordillera, Bolivia: Implications for the uplift history of the Central Andes. *Am. J. Bot.* 88:1545–57.

Gregory-Wodzicki, K. M. 2000. Relationships between leaf morphology and climate, Bolivia: Implications for estimating paleoclimate from fossil floras. *Paleobiology* 26:668–88.

Gregory-Wodzicki, K. M., W. C. McIntosh, and K. Velásquez. 1998. Climate and tectonic implications of the late Miocene Jakokkota, Bolivian Altiplano. *J. South Am. Earth Sci.* 11:533–60.

Hansen, B. C. S., E. C. Grimm, and W. A. Watts. 2001. Palynology of the Peace Creek site, Polk County, Florida. *Geol. Soc. Am. Bull.* 113:682–92.

Haug, G. H., et al. (ten coauthors). 2005. North Pacific seasonality and the glaciation of North America 2.7 million years ago. *Nature* 433:821–25.

Hoorn, C. 1993. Marine incursions and the influence of Andean tectonics on the Miocene depositional history of northwestern Amazonia: Results of a palynostratigraphic study. *Palaeogeogr. Palaeocl. Palaeoecol.* 105:267–309.

———. 1994a. An environmental reconstruction of the palaeo–Amazon River system (middle-late Miocene, NW Amazonia). *Palaeogeogr. Palaeocl. Palaeoecol.* 112:187–238.

———. 1994b. Fluvial palaeoenvironments in the intracratonic Amazonas Basin (early Miocene–early middle Miocene, Colombia). *Palaeogeogr. Palaeocl. Palaeoecol.* 109:1–54.

———. 1996. Miocene deposits in the Amazonian foreland basin. *Science* 273:122–23.

Hueber, F. M., and J. Langenheim. 1986. Dominican amber tree had African ancestors. *Geotimes* 31:8–10.

Hughes, C., and R. Eastwood. 2006. Island radiation on a continental scale: Exceptional rates of plant diversification after uplift of the Andes. *Proc. Natl. Acad. Sci. U.S.A.* 103:10334–39.

An insight into amber. 2008. *Nature* 452.677.

Ituralde-Vinent, M. A. 2001. Geology of the amber-bearing deposits of the Greater Antilles. *Carib. J. Sci.* 37:141–67.

Jackson, J. B. C., A. F. Budd, and A. G. Coates, eds. 1996. *Evolution and environment in tropical America.* University of Chicago Press, Chicago.

Janis, C. M., J. Damuth, and J. M. Theodor. 2002. The origins and evolution of the North American grassland biome: The story from the hoofed mammals. *Palaeogeogr. Palaeocl. Palaeoecol.* 177:183–98.

Jarzen, D. M., and D. L. Dilcher. 2006. Middle Eocene terrestrial palynomorphs from the Dolime Minerals and Gulf Hammock Quarries, Florida, U.S.A. *Palynology* 30:89–110.

Kemp, M. 2005. Inventing an icon: Hans Joachim Schellnhuber's map of global "tipping points" in climatic change. *Nature* 437:1238.

Kleinert, K., and M. R. Strecker. 2001. Climate change in response to orographic barrier uplift: Paleosol and stable isotope evidence from the late Neogene Santa María Basin, northwestern Argentina. *Geol. Soc. Am. Bull.* 113:728–42.

Kleiven, H. F., E. Jansen, T. Fronval, and T. M. Smith. 2002. Intensification of Northern Hemisphere glaciations in the circum Atlantic region (3.5–2.4 Ma)—ice-rafted detritus evidence. *Palaeogeogr. Palaeocl. Palaeoecol.* 184:213–23.

Knowlton, F. H. 1919. Description of a supposed new fossil species of maize from Peru. *J. Wash. Acad. Sci.* 9:134–36.

Lee, D. E., W. G. Lee, and N. Mortimer. 2001. Where and why have all the flowers gone? Depletion and turnover in the New Zealand Cenozoic angiosperm flora in relation to palaeogeography and climate. *Austral. J. Bot.* 49:341–56.

Leopold, E. B., and M. E. Denton. 1987. Comparative age of grassland and steppe east and west of the northern Rocky Mountains. *Ann. Mo. Bot. Gard.* 74:841–67.

Lewis, A. R., et al. (twelve coauthors). 2008. Mid-Miocene cooling and the extinction of tundra in continental Antarctica. *Proc. Natl. Acad. Sci. U.S.A.* 105:10676–80.

Liu, Y. S. In press. *Late Miocene flora from Gray, Tennessee, and its paleoclimatic and biogeographic significance*. Report to the National Geographic Society, 18 May 2009.

Lunt, D. J., G. L. Foster, A. M. Haywood, and E. J. Stone. 2008. Late Pliocene Greenland glaciation controlled by a decline in atmospheric CO_2 levels. *Nature* 454:1102–5.

MacFadden, B. J., Y. Wang, T. E. Cerling, and F. Anaya. 1994. South American fossil mammals and carbon isotopes: A 25 million-year sequence from the Bolivian Andes. *Palaeogeogr. Palaeocl. Palaeoecol.* 107:257–68.

Maguire, K. C., and A. L. Stigall. 2008. Paleobiogeography of Miocene Equinae of North America: A phylogenetic biogeographic analysis of the relative roles of climate, vicariance, and dispersal. *Palaeogeogr. Palaeocl. Palaeoecol.* 267:175–84.

Markgraf, V., ed. 2001. *Interhemispheric climate linkages*. Academic Press, New York.

Martin, R. A., et al. (eight coauthors). 2008. Rodent community change at the Pliocene-Pleistocene transition in southwestern Kansas and identification of *Microtus* immigration event on the central Great Plains. *Palaeogeogr. Palaeocl. Palaeoecol.* 267:196–207.

Monsch, K. A. 1998. Miocene fish faunas from the northwestern Amazonian basin (Colombia, Peru, Brazil), with evidence of marine incursions. *Palaeogeogr. Palaeocl. Palaeoecol.* 143:31–50.

Nash, J. M. 2002. El Niño, Unlocking the Secrets of the Master Weather-Maker. Warner Books, New York.

Ohlemüller, R., et al. (six coauthors). 2008. The coincidence of climatic and species rarity: High risk to small-range species from climate change. *Biol. Lett.* (The Royal Society) 4:568–72.

Palamarczuk, S., and V. Barreda. 1998. Bioestratigrafía en base a quistes de dinoflagelados de la Formación Chenque (Mioceno), Provincia del Chubut, Argentina. *Ameghiniana* 35:415–26.

Péwé, T. L., J. A. Westgate, S. L. Preece, P. M. Brown, and S. W. Leavitt. 2009. Late Pliocene Dawson Cut Forest Bed and new tephrochronological findings in the Gold Hill Loess, east-central Alaska. *Geol. Soc. Am. Bull.* 121:294–320.

Prange, M., and M. Schulz. 2004. A coastal upwelling seesaw in the Atlantic Ocean as a result of the closure of the Central American seaway. *Geophys. Res. Lett.* 31:L17207, doi: 1029/2004GL020073.

Rea, D. K., M. Leinen, and T. R. Janecek. 1985. Geologic approach to the long-term history of atmospheric circulation. *Science* 227:721–25.

Reinink-Smith, L. M., and E. B. Leopold. 2005. Warm climate in the late Miocene of the south coast of Alaska and the occurrence of Podocarpaceae pollen. *Palynology* 29:205–62.

Retallack, G. J. 2004. Late Miocene climate and life on land in Oregon within a context of Neogene global change. *Palaeogeogr. Palaeocl. Palaeoecol.* 214:97–123.

Retallack, G. J., and M. X. Kirby. 2007. Middle Miocene global change and paleogeography of Panama. *Palaios* 22:667–79.

Ribas, C. C., R. G. Moyle, C. Y. Miyaki, and J. Cracraft. 2007. The assembly of montane biotas: Linking Andean tectonics and climatic oscillations to independent regimes of diversification in *Pionus* parrots. *Proc. Biol. Sci. B* (The Royal Society) 274:2399–2408.

Riishuus, M. S., D. K. Bird, L. E. Heister, C. K. Brooks, and M. T. Hren. 2006. Tephras and soils as terrestrial paleoclimate proxies in large igneous provinces: Examples from Greenland and Iceland. Abstract. *Am. Geophys. Union*, fall meeting.

Rouchy, J. M., A. Caruso, C. Pierre, M.-M. Blanc-Valleron, and M. A. Bassetti. 2007. The end of the Messinian salinity crisis: Evidences from the Chelif Basin (Algeria). *Palaeogeogr. Palaeocl. Palaeoecol.* 254:386–417.

Rouchy, J. M., et al. (six coauthors). 2001. Paleoenvironmental changes at the Messinian-Pliocene boundary in the eastern Mediterranean (southern Cyprus basins): Significance of the Messinian Lago-Mare. *Sed. Geol.* 145:93–117.

Salzmann, U., A. M. Haywood, D. J. Lunt, P. J. Valdes, and D. J. Hill. 2008. A new global biome reconstruction and data-model comparison for the middle Pliocene. *Global Ecol. Biogeogr.* 17:432–47.

Seeing the light. 2008. *Economist*, 12 April, 88–89.

Shunk, A. J., S. G. Driese, and G. M. Clark. 2006. Latest Miocene to earliest Pliocene sedimentation and climate record derived from paleosinkhole fill deposits, Gray Fossil Site, northeastern Tennessee, U.S.A. *Palaeogeogr. Palaeocl. Palaeoecol.* 231:265–78.

Strömberg, C. A. E. 2002. The origin and spread of grass-dominated ecosystems in the late Tertiary of North America: Preliminary results concerning the evolution of hypsodonty. *Palaeogeogr. Palaeocl. Palaeoecol.* 177:59–75.

———. 2004. Using phytolith assemblages to reconstruct the origin and spread of grass-dominated habitats in the Great Plains of North America during the late Eocene to early Miocene. *Palaeogeogr. Palaeocl. Palaeoecol.* 207:239–75.

———. 2005. Decoupled taxonomic radiation and ecological expansion of open-habitat grasses in the Cenozoic of North America. *Proc. Natl. Acad. Sci. U.S.A.* 102:11980–84.

Strömberg, C. A. E., L. Werdelin, E. M. Friis, and G. Saraç. 2007. The spread of grass-dominated habitats in Turkey and surrounding areas during the Cenozoic: Phytolith evidence. *Palaeogeogr. Palaeocl. Palaeoecol.* 250:18–49.

Takeuchi, A., P. B. Larson, and K. Suzuki. 2007. Influence of paleorelief on the mid-Miocene climate variation in southeastern Washington, northeastern Oregon, and western Idaho, USA. *Palaeogeogr. Palaeocl. Palaeoecol.* 254:462–76.

Tauber, A. A. 1997a. Bioestratigrafía de la Formación Santa Cruz (Mioceno Inferior) en el extreme sudeste de la Patagonia. *Ameghiniana* 34:413–26.

———. 1997b. Paleoecología de la Formación Santa Cruz (Mioceno Inferior) en el extreme sudeste de la Patagonia. *Ameghiniana* 34:517–29.

Tedford, R. H., and C. R. Harington. 2003. An Arctic mammal fauna from the early Pliocene of North America. *Nature* 425:388–90.

Verzi, D. H., and C. A. Quintana. 2005. The caviomorph rodents from the San Andrés Formation, east-central Argentina, and global late Pliocene climate change. *Palaeogeogr. Palaeocl. Palaeoecol.* 219:303–20.

Wallace, S. C., and X. Wang. 2004. Two new carnivores from an unusual late Tertiary forest biota in eastern North America. *Nature* 431:556–59.

Wen, J. 1999. Evolution of eastern Asian and eastern North American disjunct distributions in flowering plants. *Ann. Rev. Ecol. Syst.* 30:421–55.

Wendt, T. 1993. Composition, floristic affinities, and origins of the canopy tree flora of the Mexican Atlantic slope rain forests. In *Biological diversity of Mexico: Origins and distributions,* ed. T. P. Ramamoorthy, R. Bye, A. Lot, and J. Fa, 595–680. Oxford University Press, Oxford.

Wijninga, V. M. 1995. A first approximation of montane forest development during the late Tertiary in Colombia. In *Biodiversity and conservation of neotropical montane forests,* ed. S. P. Churchill, H. Balslev, E. Forero, and J. L. Luteyn, 23–34. New York Botanical Garden, New York.

———. 1996a. Neogene ecology of the Salto de Tequendama site (2475 m altitude, Cordillera Oriental, Colombia): The paleobotanical record of montane and lowland forests. *Rev. Palaeobot. Palynol.* 92:97–156.

———. 1996b. Paleobotany and palynology of Neogene sediments from the High Plain of Bogotá (Colombia): Evolution of the Andean flora from a paleoecological perspective. Ph.D. diss., University of Amsterdam. [Available in vol. 21 of the reprint series *The Quaternary of Colombia / El Cuaternario de Colombia;* various parts also published in other works by the author cited here.]

———. 1996c. Palynology and paleobotany of the early Pliocene section Río Frio 17 (Cordillera Oriental, Colombia): Biostratigraphical and chronostratigraphical implications. *Rev. Palaeobot. Palynol.* 92:329–50.

———. 1996d. A Pliocene *Podocarpus* forest mire from the area of the High Plain of Bogotá. *Rev. Palaeobot. Palynol.* 92:157–205.

Wijninga, V. M., and P. Kuhry. 1990. A Pliocene flora from the Subachoque Valley (Cordillera Oriental, Colombia). *Rev. Palaeobot. Palynol.* 62:249–90.

Williams, C. J., et al. (five coauthors). 2008. Paleoenvironmental reconstruction of a middle Miocene forest from the western Canadian Arctic. *Palaeogeogr. Palaeocl. Palaeoecol.* 261:160–76.

Wolfe, J. A., and T. Tanai. 1980. *The Miocene Seldovia Point flora from the Kenai Group, Alaska.* U.S. Geological Survey Professional Paper 1105. U.S. Department of the Interior, Washington, D.C.

Zubakov, V. A., and I. I. Bozenkova. 1990. *Global palaeoclimate of the Late Cenozoic.* Developments in Palaeontology and Stratigraphy 12. Elsevier Science Publishers, Amsterdam.

Additional Readings and Updates

Billups, K. 2005. Snow maker for the ice ages. *Nature* 433:809–10. [Summary of Haug et al. 2005.]

Bond, W. J. 2008. What limits trees in C4 grasslands and savannas? *Annual Rev. Ecol. Evol. Syst.* 39:641–59.

Buechler, W. K., M. T. Dunn, and W. C. Rember. 2007. Late Miocene Pickett Creek flora of Owyhee County, Idaho. *Contrib. Mus. Paleon. Univ. Mich.* 31:305–62.

Culver, S. J., et al. (ten coauthors). 2008. Micropaleontologic record of late Pliocene and Quaternary paleoenvironments in the northern Albermarle Embayment, North Carolina, U.S.A. *Palaeogeogr. Palaeocl. Palaeoecol.* 264:54–77.

Florindo, F., A. Nelson, and A. M. Haywood, eds. 2008. *Antarctic cryosphere and Southern*

Ocean climate evolution (Cenozoic-Holocene). Special issue, *Palaeogeogr. Palaeocl. Palaeoecol.* 260:1–298. [Papers from the European Geosciences Union General Assembly 2009.]

Fox, D. L., and P. L. Koch. 2003. Tertiary history of C_4 biomass in the Great Plains, USA. *Geology* 31:809–12.

Grimsson, F., and T. Denk. 2007. Floristic turnover in Iceland from 15 to 6 Ma—extracting biogeographic signals from fossil floral assemblages. *J. Biogeogr.* 34:1490–1504.

Hoppe, K. A., R. Amundson, M. Vavra, M. P. McClaran, and D. L. Anderson. 2004. Isotopic analysis of tooth enamel carbonate from modern North American feral horses: Implications for paleoenvironmental reconstructions. *Palaeogeogr. Palaeocl. Palaeoecol.* 203:299–311.

Kirby, M. X., D. S. Jones, and B. J. MacFadden. 2008. Lower Miocene stratigraphy along the Panama Canal and its bearing on the Central American peninsula. *PLoS ONE* 3 (7): e2791, doi 10.1371/journal.pone.0002791.

Landau, B., G. Vermeij, and C. M. da Silva. 2008. Southern Caribbean Neogene palaeobiogeography revisited: New data from the Pliocene of Cubagua, Venezuela. *Palaeogeogr. Palaeocl. Palaeoecol.* 257:445–61.

Lindars, E. S., et al. (five authors). 2001. Phosphate $\delta^{18}O$ determination of modern rodent teeth by direct laser fluorination: An appraisal of methodology and potential application to palaeoclimate reconstruction. *Geoch. Cosmo. Act.* 65:2535–48.

MacFadden, B. J. 2008. Geographic variation in diets of ancient populations of 5-million-year-old (early Pliocene) horses from southern North America. *Palaeogeogr. Palaeocl. Palaeoecol.* 266:83–94.

McKown, A. D., R. A. Stockey, and C. E. Schwegert. 2002. A new species of *Pinus* subgenus *Pinus*, section *Contortae* from Pliocene sediments of Ch'ijee's Bluff, Yukon Territory, Canada. *Int. J. Plant Sci.* 163:687–97. [*Pinus matthewsii*.]

Meyer, H. W. 2005. *Metasequoia* in the Oligocene Bridge Creek flora of western North America: Ecological implications and the history of research. In *The geobiology and ecology of Metasequoia*, ed. B. A. LePage, C. J. Williams, and H. Yang, 159–86. Springer Verlag, Berlin.

Naish, T., et al. (fifty-five coauthors). 2009. Obliquity-paced Pliocene West Antarctic ice sheet oscillations. *Nature* 458:322–28.

Olivier, J., et al. (five coauthors). 2009. First macrofossil evidence of a pre-Holocene thorny bamboo cf. *Guadua* (Poaceae: Bambusoideae: Bambuseae: Guaduinae) in south-western Amazonia (Maria de Dios-Peru). *Rev. Palaeobot. Palynol.* 153:1–7.

Penney, D. 2008. *Dominican amber spiders: A comparative palaeontological-neontological approach to identification, faunistics, ecology and biogeography.* Siri Scientific Press, Totnes, Devon, U.K.

Pigg, K. B., S. M. Ickert-Bond, and J. Wen. 2004. Anatomically preserved *Liquidambar* (Altingiaceae) from the middle Miocene of Yakima Canyon, Washington State, USA, and its biogeographic implications. *Am. J. Bot.* 91:409–509. [*Liquidambar changii*.]

Pigg, K. B., and G. W. Rothwell. 2001. Anatomically preserved *Woodwardia virginica* (Blechnaceae) and a new filicalean fern from the middle Miocene Yakima Canyon flora of central Washington, USA. *Am. J. Bot.* 88:777–87.

Pollard, D., and R. M. DeConto. 2009. Modeling West Antarctic ice sheet growth and collapse through the past five million years. *Nature* 458:329–32.

Pons, D., and D. De Franceschi. 2007. Neogene woods from western Peruvian Amazon and palaeoenvironmental interpretation. *Bull. Geosci.* (Czech Geological Survey, Prague) 82:343–54. [Middle Miocene to Pliocene.]

Ramamoorthy, T. P., R. Bye, A. Lot, and J. Fa., eds. 1993. *Biological diversity of Mexico, origins and distribution*. Oxford University Press, Oxford.

Ramírez, S. R., B. Gravendeel, R. B. Singer, C. R. Marshall, and N. E. Pierce. 2007. Dating the origin of the Orchidaceae from a fossil orchid with its pollinator. *Nature* 448:1042–45. [A report on orchid pollen and stingless bee from the Dominican amber 12–20 Ma concludes from this relatively recent fossil record, based on molecular evidence, that "the most recent common ancestor of extant orchids lived in the Late Cretaceous (76–84 Ma)."]

Raymo, M. E., and P. Huybers. 2008. Unlocking the mysteries of the ice ages. *Nature* 451:284–85.

Raymo, M. E., L. Lisiecki, and K. Nisancioglu. 2006. Plio-Pleistocene ice volume, Antarctic climate, and the global $\delta^{18}O$ record. *Science* 313:492–95.

Santiago-Blay, J. A., and J. B. Lambert. 2007. Amber's botanical origin revealed. *Am. Sci.* 95:148–57.

Santos, J. C., et al. (five coauthors). 2009. Amazonian amphibian diversity is primarily derived from late Miocene Andean lineages. *PLoS ONE* 7 (3): e1000056, doi:10.1371/journal .pbio.1000056.

Strömberg, C. A. E. 2006. Evolution of hypsodonty in equids: Testing a hypothesis of adaptation. *Paleobiology* 32:236–58.

Strömberg, C. A. E., and R. S. Feranec, eds.. 2004. *Evolution of grass-dominated ecosystems during the Late Cenozoic*. Special issue, *Palaeogeogr. Palaeocl. Palaeoecol.* 207:199–424. [Papers from the Evolution of Grass-dominated Ecosystems during the Late Cenozoic Session at the 2001 North American Paleontological Convention.]

Svenning, J.-C., S. Normand, and F. Skov. 2009. Plio-Pliocene climate change and geographic heterogeneity in plant diversity-environment relationships. *Ecography* 32:13–21.

Verzi, D. H., and C. I. Montalvo. 2008. The oldest South American Cricetidae (Rodentia) and Mustelidae (Carnivora): Late Miocene faunal turnover in central Argentina and the great American biotic interchange. *Palaeogeogr. Palaeocl. Palaeoecol.* 267:284–91.

Wille, A. 1959. A new fossil stingless bee (Meliponini) from the amber of Chiapas, Mexico. *J. Paleontol.* 33:849–52.

8

Latest Touch

The Great Ice Ages

The relatively brief interval of the Quaternary, some 2.6 million years duration, belies the great interest in and the importance of this last geologic period (Pasini and Colalongo 1997; Balco et al. 2005; Gibbard 2007; Kerr 2008; Meltzer 2008). Its significance for us derives in no small part from the fact that it encompasses the span of *Homo sapiens*. Assuming the species appeared about 160,000 years ago, all the world's ecosystems had long been established, and the locales of human origin and the early routes along which they dispersed had been alternating between near-desert/ savanna/grassland and more mesic forest.

Another intriguing aspect of the Quaternary was the disappearance or major reduction in the great Pleistocene megafauna in many parts of the world. This event took place globally between about 50,000 and 10,000 years ago, it began in North America with the crossing of Beringia by hunters at around 17,000 BP, and it is documented by the finds at Murray Springs, Arizona, dating from 13,500 BP (chap. 2). The possibility of Pleistocene overkill has been suggested by Paul Martin (1982, 1984) because the hunters arrived in North America at about the same time as the extinctions. However, the importance earlier attributed to hunting has recently diminished because kill sites are relatively few (see summary by Fagan 2004, 38–41). Climate change is another factor long-regarded as playing a role in the demise of the mammoths, giant ground sloths, and the huge kangaroos of Australia (*Procoptodon*), in part, because the change also was happening at the time they disappeared. In itself, however, climate was not likely the only factor, because the fauna had survived many similar fluctuations throughout the Pleistocene. A third factor has been proposed by Chris Johnson (2002), namely, the low birth rate of many of the large plains and savanna mammals that disappeared. This supposedly rendered them vulnerable to the combined impacts of climate and hunting. Most of those that did survive were, in fact, arboreal, nocturnal, or lived in dense forests. As noted elsewhere, it is reassuring that rather than interminably perpetuating the one-cause-versus-another mode of discussion, a number of plausible factors—overkill, climate, reproductive rate—are offering a multifaceted explanation approaching the complexity of the event.

A further hallmark of the Quaternary period is that the dramatic changes in the environment, their causes, and the biotic responses can be traced with a relatively high degree of precision. This raises a question about whether the myriad apparent tipping points and fluctuations in the Quaternary really were occurring at a pace faster than in the Cretaceous and Tertiary, or if their cluster is just an artifact of the higher resolution possible for this most recent period. There certainly were reversals in the Tertiary, such as the cool interval of 1–1.5 million years at the Paleocene-Eocene boundary in the northern Rocky Mountains (Wing 1996), the four warm spells following that boundary in New Zealand (Newton 2007; Nicolo et al. 2007), and the well-documented warm periods of the middle Miocene and middle Pliocene. Nonetheless, the evidence is convincing from Greenland ice cores taken in the 1980s that the past 150,000 years was a wild if not crazy time of rapid "switch and dial" climate changes (Alley 2000). There was usually a relatively gradual buildup to glacial conditions, then a precipitous and chaotic fall into interglacial climates. After the LGM circa 18,000 BP, Greenland warmed by 12°C–15°C in a matter of several decades. Within the overall warming trend, there were reversals, and one of the most prominent was the Younger Dryas (named after the European alpine plant *Dryas octopetala*). It was a cold interval between 12,800 and 11,500 BP and included a decrease of 4.5°C ± 1.3°C in surface marine water temperatures (e.g., Corrège et al. 2004). There was another drop in temperature at

8200 BP. (Dean et al. 2002; Ellison et al. 2006). Warming periods include the PGCO (Postglacial Climatic Optimum, hypsithermal interval, or xerothermic period) between 8000 and 6000 BP (+2°C–3°C warmer than at present), and a Medieval Warm Period at 800–1200 CE (+1°C warmer) in the North Atlantic region. As recounted in chapter 2, the Medieval Warm Period was when the Vikings made their extended sea voyages into the New World. It was also the last time large mobile sand dunes formed in the Nebraska Sand Hills (Sridhar et al. 2006). The warm period was followed by the Little Ice Age between 1300 and 1850 CE, when the Vikings abandoned their sites. An early sign of near-glacial or impending glacial climates is the formation of sea ice along the shore (fig. 8.1), which leaves telltale evidence in coastal sediment composition. Short-term glacial fluctuations include Heinrich events of a few to several thousand years' duration, representing the culmination of cold periods when armadas of icebergs are calved from the edge of the continents into the ocean (fig. 8.2).

The physical evidence for Heinrich events is layers of gravel on the ocean floor released from icebergs as they melt. The latest H1 event, of a possible thirteen (Hiscott et al. 2001), occurred about 15,000 BP. These short-term

Figure 8.1 Sea ice, Scoresby Sound, East Greenland. Photograph by Shirley A. Graham.

Figure 8.2 Icebergs in the North Atlantic. Armadas of ice fragments are calved into the ocean during cold intervals, and the debris left as layers of gravel on the ocean floor marks these Heinrich events. Photograph by Shirley A. Graham.

fluctuations in climate were likely caused by interference with ocean circulation (Schmittner and Galbraith 2008), for example, by input of freshwater from melting glaciers into the North Atlantic Ocean (Tarasov and Peltier 2005; Broecker 2003, 2006). Dansgaard-Oeschgen (D-O) events operate on an even finer time scale of several hundred to a few thousand years and may result from fluctuations in solar input of heat that also influences the rate of freshwater flow into the North Atlantic.

With increased resolution comes the possibility of incorporating historical factors in statistical/modeling efforts to explain the current status of communities. One such effort is the Community Assembly Rules (CARs) proposed by Jared Diamond (Cody and Diamond 1975) to help understand the composition and spatial arrangement of components within avian populations on islands. Initially, competition was seen as preeminent, but this conclusion was later expanded to include niche diversity, dispersal ability, the nature of the surrounding reservoir from which the component species are derived, and environmental dynamics. Climate change, the nature of the reservoir, and the sequence of appearance of different species can be detected from vegetation history studies when resolution is sufficiently high. The "rules" or hypotheses are now being refined as part of various studies, for example, of packrat middens in the arid southwest, designed to identify Holocene refugia in the southern Sonoran Desert during cold intervals (Camille Holmgren, work in progress).

One of the hallmarks of recent paleoclimate studies has been to demonstrate that these changes in temperature, identified first in the North Atlantic region, also extended into the lower latitudes. The first large-scale attempts to reconstruct Quaternary climates were the Climate Long-Range Investigation and Mapping Program (CLIMAP 1976, 1981, 1984) and the Cooperative Holocene Mapping Project (COHMAP 1988). These early efforts suggested that while climates cooled significantly in the north (12°C–14°C), they waned toward the lower latitudes and cooled there only by as much as 2°C. Guilderson and others (1994) have shown from oxygen isotope and Sr/Ca thermometry in Barbados corals (13°11′N) that temperatures there were 5°C cooler at the LGM, and in the lowlands of Brazil they were 5°C–6°C cooler (Stute et al. 1995; Webb et al. 1997; Aeschbach-Hertig et al. 2000). The Younger Dryas, Heinrich events, and the Medieval Warm Period are now recognized from several sites in the equatorial regions, and the results document that the tropics experienced a muted but nonetheless dynamic Quaternary climate. The related precipitation patterns were the result of the meanderings of the Intertropical Convergence Zone and the pulsating occurrence of El Niños. To the far south, similar Quaternary changes in climate are also preserved in the 3623-meter Vostok and other ice cores from Antarctica and in marine cores from the Southern Ocean. At times the temperature shifts in the two hemispheres were out of phase, leading to the idea of a bipolar seesaw, whereby heat was reciprocally transferred via the ocean conveyor belt. If this were the case, however, the pace of temperature changes in the two regions should have been more or less equal, but those in the Southern Hemisphere appear to have been slower. Barker and colleagues 2009 now present evidence for the past 26,000 years that, at least for surface waters during the last deglaciation, the changes in the two hemispheres were at times opposite and abrupt (see summary by Severinghaus 2009).

Evidence that major glacial advances and changes in climate occurred in the relatively recent past was building during the nineteenth century, based in part on the presence of boulder erratics on the plains and in the mountains of Europe. These rocks, some the size of small houses, had a composition different from the underlying substrate and had obviously been transported some distance by powerful forces. Among the early expla-

Figure 8.3 Agassiz in concrete. The statue toppled from the Zoology Building on the Stanford University campus during the 1906 earthquake. From the Web site of the Stanford University Quake '06 Centennial Alliance, http://quake06.stanford.edu. Photograph courtesy of the Stanford University Archives, Stanford, CA.

nations was movement by water during the biblical flood, but Louis Agassiz for one, in his *Etudes sur les Glaciers* (1840), showed that the rocks had been transported on and within valley and continental glaciers and were left stranded on distant surfaces as the ice melted. Agassiz (1807–73) is an interesting figure in the history of science because as a prominent zoologist, geologist, and founder of the Museum of Comparative Zoology at Harvard University, he was a vigorous opponent of Darwin and the concept of evolution (fig. 8.3). His arguments that glaciers covered vast areas of the world in relatively recent times were compelling, however, and they implied the monumental changes in climate and ecosystems now recognized for the past 2.6 million years.

During the 1960s, conventional wisdom was that the Quaternary encompassed approximately the last million years, and that there had been four major glacial advances—starting with the Nebraskan, through the Kansan, Illinoian, and Wisconsinian, and four glacial retreats or interglacials, from the Aftonian through the Yarmouth, Sangamon, and the present Holocene or Recent—each lasting about 175,000 years. One implication was that the present interglacial, beginning 11,000 to 12,000 years ago, would continue for another 165,000 years or so. Another was that the ecosystems that characterize each interval last for about 175,000 years, and that their present arrangement would also continue for the next 165,000 years. Then the Serbian astronomer and mathematician Milutin Milankovitch (1920) discovered that the pace of long-term climatic change had been quite different. Milankovitch documented that cycles occur in the position of the Earth relative to the sun (fig. 8.4). Any alteration in distance or orientation would affect the amount of heat reaching the Earth's surface, and if there was a periodicity in the fluctuations, this could be a factor in determining patterned

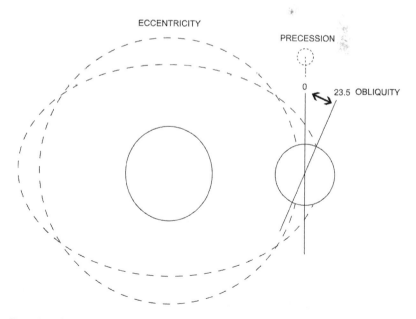

Figure 8.4 The three Milankovitch variations: eccentricity, precession, and obliquity.

climatic events such as glaciations. There is such a pattern, and the spatial relationships between the Earth and sun vary in three ways. One is in the shape of the Earth's orbit, or eccentricity, which changes from nearly circular to elliptical. At present, the Earth receives about 6 percent more heat at its closest point (the perihelion) than at its farthest point (the aphelion), but at maximum eccentricity the difference is about 27 percent. The change in orbit from most circular to most elliptical takes around 100,000 years, with a longer eccentricity supercycle of approximately 405,000 years.

The second alteration is in the obliquity or tilt of the Earth's axis. At present, the tilt is 23.5° from the vertical, but it varies from 22.1° to 24.5°. The period of the tilt variation is around 41,000 years, and the effect is to alternately minimize and maximize seasonal differences, especially toward the poles, which also affects the intensity of ocean circulation (Lisiecki et al. 2008).

The third alteration is precession. As the Earth spins, the direction to which the polar axes point changes slightly, much like that of a spinning top as it slows down. If this movement were projected onto a distant background it would inscribe a circle. At present the north polar axis points toward Polaris (the end star in the handle of the Little Dipper), and in about 12,000 it will point toward Vega. The time it takes to inscribe a full imaginary circle is about 23,000. These three variations are commonly called stretch, tilt, and wobble, and the causes derive from differences in the gravitational pull on the Earth's equatorial bulge by the moon, sun, and planets, notably Jupiter and Saturn.

After the Milankovitch variations were proposed, a wealth of supporting evidence was forthcoming, including the pattern of varves (layers) in ice from the Greenland (Alley 2000; Landais et al. 2006) and Antarctic ice core projects, layering in ocean sediments, and the movement of tree line along mountain slopes where the vegetation is sharply zoned and highly responsive to climate change. There is no reason to believe that the Milankovitch variations, based on orbital and gravitational forces, should have operated only since 2.6 Ma, and indeed, evidence exists for much older cycles in the Triassic of the Newark Basin, from Germany (Vollmer et al. 2008), and from the Paleocene of the northwestern Pacific Ocean (Westerhold et al. 2008). It is likely the cycles are less evident prior to about 2.6 Ma because of altered sediments, poorer resolution, and because high CO_2 concentration exerted an overriding influence on temperatures. With the waning of CO_2 concentration in the later part of the Neogene, the Milankovitch variations become more evident. Thus, among the principal causes for large-scale changes in past climate are (1) the Milankovitch variations, (2) fluctuations in atmospheric CO_2 concentration, and (3) changes in ocean circulation. Regarding CO_2 and climate, modeling experiments using BIOME3 simulations predict that temperatures warm with high levels of atmospheric CO_2 (Cowling 1999). The Antarctic ice cores show that glacial intervals are correlated with decreases in atmospheric CO_2 concentration from about 300 to 180 ppmv and in methane concentration from 770 to 320 ppbv.

In addition to these factors, there is a complex system of positive and negative feedbacks affecting climates and biomes (see fig. 3.5). For example, as climates warm, increasing amounts of carbon are released from soil and water that further contributes to warming. Also, global and patterned cycles of climate change are modified by regional and less patterned events, such as alterations in vegetation cover, albedo, solar variations, mountain uplift, volcanic eruptions, and catastrophes such as the superfloods that carved out the channeled scablands of eastern Washington (Baker 2002; Shaw et al. 1999).

The unique landscape of the scablands, covering 45,000 km^2, is due to repeated colossal floods during the last glacial cycle. Between 25,000 and 10,000 BP, the Purcell Trench lobe of Cordilleran ice sheet advanced to the Bitterroot Range of northern Idaho, blocking the Clark Fork River and creating glacial Lake Missoula. The volume of water in the lake was about 2500 km^3, or about six times that of modern Lake Erie. When the depth of the lake reached around 600 feet, the ice became buoyant, and this, along with melting, tunneling, and collapse of the ice dam, caused vast quantities of water to be released to the west. Ripple marks with wavelengths of 150 m, and boulders 10 m in diameter, were carried up to 3.5 km. The estimated discharge was 17×10^6 m^3/s, or ten times the flow of all the rivers in the world. As the water drained from the lake, ice reformed and settled back to the bottom. The cycle repeated some sixty times between 15,300 and 12,000 BP and scoured soil and vegetation from the southeastern quarter of the state of Washington. The first person to attribute the scablands to huge floods, also called "jökulhlaups," from Icelandic for "outbursts," was J Harlen Bretz, of interest, among other reasons, for the fact that his first name is not an abbreviation, but the letter J. It is taken as a mark of being "in" on this literature when no period is used in citing his name.

Among the many implications of catastrophes such as jökulhlaups, and past volcanic eruptions like Mount St. Helens, is that the events affect the regional spore and pollen rain onto bog and lake surfaces. Consequently, the effects of disturbance, and subsequent vegetation succession from barren ground, are superimposed on climate change and must be considered when interpreting spore and pollen diagrams.

Another forcing mechanism for climate change is the presence of the ice itself. At the maximum of the last glaciation, the two centers of continental ice, the Laurentide ice sheet centered over Hudson Bay and the Cordilleran ice sheet in the northern Rocky Mountains, fused to form a dome of ice more than a mile thick. This divided the polar jet stream into a northern and southern arm, with the latter displacing the high pressure system formed by the descending arm of the Hadley circulation cell over southwestern North America. There was also a southern shift of another high pressure system, the Bermuda-Azores High, and the result of both movements was a change from arid climates in the southwest to cooler and moister pluvial regimes.

The geological and biological records show the expected eighteen to twenty cycles of glaciation, on a periodicity of 100,000 years for about the past 780,000 years. Prior to that time, glaciation followed the tilt cycle of 43,000 years, which actually results in greater differences in solar insolation. The cause for the change in timing is not fully understood, but amplification of the eccentricity cycle is thought to be modulated by the diminishing trend in atmospheric CO_2 concentration until recent times. There are three implications of these patterned changes in climate. One is that for about 90,000 years of each eccentricity cycle, it was getting cold, it was very cold, or it was still cold, while in only about 10,000 years of the cycle was it warm like the present. This means that in areas of western North America, the desert vegetation, so characteristic of today, is the exception, and that for about 90 percent of the last 100,000 years the typical ecosystem was a more mesic shrubland/chaparral-woodland-savanna. The second realization is that any change in climate or vegetation during the last glacial/postglacial interval was repeated at least eighteen to twenty times since about 2.6 Ma. For example, if it is determined that recent fluctuations in Quaternary climate affected the lowland neotropical rain forest, as it did, then it must also be assumed that there were similar effects multiple times during the Quaternary. In other words, ecosystems are revealed as far more dynamic than if there were only four glaciations lasting 175,000 years and affecting only the high latitudes. This is another important paradigm shift in our view of recent climatic change and biotic history.

The third implication is that if ecosystems experienced such profound and frequent changes in range and composition during glacial intervals, it is unlikely they would regain the exact same range and composition after each climatic reshuffling (Davis 1981, 1986). During each time segment within a fluctuating climate, the prevailing ecosystems in a region are composed of ecologically compatible species that have had the opportunity to assemble depending on the presence or absence of barriers and the dispersal potential of the propagules. Ecosystems are not composed of the same species in the same proportions over interminable spans of time, and within limits, they have had different composition, and may still be expanding their range since the last of the eighteen to twenty cycles.

Although direct paleobotanical evidence is rare for Quaternary intervals beyond the last glaciation, the North Greenland Ice Core Project Members (NGRIP 2004) has recovered a record back 123,000 years into the last interglacial. Climates were more stable in the interglacials than in the glacial and glacial/interglacial transitions, and it was about 5°C warmer than today. All this information, collectively, establishes a valuable context for interpreting the Quaternary history of plant communities and ecosystems.

METHODOLOGIES

There are several ways of collecting material for Quaternary spore and pollen analysis, but the most common is to drill cores through peat bogs, lake sediments, or near-shore coastal deposits. Samples are typically collected every 10 cm along the core and processed by the techniques described in chapter 4. Counts of at least 300 microfossils are usually made at each level, and the percentage of each taxon is tabulated. Identifications are made mostly to genus, occasionally to family (e.g., for the grasses, sedges, and Chenopodiaceae-Amaranthaceae, recorded as chenoams); or to groups of families (e.g., for some ferns and for the Taxodiaceae-Cupressaceae-Taxaceae, recorded as t-c-t). The conventional method for presenting the results is the spore and pollen diagram (fig. 8.5). There are many of these high-resolution records available, and they are now being databased to cover extensive areas of, for example, North America and Greenland (Whitmore et al. 2005).

Most diagrams include certain standard information. The depth in centimeters, radiocarbon ages, and sediment type (peat, sand, charcoal, oxidized, or artifact layers) are plotted along the vertical axis, and the percentage of each spore or pollen type is given on the horizontal axis. The change through time for an individual taxon is called a profile; all the types for any one level constitute a spectrum; and the two together constitute the spore and pollen diagram. The ratio of arboreal (tree) pollen (AP) to nonarboreal (herbaceous) pollen (NAP) gives an indication of closed, dense forest (warm, moist conditions) versus open herbaceous vegetation, which often reflects either colder, drier conditions, perhaps with tundra or grassland, or disturbances caused by fire, epidemic disease, or human presence. Pol-

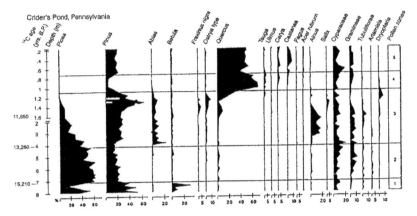

Figure 8.5 Quaternary pollen diagram from Crider's Pond, Pennsylvania. Modified from Ritchie 1987. Used with permission from Cambridge University Press, Cambridge.

len zones denote intervals characterized by particular pollen types and the inferred climate. Zone 1, for example, indicates the cold, moist climate of a boreal forest, with representative *Abies, Picea,* and *Betula;* zone 2, a warm, dry hypsithermal interval, with pine maximum; and zone 3, the cool, moist conditions for a deciduous forest, with *Acer, Castanea, Fagus,* and *Quercus.* There may be supplemental diagrams for fungal spores, seeds, diatoms, phytoliths, macrofossils (e.g., pine needles), bryophytes, and insect remains. Modern pollen rain studies are made to better associate a particular spectrum with the arrangement and quantitative representation of plants in a modern community (Gosling and Bunting 2008; Minckley et al. 2008). For example, at Sicamous Creek Lake in British Columbia, *Abies lasiocarpa,* which constitutes 50 percent of the vegetation, is 10 percent of the pollen rain (Minckley and Whitlock 2000); and on southern Vancouver Island, needles document the presence of *Abies* where the pollen is 1–2 percent (Heinricks et al. 2002). These and other techniques provide ever-improving estimates about the composition and spatial arrangement of Quaternary vegetation and the implied paleoenvironments based on spore and pollen diagrams (Delcourt and Delcourt 1991; Birks 1993; Birks and Gordon 1985). Even so, the diagrams reflect mostly major changes in the dominant vegetation. To prevent interpretations from exceeding the sensitivity of the method, a wide array of ancillary information must be incorporated from a knowledge of the structure, ecology, and composition of local communities, catastrophic events, glacial geology, isotope records, water chemistry, solar variations, and archeology. When this is done, a remarkably detailed and consistent history emerges for New World vegetation since 2.6 Ma.

REGIONAL HISTORIES

I have been emphasizing the origin, sequence of appearance, and causal factors involved in the development of ecosystems. By the Quaternary, the current ecosystems of the New World had long been established, so the discussions below are intended to (1) give some selective examples of recent modifications in their range and composition; (2) demonstrate how various forcing mechanisms for ecosystem change are being revealed; (3) show how these mechanisms, originally demonstrated mostly for the high polar regions, are now known to extend into the tropical and southern latitudes; and (4) explain how methods and interpretations have changed over the past few decades. Some selectivity is necessary because compared to the fewer than twenty Oligocene floras studied relatively recently and in detail for the New World, there are a thousand or more for the Quaternary (I, chap. 8; II, chaps. 4–7).

North America (North of Mexico)

In northeastern and upper midwestern North America, the biotic record within the glacial boundary, along the southern margin of the ice sheet, and in interlobate areas of the region begins around 12,000–15,000 years ago. This is when the ice retreated and peat and lake sediments began filling basins created by the glaciers, and when vegetation moved onto the barren soil (figs. 2.3, 8.5; Jackson et al. 2000). A composite sequence is shown in table 8.1, starting with sediments at the bottom containing a prominent representation of pollen from herbaceous plants that suggests a treeless vegetation or tundra. Macrofossils from this "T" zone are important because the zone usually includes some tree pollen blown in from beyond the glacial boundary (e.g., *Abies*, *Picea*). As climates warmed, soils matured, and succession progressed, a boreal forest developed at the site (zone A), followed by a pine maximum (zone B) reflecting the hypsithermal warm interval. Temperatures then cooled and various combinations of deciduous trees moved in (zone C), persisting until early agriculturalists arrived several hundred years ago (zone C-3). The last zone is marked by an increase in charcoal, weeds typically associated with cultivation (e.g., *Ambrosia*, ragweed; *Rumex*, buckwheat), and crop plants (e.g., *Zea mays*, corn or maize).

Table 8.1. Pollen zone sequence and inferred vegetation and climate for the
New England region

Pollen zones	Years BP	Vegetation	Climate
C-3	Present	Oak-chestnut-hemlock (*Ambrosia, Rumex,* other weeds)	Moister, cooler
C-2		Oak with hickory	Warm, dry
C-1	7900	Oak-hemlock, some *Ambrosia*	Warm, moist
B	7900–8100	Pine maximum with oak	Warm, dry
A-4	9100–10,200	Spruce-fir with larch, birch, alder	Cooler, moister (Younger Dryas)
A-3	10,200	Spruce-oak with pine	Warmer, drier
A-2	11,700	Spruce with oak	
A-1	11,700–12,150	Spruce, fir, birch	Cool, moist
T	12,150–14,300	Tundra	Cold, dry

Sources: Various sources from Graham (I, 289)

This is the general pattern of ecosystem change within the glacial boundary in northeastern North America.

In the Appalachian region beyond the glacial boundary, coniferous forest occupied much of the highlands during each cold interval, while many of the thermophilous deciduous elements moved onto the coastal plain and to the southwest toward the Ozark and Edwards plateaus. Thus, in referring to the Appalachian highlands as an ancient (Tertiary) relictual area for the deciduous forest, it is necessary to recognize that in the Quaternary the regional ecosystems were undergoing considerable change in distribution and in composition at the level of associations. On the coastal plain of Georgia at 13–11 kyr, during the Younger Dryas, there was mesic forest and reduced pine, while after about 4.5 kyr drier conditions and fire resulted in an increase of southern pine (LaMoreaux et al. 2009). Farther beyond the glacial boundary, a pollen diagram from Lake Tulane in south-central Florida shows five peaks in pine pollen that correspond to Heinrich events H1–H5 (Grimm et al. 1993; Watts and Hansen 1994). It raises the question that if Quaternary climate and vegetation changes extended this far beyond the glacial boundary, how much farther south did they extend? As noted earlier, the answer is across the equator and all the way from pole to pole.

To the far northwest at Quartz Creek in the Yukon part of Beringia at the LGM (around 24 kyr), plant fossils have been recovered from a rodent nest, the stomach contents of an extinct horse, and an alluvial peat (Zazula et al. 2003). The vegetation included prairie sage (*Artemisia frigida*), grasses (*Poa, Elymus*), sedges, rushes (*Juncus/Luzula*), and other plants that formed a herb-tundra ecosystem (with some small, scattered trees) on which grazed mammoths, horses, and bison in a cold dry climate.

In the southwest, a study of stalagmites in caves of the Guadalupe Mountains of Texas and New Mexico show a mid-Holocene dry period with modern conditions established some 4000 years ago (Polyak and Asmerom 2001). In the broader region of the western United States, the Medieval Warm Period (900–1300 CE) is evident (Cook et al. 2004). In the Mojave Desert, *Neotoma* (packrat) middens spanning the past 24,000 years reveal that the present arid desert of *Larrea tridentata* (creosote bush) was preceded above 1000 m by a moister pluvial vegetation of *Pinus monophylla* (piñon pine), *Juniperus osteosperma* (Utah juniper), *Purshia mexicana* (bitterbush), *Cercocarpus ledifolius* (mountain mahogany), and *Prunus fasciculata* (desert almond; Koehler et al. 2005).

Studies of two rare peat bogs in central Texas provide an interesting scenario of how accounts of Quaternary biogeography have improved over time. When John Potzger and B. C. Tharp (1943, 1947, 1954) made the first investigations of these sites, Potzger identified pollen of *Abies* and *Picea* in combined amounts of 11 percent in the lowermost levels. Neither fir nor spruce grows within several hundred miles of central Texas (the present vegetation is oak-hickory woodland and oak savanna). Earlier, Brown (1938) had reported cones, twigs, and wood of the boreal *Larix* (larch), *Picea*, and *Thuja* (northern white cedar) from Pleistocene deposits in Louisiana. The combined records were included in the monumental "Biogeography of the Pleistocene," a 1949 review by E. S. Deevey running more than a hundred pages. With the data then available, it was plausible to conclude that at the LGM the boreal forest, represented by *Abies, Larix, Picea*, and *Thuja*, extended all the way to the Gulf Coast and remained there until about 9 kyr (the age of the lowermost sediments in the Texas bogs). It was thus also reasonable to assume that the deciduous forest had been forced even farther south. Today there are about fifty species of plants from eastern North America disjunct in eastern Mexico, separated by the Chihuahuan Desert of south Texas and northern Mexico, in an altitudinal zone between 1000–2000 m along the eastern slopes of the Sierra Madre Oriental, including *Acer, Alnus, Carpinus, Carya, Cornus, Diospyros, Fagus, Fraxinus, Juglans, Liquidambar, Magnolia, Ostrya, Platanus, Populus, Tilia*, and *Ulmus*.

There are also amphibians and reptiles that show the same pattern. What better time for them to have arrived than during the cold intervals of the Pleistocene, only to be left stranded when climates warmed and the desert was reconstituted during the last interglacial? From the present perspective, this implies that potentially eighteen to twenty interchanges occurred in the recent geologic past, and that this continuity, or near continuity, existed for some 90 percent of the time.

Four studies later provided amendments to these early interpretations of Quaternary biogeography in the south-central United States, with broader implications for the presence of temperate elements in eastern Mexico and farther south. First, it was observed that affinities among the amphibians, reptiles, and to some extent the trees and shrubs between southeastern United States and eastern Mexico exist mostly at the generic level (Martin 1958; Martin and Harrell 1957), implying a long period of geographic and reproductive separation. Second, a reexamination of the Texas material showed an absence of *Abies* and a much smaller presence of *Picea* than previous calculations, only 1.5 percent rather than 11 percent (Graham and Heimsch 1960). Three pollen grains of spruce out of every 200 would easily be within wind transport of stands growing further south than at present during the colder intervals of the Quaternary without requiring a full-fledged boreal forest along the Gulf Coast. Moreover, macrofossils of boreal trees like larch and spruce frequently drift down the Mississippi River and are deposited in backswamp sediments during floods. Third, Delcourt and Delcourt (1977) further showed that the macrofossils in Louisiana identified as *Thuja occidentalis* (northern white cedar) were actually the local *Chamaecyparis thyoides* (southern yellow cedar). Fourth, my study on the Paraje Solo Formation of southeastern Mexico (Graham 1976), mentioned in chapter 7, documents that *Abies, Picea, Alnus, Celtis, Ilex, Juglans, Liquidambar, Populus, Quercus,* and *Ulmus* were already established in eastern Mexico by at least the middle Pliocene. The most likely time for their principal early introduction was around the middle Miocene, that is, with the third of the major paleotemperature declines (see fig. 9.3 below).

Mexico

Periodic displacement of the jet streams and the Hadley/Bermuda-Azores convection cells southward during the Quaternary brought alternating periods of warm-dry and cool-moist conditions to the arid southwest. The latest of these sequences is revealed by cave speleothems in Carlsbad Caverns, New Mexico (Brook et al. 2006), and by packrat middens (Van Devender

1990), dune formation (Murillo de Nava et al. 1999), and diatom, spore, and pollen diagrams (Metcalfe 2006; Metcalfe et al. 1997; Minckley and Jackson 2008) from northern Mexico. The estimated change during glacial times was a MAT lower by about 5°C–6°C and about twice the winter rain (Rhode 2002). Through interglacial times, the vegetation sequence was from a juniper–piñon pine woodland at 18–12 kyr (*Juniperus scopulorum, J. coahuilensis, Pinus edulis*), to a drier oak-juniper woodland around 9 kyr, and ultimately to the modern Chihuahuan Desert circa 4 kyr (*Larrea tridentata, Fouquieria splendens*). The moist glacial intervals in the region are also reflected in the faunal record (Shaw and McDonald 1987). A metacarpal of the giant anteater (*Myrcemophaga tridactyla*) was discovered in early Pleistocene deposits from Sonora, Mexico. Its present northern limit is 3000 km to the south in the more moist climates of Central and South America. The dynamic nature of the Quaternary environment in the Basin of Mexico is shown by evidence for the Younger Dryas, a mid-Holocene dry period, and the Little Ice Age (Lozano-García and Ortega-Guerrero 1998; Metcalfe et al. 2007). The paleobotanical and other records show once again that in relatively recent time seemingly moderate changes in climate produced profound changes in the established ecosystems.

In the porous limestone karst topography of the Yucatán Peninsula of Mexico, Belize, and northern Guatemala, moisture is a limiting factor. The average elevation today is around 200 m; during the glacial maxima, sea levels were lower by about 120 m, and water tables dropped by 26–40 m. Lakes and cenotes in the lowland Maya region have been studied by spore and pollen analysis, sediment chemistry, stable isotope records (Leyden et al. 1998; Hodell et al. 2005), and by carbon and oxygen stable isotope records from stalagmites (Webster et al. 2007). There is evidence for (1) Milankovitch-paced forcing of the climate, (2) lower water tables and dry conditions at the LGM, and (3) a particularly dry period between 1785 and 930 BP corresponding to the collapse of the lowland Maya civilization (Haug et al. 2003).

The Antilles

In the vicinity of the Greater Antilles, an interesting discovery from the Sargasso Sea is that sea surface temperatures (SSTs) were warmer by about 1°C during the Medieval Warm Period, and cooler by about 1°C during the Little Ice Age (Keigwin 1996). In another study, ratios of Mg/Ca and coral ^{18}O isotopes off southwestern Puerto Rico at about 18°N show that SSTs were 2°C cooler during the Little Ice Age (Watanabe et al. 2001). Pollen profiles from Lake Miragoane on the north side of the southern peninsula of Haiti record two periods of forest decline in the past 500 years (Binford et al. 1987; Brenner and Binford 1988; Higuera-Gundy, 1989). One was between 1500 and 1700 CE and corresponds to the Spanish occupation of the island. The other was between 1700 and 1800 CE marking the time of French occupation. Pollen records for later times show forest regeneration when Haiti became independent in 1804 and the large colonial plantations were abandoned in favor of small subsistence farms. The upper 6 cm of the core documents the virtual destruction of Haiti's natural ecosystems in recent times.

Central America

An example of the inferred climate, vegetation change, and ecosystem adjustments is provided by the sediments at La Chonta and La Trinidad in Costa Rica around 9°N. At the LGM, the tree line moved down from about 3000–3500 m to about 2000 m. In the warmer parts of the late glacial, oak pollen reached 80 percent in some samples, the tree line moved upward, then temperatures fell by about 2°C for 500–600 years and the tree line descended to 2400 m at a time corresponding to the Younger Dryas (Hooghiemstra et al. 1992; Islebe and Hooghiemstra 1997). In Panama, the classic study of Alexandra Bartlett and Elso Barghoorn (1973) on sediments from Gatún Lake, as well as studies from other tropical lakes (e.g., Bush et al. 1992), revealed considerable changes in sea level as measured by the coming and going of mangrove pollen. Disturbance of the vegetation by human activity in southern Central America is shown by the increased carbon content of the sediments, charred phytoliths, pollen of cultivars, and starch grains of maize and manioc on grinding stones dated at 6000–7000 BP (Perry et al. 2007; Piperno 2006; Piperno et al. 1985; Piperno et al. 1990; Piperno et al. 2000; Piperno and Pearsall 1998). As a generalization, the interpretation of spore and pollen diagrams from Latin America must take into account modification of the vegetation by human disturbance after 6000 BP.

South America

There have been numerous studies of Quaternary vegetation and environments in South America, but the information is dispersed over a vast area and interpretations are complicated by the great diversity of the vegetation. One of the most important findings has been evidence for a LGM interval cooler by 5°C–6°C in the Amazon lowlands (Webb et al. 1997; Stute et al. 1995; Aeschbach-Hertig et al. 2000), associated with significantly drier climates (Muhs and Zárate 2001). This is shown not only in the pollen profiles, but also by a wealth of information from sedimentology and faunal evidence (II, chap. 7). For example, Wüster and others (2005a, b) used three mitochondrial genes in the dry-habitat *Crotalus durissus* (neotropical rattlesnake) from populations to the north and south of Amazonia to infer divergence (separation) in the mid-Pleistocene because of drier conditions in the basin. They conclude that "the presence of *C. durissus* on both sides of the Amazon Basin does represent evidence of profound changes in the distribution of rain forests in the Pleistocene and cannot be explained by relatively minor changes in rain forest community composition" (Wüster 2005b, 3619). A similar opinion is expressed by Anhuf and others: "The point of view that claims that the Amazon lowland forests were not replaced by savanna during the LGM can no longer be supported by the data" (2006, 518) There is such a view, still vigorously defended by Paul Colinvaux and colleagues (Colinvaux 2007; Colinvaux et al. 2000); but as we shall see in chapter 9, it now seems to be drifting toward a minority opinion. It maintains that although the lowland Amazon Basin experienced coolness and dryness at the LGM, and presumably at least eighteen to twenty times during the Quaternary, the changes were not sufficient to disrupt the lowland neotropical rain forest (e.g., Colinvaux et al. 2000). This grades into the subject of refugia that will also be discussed in chapter 9. In other areas of northern lowland Brazil, for example, in the state of Pará at about 5°S, there is evidence for a Little Ice Age (Cohen et al., 2005); and elsewhere in Brazil the Younger Dryas, Heinrich, and D-O events are recognized.

The importance of these climatic events is that they are relatively recent, short-term, rapid-paced occurrences that had the potential to fragment the biota into reproductively isolated populations. Thus, climatic change becomes one of several forcing mechanisms for speciation from a systems view. None of the events individually constitute "the" explanation, and as other debates have shown, it is counterproductive to discuss them in a "versus" format. Biodiversity is difficult enough to explain without overemphasizing one school of thought (literally or figuratively) when multiple factors are available.

In addition to climate change, certain geologic events provide an additional mechanism for explaining biodiversity. Subdivision of the Amazon Basin floor into parcels has evolutionary and speciation consequences. Alfred Russel Wallace, a contemporary of Charles Darwin whose paper on the theory of evolution was copresented at the Linnean Society of London in July 1858, developed a biogeographic theory now called the riverine hypothesis (Wallace 1849, 1876). The theory was based primarily on Wallace's travels in Amazonia, and it posited that the distribution of species was strongly influenced by the pattern of drainage systems, and by former changes in these patterns. This has proven true for particular groups of mammals; but as expected, it is neither true for all organisms, nor is it the only factor operating within an individual group of organisms. Some tamarin populations, for example, have been identified as having different mtDNA profiles even though there were no morphological differences between them (Patton and da Silva 1998, 2005; Patton et al. 1994). Also, the molecularly defined segregates in these studies were not separated on either side of the obvious barrier (the Juruá River), but rather between the upper reaches and the mouth of the river: "What is really striking . . . is that all 11 [of the 17 species] are separated at almost the same geographical point on the river, although there is nothing remarkable about the spot—no bend, no hill, no valley" (Patton, quoted in Morell 1996, 1497). The separation among these groups of tamarins was, rather, the Iquitos Arch, one of several arches created by uplift of the Andes Mountains that subdivided Amazonia into subbasins. Over 4000 feet of sediment now mostly obscure the arches, but when they formed, and while they lasted, the distribution of some, but not all, of the biotic components in Amazonia were fragmented to varying degrees depending on the dispersal ability of the organisms. Evidence still remains in the molecular signature of the tamarins, and DNA analysis suggests the divergence dates to 3–1 Ma.

On the subject of the interaction between recent geologic events and speciation in neotropical ecosystems, there is another intriguing example from the Galápagos Islands. Isla Isabela has five taxa of the large Galápagos tortoise, one each on its five volcanoes. One taxon, *Geochelone nigra vanadenburghi*, on Volcano Alecdo, has three to five times less matrilineal diversity than the others as revealed by mtDNA. All the volcanoes originated at about the same time (500,000 years ago), but Alecdo was the most recent to erupt (100,000 years ago), causing a significant reduction in the biota. The hypothesis is that this decrease resulted in a genetically reduced parent population from which the modern lineage is derived. According to Beheregaray and colleagues, these data "emphasize the value of modern molecular population approaches in obtaining historical demographic information that cannot be discerned based on contemporary scenarios" (2003, 75).

In Colombia, the High Plain of Bogotá is located in the eastern cordillera at an elevation of 2550 m. It formed between 6 and 3 Ma with uplift of the mountains, and has been subsiding since the end of the Pliocene. It is now filled with an almost continuous sequence of sediments and plant microfossils dating to 3.2 Ma. For more than fifty years, this exceptional trove of vegetation and environmental information has been studied by Thomas van der Hammen, and now by Henry Hooghiemstra, A. M. Cleef, and others at the University of Amsterdam and in Colombia (Hooghiemstra 1984; Hooghiemstra et al. 1992; Hooghiemstra et al. 2006). Among the results is the demonstration that páramo and Andean forest zones (fig. 2.42) moved up and down slope in concert with climatic and glacial cycles. Also, the major changes followed a periodicity of about 100,000 years. In the glacial intervals, temperatures are estimated to have been about 8°C colder at the high elevations, and this extrapolates to cooler conditions in the lowlands now documented at around 5°C–6°C. In fact, the early studies by van der Hammen were among the first to suggest that the idea of a stable and unchanging lowland neotropical rain forest was a myth. Early human influence on the vegetation is evident between 4810 and 3800 BP, and between 3800 and 2470 BP. There is crop cultivation (*Zea mays*) and abundant charcoal in the sediments (Gómez et al. 2007).

Human occupation

To the south in Peru, new information on human occupation and coastal sediments also reveal the antiquity and impact of El Niños on ancient civilizations in the region, and a wealth of intriguing geological and biological data are additionally preserved in rapidly melting ice caps. The earliest and most certain evidence of humans in Peru is from the lowlands at 12,000–11,000 BP. The record of agriculture can be traced back about 4000 years from starch grains and phytoliths of *Zea mays* and *Maranta arundinacea* (arrowroot) found on tools and in soils at a preceramic house at Waynuma, southern Peru (Perry et al. 2007). This extends the evidence for cultivation from previous studies back by more than 1000 years. As noted, accurate information on the appearance of farming in an area, as opposed to subsistence by hunting and gathering, is important because like catastrophes (e.g., megafloods, volcanism) it represents modification of vegetation by a factor other than climate. Agriculture was established in many places in South America by 6000 BP, and after that time the interpretation of spore and pollen diagrams must take into account possible disturbance from human activity.

El Niños

The beginning of El Niños in coastal South America is difficult to date accurately because sediments from flood-induced landslides are similar to those from tectonic-induced earthquakes. In Ecuador there are debris flows possibly dating as far back as 38,000 BP, and certainly to 15,000 BP. In Peru the record suggests mild El Niños between 8000 and 5800 BP, then periodic mega–El Niños afterward (Sandweiss 2003; Sandweiss et al. 2001; Keefer et al. 2003; Keefer et al. 1998). This record is important for cultural history, and indirectly for vegetational history, because grand-scale monument and ceremonial construction often coincides with the good times provided by moderate rain, and abandonment of the sites often coincides with the bad times created by excessive rain and flooding. Exceptionally strong El Niños occurred at the end of the Younger Dryas 12,000 years ago, when the Quebrada Tacahuay site in southernmost coastal Peru was deserted. The latest cycle of intense and regular El Niños beginning around 5800 BP corresponds to active temple building on the coast; while the onset of even stronger ones after 3200–2800 BP marks their abandonment.

The Quelccaya ice cap in Peru is at 13°56′S and at 5670 m (18,602 ft) elevation. The ice shows annual accumulations of winter snow (white layers) and summer debris (grey layers). Volcanic eruptions, cold and dry intervals, the beginnings of agriculture (greater amounts of dust), and alterations in atmospheric chemistry (smelting to produce bronze, beginning of the industrial revolution, addition of lead to gasoline, introduction of PCBs, use of DDT) are not only recorded in the ice, but they can be dated nearly to the year. El Niños can be detected because moisture in the region is reduced by 30 percent, and in the unusually strong event of 1982–83, the surface of the ice was covered with a blanket of dust. The herculean efforts required to retrieve deep ice cores at this elevation, and to maintain them in frozen condition for future study when the ice caps no longer exist, has been the work of Lonnie Thompson of the Byrd Polar Research Center at Ohio State University (Thompson and Moseley-Thompson 1987, 1989; Thompson et al. 1979; Thompson et al. 1988; Thompson et al. 2005). An account of these efforts is given by J. Madeleine Nash in her 2002 book, El Niño. As she notes, "The Quelccaya Ice Cap was like a weather station that had been in continuous operation since A.D. 500. In its layers archaeologists would find evidence of the El Niño floods that were said to have devastated the great coastal city of Chan Chan around A.D. 1100, they would find evidence of the decades-long drought that decimated the pre-Incan civilization of Tiwanaku at around the same time. And they would also find evidence of centuries-long El Niño– and La Niña–like periods that eerily parallel the rise and fall of these and other pre-Incan cultures" (2002, 84). A finding of special interest for Holocene ecosystem history is that between the annual layers of 1490 and 1880 there is evidence of a cold period corresponding to the Little Ice Age even at this equatorial latitude. Similar cores have been retrieved from Tibet and Kilimanjaro that give a global perspective to the findings. Today, there are about 160,000 mountain glaciers, and multispectral satellite data from the LandSat Thematic Mapper (TM) show that between 1985 and 1999, a set of 930 glaciers in the Swiss Alps lost 18 percent of their former area at a rate seven times higher than between 1850 and 1973 (Paul et al. 2004).

Events at the transition between the late Pliocene and the early to middle Pleistocene are recorded in the Canoa Formation of coastal central Ecuador from stable isotope studies of fossil fish, otoliths, mollusks shells, benthic foraminifera, and shark teeth (Pellegrini and Longinelli 2008). An interesting hypothesis is that there was a north to south migration of cold temperate fish from southern North America (e.g., California) to coastal Ecuador as a result of the closing of the Panama seaway. The idea is that the closure caused coastal upwelling of cold waters, like that noted in chapter 7 for the middle Pliocene Paraje Solo Formation of southeastern Mexico, and this afforded a migratory pathway of suitable water temperatures for cool to cold-water fish.

A number of innovative studies are being made in the high Central Andes (Betancourt et al. 2000; Baker et al. 2001; Grosjean et al. 2003; Placzek et al. 2006; Placzek et al. 2009; Theissen et al. 2008). They show that the dry lake beds on the Altiplano of Bolivia, for example, the Salar de Uyuni (fig. 8.6), have repeatedly filled up and dried out since 120 kyr. Between 120 and 98 kyr, the now dry lakes were 80 m deep; at 95–20 kyr there was intermittent shallowing; then between 18.1 and 14.1 kyr there developed "the deepest and largest lake in the basin" (140 m; Placzek et al. 2006). If such profound changes were happening during the last glacial-interglacial cycle, surely they must have occurred on multiple occasions during the past 2.6 million years.

In Argentina and Chile, the Quaternary vegetation and climatic history often involves subtle shifts in shrubs, herbs (especially grass), and trees (especially *Nothofagus*) under the influence of the meandering Antarctic anticyclone high pressure system. At the glacial maxima, this system covered an area about double that of the interglacials and shifted northward by around 10° latitude. There was a Younger Dryas cold interval, a mid-Holocene warm/dry period, and a Little Ice age.

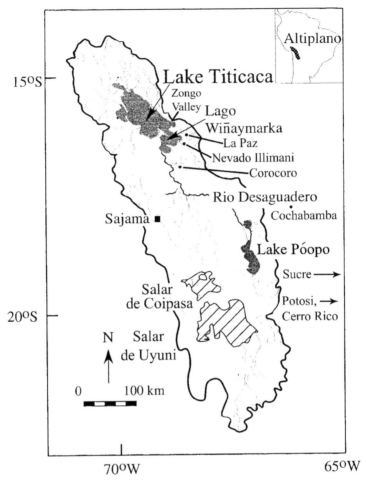

Figure 8.6 Place-names and principal salars of the Central Andes Altiplano.

These findings on the Quaternary of South America put some conclud-
ing touches on the ecosystem history of the New World. They are of considerable importance for reconstructing the local vegetation and environment
(e.g., Heusser 2003), but collectively the information has broader applications. The findings offer valuable analogs for reading the more ancient
and less clear history of environments and biotas, and they add more times
and more places supporting the generalization that neither from lowlands
to páramo, nor from pole to pole, have Neogene and Quaternary climates
and biotas been stable for geologically long periods of time. We know en-

vironments also varied, albeit at a slower pace, in the Cretaceous and Paleogene, and this removes the earlier misconception of long-term stability that plagued evolutionary biologists and biogeographers in their efforts to discover mechanisms to explain the origin of biodiversity and patterns of distribution—a rather nice contribution from "historic demographic information that cannot be discerned based on contemporary scenarios" (Beheregaray et al. 2003, 75).

: : **REFERENCES** : :

Aeschbach-Hertig, W., F. Peeters, U. Beyerle, and R. Kipfer. 2000. Palaeotemperature reconstruction from noble gases in ground water taking into account equilibration with entrapped air. *Nature* 405:1040–44.

Agassiz, L. 1840. *Etudes sur les glaciers*. Privately published, Neuchâtel.

Alley, R. B. 2000. *The two-mile time machine, ice cores, abrupt climate change, and our future.* Princeton University Press, Princeton.

Anhuf, D., et al. (twelve coauthors). 2006. Paleo-environmental change in Amazonian and African rainforest during the LGM. *Palaeogeogr. Palaeocl. Palaeoecol.* 239:510–27.

Baker, P. A., et al. (six coauthors). 2001. Tropical climate changes at millennial and orbital timescales on the Bolivian Altiplano. *Nature* 409:698–701.

Baker, V. R. 2002. The study of superfloods. *Science* 295:2379–80.

Balco, G., C. W. Rovey II, and J. O. H. Stone. 2005. The first glacial maximum in North America. *Science* 307:222. [Atlanta and Whippoorwill tills of central Missouri, dated respectively at 2.41 Ma (plus or minus 0.14 million years) and at 1.8–1.6 Ma.]

Barker, S., et al. (six coauthors). 2009. Interhemispheric Atlantic seesaw response during the last deglaciation. *Nature* 457:1097–1103.

Bartlett, A. S., and E. S. Barghoorn. 1973. Phytogeographic history of the Isthmus of Panama during the past 12,000 years: A history of vegetation, climate, and sea-level change. In *Vegetation and vegetational history of northern Latin America*, ed. A. Graham, 203–99. Elsevier Science Publishers, Amsterdam.

Beheregaray, L. B., et al. (five coauthors). 2003. Genes record a prehistoric volcano eruption in the Galápagos. *Science* 302:75.

Betancourt, J. L., C. Latorre, J. A. Rech, J. Quade, and K. A. Rylander. 2000. A 22,000-year record of monsoonal precipitation from northern Chile's Atacama Desert. *Science* 289:1542–46.

Binford, M. W., et al. (five coauthors). 1987. Ecosystems, paleoecology and human disturbance in subtropical and tropical America. *Quat. Sci. Rev.* 6:115–28.

Birks, H. J. B. 1993. Quaternary palaeoecology and vegetation science—current contributions and possible future developments. *Rev. Palaeobot. Palynol.* 79:153–77.

Birks, H. J. B., and A. D. Gordon. 1985. *Numerical methods in Quaternary pollen analysis.* Academic Press, New York.

Brenner, M., and M. W. Binford. 1988. A sedimentary record of human disturbance from Lake Miragoane, Haiti. *J. Paleolimn.* 1:85–97.

Broecker, W. S. 2003. Does the trigger for abrupt climate change reside in the ocean or in the atmosphere? *Science* 300:1519–22.

———. 2006. Was the Younger Dryas triggered by a flood? *Science* 312:1146–48.

Brook, G. A., B. B. Ellwood, L. B. Railsback, and J. B. Cowart. 2006. A 164 ka record of environmental change in the American Southwest from a Carlsbad Cavern speleothem. *Palaeogeogr. Palaeocl. Palaeoecol.* 237:483–507.

Brown, C. A. 1938. The flora of Pleistocene deposits in the western Florida Parishes, west Feliciana Parish, and east Baton Rouge Parish, Louisiana. In *Contributions to the Pleistocene history of the Florida parishes of Louisiana*, Geological Bulletin 12, 59–96. Department of Conservation, Louisiana Geological Survey, New Orleans.

Bush, M. B., et al. (six coauthors). 1992. A 14,300-yr paleoecological profile of a lowland tropical lake in Panama. *Ecol. Monogr.* 62:251–75.

CLIMAP. 1976. The surface of the ice-age Earth. *Science* 191:1131–37.

———. 1981. *Seasonal reconstruction of the Earth's surface at the Last Glacial Maximum.* Map Chart Series MC-36. Geological Society of America, Boulder.

———. 1984. *The last interglacial ocean. Quat. Res.* 21:123–224.

Cody, M. L., and J. M. Diamond, eds. 1975. *Ecology and evolution of communities.* Belknap Press, Harvard University, Cambridge, MA.

Cohen, M. C. L., H. Behling, and R. J. Lara. 2005. Amazonian mangrove dynamics during the last millennium: The relative sea-level and the Little Ice Age. *Rev. Palaeobot. Palynol.* 136:93–108.

COHMAP. 1988. Climatic changes of the last 18,000 years: Observations and model simulations. *Science* 241:1043–52.

Colinvaux, P. A. 2007. *Amazon expeditions: My Quest for the ice-age Equator.* Yale University Press, New Haven, CT. [See review by W. E. Duellman, *Quart. Rev. Biol.*, September 2008.]

Colinvaux, P. A., P. E. De Oliveira, and M. B. Bush. 2000. Amazonian and neotropical plant communities on glacial time-scales: The failure of the aridity and refuge hypothesis. *Quat. Sci. Rev.* 19:141–69.

Cook, E. R., C. A. Woodhouse, C. M. Eakin, D. M. Meko, and D. A. Stahle. 2004. Long-term aridity changes in the western United States. *Science* 306:1015–18.

Corrège, T., et al. (five coauthors). 2004. Interdecadal variation in the extent of South Pacific tropical waters during the Younger Dryas event. *Nature* 428:927–29.

Cowling, S. A. 1999. Simulated effects of low atmospheric CO_2 on structure and composition of North American vegetation at the Last Glacial Maximum. *Global Ecol. Biogeogr.* 8:81–93.

Davis, M. B. 1981. Quaternary history and the stability of forest communities. In *Forest Succession: Concepts and Applications*, ed. D. C. West, H. H. Shugart, and D. B. Botkin, 132–53. Springer Verlag, Berlin.

———, ed. 1986. *Vegetation-climate equilibrium.* Special issue, *Vegetatio* 67:1–141.

Dean, W. E., R. M. Forester, and J. P. Bradbury. 2002. Early Holocene change in atmospheric circulation in the northern Great Plains: An upstream view of the 8.2 ka cold event. *Quat. Sci. Rev.* 21:1763–75.

Deevey, E. S. Jr. 1949. Biogeography of the Pleistocene. *Geol. Soc. Am. Bull.* 60:1315–1416.

Delcourt, H. R., and P. A. Delcourt. 1991. *Quaternary ecology: A paleoecological perspective.* Chapman and Hall, New York.

Delcourt, P. A., and H. R. Delcourt. 1977. The Tunica Hills, Louisiana-Mississippi: Late glacial locality for spruce and deciduous forest species. *Quat. Res.* 7:218–37.

Ellison, C. R. W., M. R. Chapman, and I. R. Hall. 2006. Surface and deep ocean interactions during the cold climate event 8200 years ago. *Science* 312:1929–32.

Fagan, B. 2004. *The long summer: How climate changed civilization.* Basic Books, Cambridge, Mass.

Gibbard, P. 2007. Annual report of the Subcommission on Quaternary Stratigraphy, in the 2007 annual report of the International Commission for Stratigraphy. http://www .quaternary.stratigraphy.org.uk/about/annualreports/.

Gómez, A., J. C. Berrío, H, Hooghiemstra, M. Becerra, and R. Marchant. 2007. A Holocene pollen record of vegetation change and human impact from Pantano de Vargas, an intra-Andean basin of Duitama, Colombia. *Rev. Palaeobot. Palynol.* 145:143–57.

Gosling, W. D., and J. M. Bunting, eds. 2008. *The paleoecological record from mountain regions, Open Science Conference, Global Change in Mountain Regions.* Special issue, *Palaeogeogr. Palaeocl. Palaeoecol.* 259:1–92. [Papers on Northern Andes (Ecuador) and tropical Andes (Bolivia, Peru).]

Graham, A. 1976. Studies in neotropical paleobotany. Part 2, The Miocene communities of Veracruz, Mexico. *Ann. Mo. Bot. Gard.* 63:787–842.

Graham, A., and C. Heimsch. 1960. Pollen studies of some Texas peat deposits. *Ecology* 41:785–90.

Grimm, E. C., G. L. Jacobson Jr., W. A. Watts, B. C. S. Hansen, and K. A. Maasch. 1993. A 50,000-year record of climate oscillations from Florida and its temporal correlation with the Heinrich events. *Science* 261:198–200.

Grosjean, M., I. Cartajena, M. A. Geyh, and L. Núñez. 2003. From proxy data to paleoclimate interpretations: The mid-Holocene paradox of the Atacama Desert, northern Chile. *Palaeogeogr. Palaeocl. Palaeoecol.* 194:247–58.

Guilderson, T. P., R. G. Fairbanks, and J. L. Rubenstone. 1994. Tropical temperature variations since 20,000 years ago: Modulating interhemispheric climate change. *Science* 263:663–68.

Haug, G. H., et al. (five coauthors). 2003. Climate and the collapse of Maya civilization. *Science* 299:1731–35.

Heinricks, M. L., J. A. Antos, R. J. Hebda, and G. B. Allen. 2002. *Abies lasiocarpa* (Hook.) Nutt. in the late-glacial and early-Holocene vegetation of British Colombia, Canada, and adjacent regions in Washington, USA. *Rev. Palaeobot. Palynol.* 120:107–22.

Heusser, C. J. 2003. *Ice age Southern Andes: A chronicle of paleontological events.* Elsevier Science Publishers, Amsterdam.

Higuera-Gundy, A. 1989. Recent vegetation changes in southern Haiti. In *Biogeography of the West Indies,* ed. C. A. Woods, 191–200. Sand Hill Crane Press, Gainesville.

Hiscott, R. N., A. E. Aksu, P. J. Mudie, and D. F. Parsons. 2001. A 340,000 year record of ice rafting, palaeoclimatic fluctuations, and shelf-crossing glacial advances in the southwestern Labrador Sea. *Global Planet. Change* 28:227–40. [The record extends to Marine Isotope Stage 9 and the fluctuations include 13 cold Heinrich events (H1–H13).]

Hodell, D. A., M. Brenner, and J. H. Curtis. 2005. Terminal classic drought in the northern

Maya lowlands inferred from multiple sediment cores in Lake Chichancanab (Mexico). *Quat. Sci. Rev.* 24:1413–27.

Hooghiemstra, H. 1984. Vegetational and climatic history of the High Plain of Bogotá, Colombia: A continuous record of the last 3.5 million years. *The Quaternary of Colombia / El Cuaternario de Colombia* 10:1–368. [Reprinted from *Dissertationes botanicae*, vol. 79 (J. Cramer, Vaduz, Germany).]

Hooghiemstra, H., A. M. Cleef, G. W. Noldus, and M. Kappelle. 1992. Upper Quaternary vegetation dynamics and palaeoclimatology of the La Chonta bog area (Cordillera de Talamanca, Costa Rica). *J. Quat. Sci.* 7:205–25.

Hooghiemstra, H., V. M. Wijninga, and A. M. Cleef. 2006. The paleobotanical record of Colombia: Implications for biogeography and biodiversity. In *Latin American biogeography—causes and effects*. Proceedings of the 51st Annual Systematics Symposium of the Missouri Botanical Garden (Alan Graham, organizer). *Ann. Mo. Bot. Gard.* 93:297–324.

Islebe, G. A., and H. Hooghiemstra. 1997. Vegetation and climate history of montane Costa Rica since the last glacial. *Quat. Sci. Rev.* 16:589–604.

Jackson, S. T., et al. (six coauthors). 2000. Vegetation and environment in eastern North America during the Last Glacial Maximum. *Quat. Sci. Rev.* 19:489–508.

Johnson, C. N. 2002. Determinants of loss of mammal species during the late Quaternary "megafauna" extinctions: Life history and ecology, but not body size. *Proc. R. Soc. Lond. B* 269:2221–27.

Keefer, D. K., M. E. Moseley, and S. D. deFrance. 2003. A 38,000-year record of floods and debris flows in the Ilo region of southern Peru and its relation to El Niño events and great earthquakes. *Palaeogeogr. Palaeocl. Palaeoecol.* 194:41–77.

Keefer, D. K., et al. (five coauthors). 1998. Early maritime economy and El Niño events at Quebrada Tacahuay, Peru. *Science* 281:1833–35.

Keigwin, L. D. 1996. The Little Ice Age and Medieval Warm Period in the Sargasso Sea. *Science* 274:1504–8.

Kerr. R. A. 2008. A time war over the period we live in. *Science* 319:402–3.

Koehler, P. A., R. S. Anderson, and W. G. Spaulding. 2005. Development of vegetation in the central Mojave Desert of California during the late Quaternary. *Palaeogeogr. Palaeocl. Palaeoecol.* 215:297–11.

LaMoreaux, H. K., G. A. Brook, and J. A. Knox. 2009. Late Pleistocene and Holocene environments of the southeastern United States from the stratigraphy and pollen content of a peat deposit on the Georgia coastal plain. *Palaeogeogr. Palaeocl. Palaeoecol.* 280:300–312.

Landais, A., et al. (eight coauthors). 2006. The glacial inception as recorded in the NorthGRIP Greenland ice core: Timing, structure and associated abrupt temperature changes. *Clim. Dynam.* 26:273–84.

Leyden, B. W., M. Brenner, and B. H. Dahlin. 1998. Cultural and climatic history of Coba, a lowland Maya city in Quintana Roo, Mexico. *Quat. Res.* 49:111–22.

Lisiecki, L. E., M. E. Raymo, and W. B. Curry. 2008. Atlantic overturning responses to late Pleistocene climate forcings. *Nature* 456:85–88.

Lozano-García, M. S., and B. Ortega-Guerrero. 1998. Late Quaternary environmental changes of the central part of the Basin of Mexico: Correlation between Texcoco and Chalco basins. *Rev. Palaeobot. Palynol.* 99:77–93.

Martin, P. S. 1958. *A biogeography of reptiles and amphibians in the Gomez Farías Region,*

Tamaulipas, Mexico. Miscellaneous Publications 101. Museum of Zoology, University of Michigan, Ann Arbor.

———. 1982. The pattern and meaning of Holarctic mammoth extinction. In *Paleoecology of Beringia*, ed. D. M. Hopkins, J. V. Matthews Jr., C. E. Scheweger, and S. B. Young, 399–408. Academic Press, New York.

———. 1984. Pleistocene overkill. In *Quaternary Extinctions: A prehistoric revolution*, ed. P. S. Martin and R. G. Klein. University of Arizona Press, Tucson.

Martin, P. S., and B. E. Harrell. 1957. The Pleistocene history of temperate biotas in Mexico and eastern United States. *Ecology* 38:468–80.

Meltzer, D. J. 2008. Letters: A Quaternary question. *Science* 320:177–78. [See W. A. Berggren's response in the same issue.]

Metcalfe, S. E. 2006. Late Quaternary environments of the northern deserts and central Volcanic Belt of Mexico. In *Latin American biogeography—causes and effects*. Proceedings of the 51st Annual Systematics Symposium of the Missouri Botanical Garden (Alan Graham, organizer). *Ann. Mo. Bot. Gard.* 93:258–73.

Metcalfe, S. E., A. Bimpson, A. J. Courtice, S. L. O'Hara, and D. M. Taylor. 1997. Climate change at the monsoon/westerly boundary in northern Mexico. *J. Paleolimn.* 17:155–71.

Metcalfe, S. E., et al. (six coauthors). 2007. Long and short-term change in the Pátzcuaro Basin, central Mexico. *Palaeogeogr. Palaeocl. Palaeoecol.* 247:272–95.

Milankovitch, M. 1920. *Théorie mathematique des phenomènes thermiques produits par la radiation solaire.* Gauthier-Villars, Académie Yugoslave des Sciences et des Arts de Zagreb, Yugoslavia.

Minckley, T. A., and S. T. Jackson. 2008. Ecological stability in a changing world? Reassessment of the palaeoenvironmental history of Cuatrociénegas, Mexico. *J. Biogeogr.* 35: 188–90.

Minckley, T. A., and C. Whitlock. 2000. Spatial variation of modern pollen in Oregon and southern Washington, USA. *Rev. Palaeobot. Palynol.* 112:97–123.

Minckley, T. A., et al. (five coauthors). 2008. Associations among modern pollen, vegetation, and climate in western North America. *Quat. Sci. Rev.* 27:1962–91.

Morell, V. 1996. Amazonian diversity: A river doesn't run through it. *Science* 273:1496–97.

Muhs, D. R., and M. Zárate. 2001. Late Quaternary eolian records of the Americas and their paleoclimatic significance. In *Interhemispheric climate linkages*, ed. V. Markgraf, 183–216. Academic Press, New York.

Murillo de Nava, J. M., D. S. Gorsline, G. A. Goodfriend, V. K. Vlasov, and R. Cruz-Orozco. 1999. Evidence of Holocene climatic changes from aeolian deposits in Baja California Sur, Mexico. *Quat. Int.* 56:141–54.

Nash, J. M. 2002. *El Niño: Unlocking the secrets of the master weather-maker.* Warner Books, New York.

Newton, A. 2007. Palaeoclimatology: Multiple warmings? *Nature Geoscience*, 1 September, doi:10.1038/ngeo2007.4. Published online 6 September 2007. [See Nicolo et al. 2007 reference.]

Nicolo, M., J. G. R. Dickens, C. J. Hollis, and J. C. Zachos. 2007. Multiple early Eocene hyperthermals: Their sedimentary expression on the New Zealand continental margin and in the deep sea. *Geology* 35:699–702.

NGRIP. 2004. High-resolution record of Northern Hemisphere climate extending into the last interglacial period. *Nature* 431:147–51.

Pasini, G., and M. L. Colalongo. 1997. The Pliocene-Pleistocene boundary-stratotype at Vrica, Italy. In *The Pleistocene boundary and the beginning of the Quaternary*, ed. J. A. Van Couvering, 15–45. Cambridge University Press, Cambridge.

Patton, J. L., and M. N. F. da Silva. 1998. Rivers, refuges, and ridges: The geography of speciation of Amazonian mammals. In *Endless forms: Species and speciation*, ed. D. Howard and S. H. Berlocher, 202–13. Oxford University Press, Oxford.

———. 2005. The history of Amazonian mammals: Mechanisms and timing of diversification. In *Tropical rainforests, Past, present, and future*, ed. E. Bermingham, C. W. Dick, and C. Moritz, 107–26. University of Chicago Press, Chicago.

Patton, J. L., M. N. F. da Silva, and J. R. Malcolm. 1994. Gene genealogy and differentiation among arboreal spiny rats (Rodentia: Echimydae) of the Amazon Basin: A test of the riverine barrier hypothesis. *Evolution* 48:1314–23.

Paul, F., A Kääb, M. Maisch, T. Kellenberger, and W. Haeberli. 2004. Rapid disintegration of alpine glaciers observed with satellite data. *Geophys. Res. Lett.* 31:L21402, doi: 10.1029/2004glo 20816.

Pellegrini, M., and A. Longinelli. 2008. Palaeoenvironmental conditions during the deposition of the Plio-Pleistocene sedimentary sequence of the Canoa Formation, central Ecuador: A stable isotope study. *Palaeogeogr. Palaeocl. Palaeoecol.* 266:119–28.

Perry, L., et al. (fourteen coauthors). 2007. Starch fossils and the domestication and dispersal of chili peppers (*Capsicum* ssp. L.) in the Americas. *Science* 315:986–88.

Piperno, D. R. 2006. Quaternary environmental history and agricultural impact on vegetation in Central America. In *Latin American biogeography—causes and effects*. Proceedings of the 51st Annual Systematics Symposium of the Missouri Botanical Garden (Alan Graham, organizer). *Ann. Mo. Bot. Gard.* 93:274–96.

Piperno, D. R., M. B. Bush, and P. A. Colinvaux. 1990. Paleoenvironmental and human settlement in late-glacial Panama. *Quat. Res.* 33:108–16.

Piperno, D. R., K. Husum-Clary, R. G. Cooke, A. J. Ranere, and D. Weiland. 1985. Preceramic maize in central Panama: Evidence from phytoliths and pollen. *Am. Anthropo.* 87:871–78.

Piperno, D. R., and D. M. Pearsall. 1998. *The origins of agriculture in the lowland neotropics*. Academic Press, New York.

Piperno, D. R., A. J. Ranere, I. Holst, and P. Hansell. 2000. Starch grains reveal early root crop horticulture in the Panamanian tropical forest. *Nature* 407:894–97.

Placzek, C. J., J. Quade, and P. J. Patchett. 2006. Geochronology and stratigraphy of late Pleistocene lake cycles on the southern Bolivian Altiplano: Implications for causes of tropical climate change. *Geol. Soc. Am. Bull.* 118:515–32.

Placzek, C. J., et al. (eight coauthors). 2009. Climate in the dry, central Andes over geologic, millennial, and interannual timescales. *Ann. Mo. Bot. Gard.* 96:386–97.

Polyak, V. J. and Y. Asmerom. 2001. Late Holocene climate and cultural changes in the southwestern United States. *Science* 294:148–51.

Potzger, J. E., and B. C. Tharp. 1943. Pollen record of Canadian spruce and fir from Texas bog. *Science* 98:584–85.

———. 1947. Pollen profile from a Texas bog. *Ecology* 28:274–80.

———. 1954. Pollen study of two bogs in Texas. *Ecology* 35:462–66.

Rhode, D. 2002. Early Holocene juniper woodland and chaparral taxa in the central Baja California Peninsula, Mexico. *Quat. Res.* 57:102–8.

Ritchie, J. C. 1987. *Postglacial vegetation of Canada*. Cambridge University Press, Cambridge.

Sandweiss, D. H. 2003. Terminal Pleistocene through mid-Holocene archaeological sites as paleoclimate archives for the Peruvian coast. *Palaeogeogr. Palaeocl. Palaeoecol.* 194:23–40.

Sandweiss, D. H. (five coauthors). 2001. Variation in Holocene El Niño frequencies: Climate records and cultural consequences in ancient *Peru. Geol.* 29:603–6.

Severinghaus, J. P. 2009. Southern see-saw seen. *Nature* 457:1093–94.

Schmittner, A., and E. D. Galbraith. 2008. Glacial greenhouse-gas fluctuations controlled by ocean circulation changes. *Nature* 456:373–76.

Shaw, C. A., and H. G. McDonald. 1987. First record of giant anteater (*Xenarthra*, Myrmecophagidae) in North America. *Science* 236:186–88.

Shaw, J., et al. (seven coauthors). 1999. The Channeled Scabland: Back to Bretz? *Geology* 27:605–8.

Sridhar, V., et al. (five coauthors). 2006. Large wind shift on the Great Plains during the Medieval Warm Period. *Science* 313:345–47.

Stute, M., et al. (seven coauthors). 1995. Cooling of tropical Brazil (5°C) during the Last Glacial Maximum. *Science* 269:379–83.

Tarasov, L., and W. R. Peltier. 2005. Arctic freshwater forcing of the Younger Dryas cold reversal. *Nature* 435:662–65.

Theissen, K. M., R. B. Dunbar, H. D. Rowe, and D. A. Mucciarone. 2008. Multidecadal- to century-scale arid episodes on the northern Altiplano during the middle Holocene. *Palaeogeogr. Palaeocl. Palaeoecol.* 257:361–76.

Thompson, L. G., M. E. Davis, E. Moseley-Thompson, and K.-b. Liu. 1988. Pre-Incan agricultural activity in dust layers in two tropical ice cores. *Nature* 336:763–65.

Thompson, L. G., L. Hastenrath, and B. Arnao. 1979. Climate ice core records from the tropical Quelccaya ice cap. *Science* 203:1240–43.

Thompson, L. G., and E. Moseley-Thompson. 1987. Evidence of abrupt climatic change during the last 1500 years recorded in ice cores from the tropical Quelccaya ice cap. In *Abrupt climatic change: Evidence and implications*, ed. W. Berger and L. Labeyrie, 99–110. D. Reidel Publishing Co., Dordrecht.

———. 1989. One-half millennia of tropical climate variability as recorded in the stratigraphy of the Quelccaya ice cap, Peru. In *Aspects of climate variability in the Pacific and Western Americas*, ed. D. H. Peterson, 15–31. AGU Monograph 55. American Geophysical Union, Washington, D.C.

Thompson, L. G., et al. (five coauthors). 2005. Tropical ice core records: Evidence for asynchronous glaciation on Milankovitch timescales. *J. Quat. Sci.* 20:723–33.

Van Devender, T. R. 1990. Late Quaternary vegetation and climate of the Chihuahuan Desert, United States and Mexico. In *Packrat middens, The last 40,000 years of biotic change*, ed. J. L. Betancourt, T. R. Van Devender, and P. S. Martin, 104–33. University of Arizona Press, Tucson.

Vollmer, T., et al. (five coauthors). 2008. Orbital control on upper Triassic playa cycles of the Steinmergel-Keuper (Norian): A new concept for ancient playa cycles. *Palaeogeogr. Palaeocl. Palaeoecol.* 267:1–16.

Wallace, A. R. 1849. On the monkeys of the Amazon. *Proc. Zool. Soc. Lond.* 20:107–10.

———. 1876. *The geographical distribution of animals.* Hafner, New York.

Watanabe, T., A. Winter, and T. Oba. 2001. Seasonal changes in sea surface temperature and salinity during the Little Ice Age in the Caribbean Sea deduced from Mg/Ca and $^{18}O/^{16}O$ ratios in corals. *Mar. Geol.* 173:21–35.

Watts, W. A., and B. C. S. Hansen. 1994. Pre-Holocene and Holocene pollen records of vegetation history from the Florida peninsula and their climatic implications. *Palaeogeogr. Palaeocl. Palaeoecol.* 109:163–76.

Webb, R. S., D. H. Rind, S. J. Lehman, R. J. Healy, and D. Sigman. 1997. Influence of ocean heat transport on the climate of the Last Glacial Maximum. *Nature* 385:695–99.

Webster, J. W., et al. (six coauthors). 2007. Stalagmite evidence from Belize indicating significant droughts at the time of Preclassic abandonment, the Maya hiatus, and the Classic Maya collapse. *Palaeogeogr. Palaeocl. Palaeoecol.* 250:1–17.

Westerhold, T., et al. (seven coauthors). 2008. Astronomical calibration of the Paleocene time. *Palaeogeogr. Palaeocl. Palaeoecol.* 257:377–403.

Whitmore, J., et al. (eleven coauthors). 2005. Modern pollen data from North America and Greenland for multi-scale paleoenvironmental applications. *Quat. Sci. Rev.* 24:1828–48.

Wing, S. L. 1996. Paleocene-Eocene floral and climate change in the northern Rocky Mountains, USA. Paper delivered at the International Organization of Palaeobotany Fifth Quadrennial Conference, Santa Barbara, California, June-July 1996. [Abstracts, 113.]

Wüster, W., et al. (five coauthors). 2005a. No rattlesnakes in the rainforests: Reply to Gosling and Bush. *Mol. Ecol.* 14:3619–21.

———. 2005b. Tracing an invasion: Landbridges, refugia, and the phylogeography of the neotropical rattlesnake (Serpentes: Viperidae: *Crotalus durissus*). *Mol. Ecol.* 14:1095–1108.

Zazula, G. D., et al. (seven coauthors). 2003. Ice-age steppe vegetation in east Beringia. *Nature* 423:603.

Additional Readings and Updates

Extinction of the North American Pleistocene Megafauna

Barnosky, A. D., P. L. Koch, R. S. Feranec, S. L. Wing, and A. B. Shabel. 2004. Assessing the causes of late Pleistocene extinctions on the continents. *Science* 306:70–75 ["(E)vidence suggests that the intersection of human impacts with pronounced climatic change drove the precise timing and geography of extinction in the Northern Hemisphere."]

Cardillo, M., and A. Lister. 2002. Death in the slow lane. *Nature* 419:440–41. [Reviews factors of low reproductive rates and open, grassland/savanna habitats, in addition to hunting and climatic change, in the demise of the Pleistocene megafauna. Survivors with low reproductive rates tended to be arboreal, nocturnal, or inhabitants of dense forest habitats.]

Koch, P. L. 2006. Land of the lost. *Science* 311:957. [Review of Martin 2005.]

Martin, P. S. 2005. *Twilight of the mammoths: Ice age extinctions and the rewilding of America.* University of California Press, Berkeley.

Alsos, I. G., T. Engelskjøn, L. Gielly, P. Taberlet, and C. Brochmann. 2005. Impact of ice ages on circumpolar molecular diversity: Insights from an ecological key species. *Mol. Ecol.* 14:2739–53.

Anderson, L. L., F. Sheng Hu, D. M. Nelson, R. J. Petit, and K. N. Paige. 2006. Ice-age endurance: DNA evidence of a white spruce refugium in Alaska. *Proc. Natl. Acad. Sci. U.S.A.* 103:12447–50.

Anderson, R. S., et al. (seven coauthors). 2008. Development of the mixed conifer forest in northern New Mexico and its relationship to Holocene environmental change. *Quat. Res.* 69:263–75.

Baker, R. G., et al. (five coauthors). 2002. Holocene paleoenvironments in southeastern Minnesota—chasing the prairie-forest ecotone. *Palaeogeogr. Palaeocl. Palaeoecol.* 177:103–22.

Bard, E., and R. E. M. Rickaby. 2009. Migration of the subtropical front as a modulator of glacial climate. *Nature* 460:380–83.

Barron, J. A., D. Bukry, W. E. Dean, J. A. Addison, and B. Finney. 2009. Paleoceanography of the Gulf of Alaska during the past 15,000 years: Results from diatoms, silicoflagellates, and geochemistry. *Mar. Micropaleontol.* 72:176–95.

Bintanja, R., and R. S. W. van de Wal. 2008. North American ice-sheet dynamics and the onset of 100,000-year glacial cycles. *Nature* 454:869–72. [Suggests a possible explanation for the change from the 41,000- to the 100,000-year orbital-induced climate cycle during the Quaternary.]

Blinnikov, M. S. 2005. Phytoliths in plants and soils of the interior Pacific Northwest, USA. *Rev. Palaeobot. Palynol.* 135:71–98. [Phytolith composition is demonstrated distinctive for eight kinds of regional modern communities: *Artemesia* shrubland, four types of grasslands, and three woodlands (*Pinus ponderosa*, *Abies grandis–Picea engelmanni*, *Abies lasiocarpa–Picea engelmanni*).]

Blinnikov, M. S., A. Busacca, and C. Whitlock. 2002. Reconstruction of the late Pleistocene grassland of the Columbia Basin, Washington, USA, based on phytolith records in loess. *Palaeogeogr. Palaeocl. Palaeoecol.* 177:77–101.

Brubaker, L. B., P. M. Anderson, M. E. Edwards, and A. V. Lozhkin. 2005. Beringia as a glacial refugium for boreal trees and shrubs: New perspectives from mapped pollen data. *J. Biogeogr.* 32:833–48.

Caseldine, C. 2001. Changes in *Betula* in the Holocene record from Iceland—a palaeoclimatic record or evidence for early Holocene hybridization? *Rev. Palaeobot. Palynol.* 117:139–52.

Cazenave, A. 2006. How fast are the ice sheets melting? *Science* 314:1250–52. [Since the end of the nineteenth century, tide gauges show an average annual rise of 1.8 mm over the past fifty years; satellite altimetry show 3 mm a year since 1993.]

Cess, R. D. 2005. Water vapor feedback in climate models. *Science* 310:795–96.

Chapin, F. S. III, et al. (twenty coauthors). 2005. Role of land-surface changes in Arctic summer warming. *Science* 310:657–60.

Chen, D., M. A. Cane, A. Kaplan, S. E. Zebiak, and D. Huang. 2004. Predictability of El Niño over the past 148 years. *Nature* 428:733–36.

Clague, J. J., P. T. Bobrowsky, and I. Hutchinson. 2000. A review of geological records of large tsunamis at Vancouver Island, British Columbia, and implications for hazard. *Quat. Sci. Rev.* 19:849–63.

Clark, P. U., N. G. Pisias, T. F. Stocker, and A. J. Weaver. 2002. The role of the thermohaline circulation in abrupt climate change. *Nature* 415:863–69.

Clet-Pellerin, M., and S. Occhietti. 2000. Pleistocene palynostratigraphy in the St. Lawrence Valley and middle Estuary. *Quat. Int.* 68–71:39–57.

Comes, H. P., and J. W. Kadereit. 1998. The effect of Quaternary climatic changes on plant distribution and evolution. *Trends Plant Sci.* 3:432–38.

Crucifix, M. 2008. Global change: Climate's astronomical sensors. *Nature* 456:47–48.

Cuffy, K. M. 2004. Into an ice age. *Nature* 431:133–34. ["Analyses of a new core from Green-
land yield the first high-resolution picture of the start of the last ice age in the Northern
Hemisphere, and of the onset of climate instability as our planet cooled." Summary of
NGRIP 2004.]

Davies, T. J., A. Purvis, and J. L. Gittleman. 2009. Quaternary climate change and the geographic ranges of mammals. *Am. Nat.* 174:297–307.

De Batist, M., and E. Chapron, eds. 2008. *Lake systems: Sedimentary archives of climate
change and tectonics, EGU General Assembly.* Special issue, *Palaeogeogr. Palaeocl. Palaeoecol.*
259:93–356. [Papers on the Salton Basin, California; Laguna Potrok Aike, Argentina;
Mérida Andes, Venezuela; Lago Icalmam, Chile.]

Driese, S. G., K. H. Orvis, S. P. Horn, Z.-H. Li, and D. S. Jennings. 2007. Paleosol evidence for
Quaternary uplift and for climate and ecosystem changes in the Cordillera de Talamanca,
Costa Rica. *Palaeogeogr. Palaeocl. Palaeoecol.* 248:1–23.

A dust with climate. 2008. Editors' summary. *Nature* 452.

Elliot, M., L. Labeyrie, and J.-C. Duplessy. 2002. Changes in North Atlantic deep-water formation associated with the Dansgaard-Oeschger temperature oscillations (60–10 ka). *Quat.
Sci. Rev.* 21:1153–65.

Etterson, J. R., and R. G. Shaw. 2001. Constraint to adaptive evolution in response to global
warming. *Science* 294:151–54.

Fall, P. L. 1997. Timberline fluctuations and late Quaternary paleoclimates in the Southern
Rocky Mountains, Colorado. *Geol. Soc. Am. Bull.* 109:1306–20.

Florindo, F., A. Nelson, and A. M. Haywood, eds. 2008. *Antarctic cryosphere and Southern
Ocean climate evolution (Cenozoic-Holocene).* Special issue, *Palaeogeogr. Palaeocl. Palaeoecol.*
260:1–298. [Papers from the European Geosciences Union General Assembly 2007.]

Francis, D. R., A. P. Wolfe, I. R. Walker, and G. H. Miller. 2006. Interglacial and Holocene
temperature reconstructions based on midge remains in sediments of two lakes from
Baffin Island, Nunavut, Arctic Canada. *Palaeogeogr. Palaeocl. Palaeoecol.* 236:107–24.
["2.22°C for summer water temperatures . . . 1.53°C for mean July air temperatures.
Reconstructions at both sites indicate that summer temperatures during the last
interglacial were higher than at any time in the Holocene, and 5 to 10°C higher than at
present."]

Fréchette, B., A. P. Wolfe, G. H. Miller, P. J. H. Richard, and A. de Vernal. 2006. Vegetation
and climate of the last interglacial on Baffin Island, Arctic Canada. *Palaeogeogr. Palaeocl.
Palaeoecol.* 236:91–106. [" July air temperatures of the last interglacial . . . 4 to 5°C
warmer than present on eastern Baffin Island, which was warmer than any interval within
the Holocene."]

Giles, J. 2006. How much will it cost to save the world? *Nature* 444:6–7. [*Stern Review on the
Economics of Climate Change* estimates that "tackling climate change [now] would cost
20 times less than doing nothing," or about 1 percent GDP.]

González-Carranza, Z., J. C. Berrío, H. Hooghiemstra, J. F. Duivenvoorden, and H. Behling.
2008. Changes of seasonally dry forest in the Colombian Patía Valley during the early
and middle Holocene and the development of a dry climatic record for the northernmost
Andes. *Rev. Palaeobot. Palynol.* 152:1–10. ["The data also show that relatively humid and
wet conditions were present at the Patía Valley from c. 9510 to 8600 cal yr BP, whereas
between c. 8360 to 8260 cal yr BP dry conditions prevailed in the valley. The pollen rec-

ord of Potrerillo-2 illustrates well the response—and sensitivity—of dry forest ecosystems to changes in precipitation, which is mainly related with the mean position of the ITCZ."]

Grigg, L. D., and C. Whitlock. 2002. Patterns and causes of millennial-scale climate changes in the Pacific Northwest during Marine Isotope Stages 2 and 3. *Quat. Sci. Rev.* 21:2067–83. [Early MIS 3 = 60–45 kyr; late MIS 3 = 45–27 kyr; MIS 2 = 27.5–14 kyr.]

Gonzalez, C., L. M. Dupont, H. Behling, and G. Wefer. 2008. Neotropical vegetation response to rapid climatic changes during the last glacial period: Palynological evidence from the Cariaco Basin. *Quat. Res.* 69:217–30.

Gu, L., et al. (six coauthors). 2003. Response of a deciduous forest to the Mount Pinatubo eruption: Enhanced photosynthesis. *Science* 299:2035–38.

Hansen, D. M., and M. Galetti. 2009. The forgotten megafauna. *Science* 324:42–43.

Heinrichs, M. L., R. J. Hebda, I. R. Walker, and S. L. Palmer. 2002. Postglacial paleoecology and inferred paleoclimate in the Engelmann spruce-subalpine fir forest of south-central British Columbia, Canada. *Palaeogeogr. Palaeocl. Palaeoecol.* 184:347–69.

Heusser, L. E. 2000. Rapid oscillations in western North America vegetation and climate during oxygen isotope stage 5 inferred from pollen data from Santa Barbara Basin (Hole 893A). *Palaeogeogr. Palaeocl. Palaeoecol.* 161:407–21.

Hewitt, G. M. 2004. Genetic consequences of climatic oscillations in the Quaternary. *Phil. Trans. R. Soc. B* 359:183–95.

Holmgren, C. A. 2007. Paleoenvironmental response of the northern Chijuahuan Desert to the Bølling/Allerød–Younger Dryas climatic oscillation. *Current Research in the Pleistocene* 24:7–10.

Holmgren, C. A., J. L. Betancourt, and K. A. Rylander. 2006. A 36,000-yr vegetation history from the Peloncillo Mountains, southeastern Arizona, USA. *Palaeogeogr. Palaeocl. Palaeoecol.* 240:405–22.

Holmgren, C. A., M. C. Peñalba, K. A. Rylander, and J. L. Betancourt. 2003. A 16,000 14C yr BP packrat midden series from the U.S.A.-Mexico borderlands. *Quat. Res.* 60:319–29.

Holmgren, C. A., E. Rosario, C. Latorre, and J. L. Betancourt. 2008. Late Holocene fossil rodent middens from the Arica region of northernmost Chile. *J. Arid Environ.* 72:677–96.

Holmgren, C. A., et al. (seven coauthors). 2001. Holocene vegetation history from fossil rodent middens near Arequipa, Peru. *Quat. Res.* 56:242–51.

Huang, Y., et al. (five coauthors). 2006. Climatic and environmental controls on the variation of C_3 and C_4 plant abundances in central Florida for the past 62,000 years. *Palaeogeogr. Palaeocl. Palaeoecol.* 237:428–35. [There was about a 40 percent higher input from C_4 (cool-indicating) plants during the LGM than during the Holocene, consistent with climatic trends and documenting the usefulness of carbon isotope data in reflecting these trends.]

Huybers, P., and C. Wunsch. 2005. Obliquity pacing of the late Pleistocene glacial terminations. *Nature* 434:491–94.

Kerr, R. A. 2002. Mild winters mostly hot air, not Gulf Stream. *Science* 297:2202.

———. 2004a. Sea change in the Atlantic. *Science* 303:35.

———. 2004b. A slowing cog in the North Atlantic Ocean's climate machine. *Science* 304:371–72.[Notes "disaster" scenarios in discussions of global warming; see also comments of W. S. Broecker on p. 388 of the same issue.]

———. 2005. Scary Arctic ice loss? Blame the wind. *Science* 307:203.

Kennett, D. J., and J. P. Kennett. 2000. Competitive and cooperative responses to climatic instability in coastal southern California. *Am. Antiquity* 65:379–95.

Knorr, W., I. C. Prentice, J. I. House, and E. A. Holland. 2005. Long-term sensitivity of soil carbon turnover to warming. *Nature* 433:298–301.

Krajick, K. 2002. Melting glaciers release ancient relics. *Science* 296:454–56. [Mammals, birds, rodents, human artifacts from 8300-year-old deposits in the Yukon.]

Kuch, M., et al. (five coauthors). 2002. Molecular analysis of an 11,700-year old rodent midden from the Atacama Desert, Chile. *Mol. Ecol.* 11:913–24.

Kuzmina, S., S. Ellias, P. Matheus, J. E. Storer, and A. Sher. 2008. Paleoenvironmental reconstruction of the Last Glacial Maximum, inferred from insect fossils from a tephra buried soil at Tempest Lake, Seward Peninsula, Alaska. *Palaeogeogr. Palaeocl. Palaeoecol.* 267:245–55.

Lambert, F., et al. (nine coauthors). 2008. Dust-climate couplings over the past 800,000 years from the EPICA Dome C ice core. *Nature* 452:616–19.

Latorre, C. L., J. L. Betancourt, K. A. Rylander, and J. A. Quade. 2002. Vegetation invasions into Absolute Desert: A 45,000-year rodent midden record from the Calama-Salar de Atacama Basins, Chile. *Geol. Soc. Am. Bull.* 114:349–66.

Latorre, C. L, A. Maldonado, and J. Betancourt. 2006. Late Quaternary climate and vegetation changes in the Atacama Desert, northern Chile: The rodent midden record. Abstract. *Geol. Soc. Am., Abstracts with Programs* 38:456.

Ledru, M.-P., et al. (five coauthors). 2007. Regional assessment of the impact of climatic change on the distribution of a tropical conifer in the lowlands of South America. *Diversity and Distributions* 13:761–71. [*Podocarpus brasiliensis, P. lambertii, P. sellowii* endemic to Brazil.]

Lempinen, E. W. 2006. In Arctic Alaska, climate warming threatens a village and its culture. *Science* 314:609.

LePage, B. A. 2009. Earliest occurrence of *Taiwania* (Cupressaceae) from the Early Cretaceous of Alaska: Evolution, biogeography, and paleoecology. *Proc. Acad. Nat. Sci. Phila.* 158:129–58.

Levac, E. 2001. High resolution Holocene palynological record from the Scotian Shelf. *Mar. Micropaleontol.* 43:179–97.

Lovett, R. A. 2000. Mount St. Helens, revisited. *Science* 288:1578–79.

Lowenstein, T. K., et al. (six coauthors). 1999. 200 k.y. paleoclimate record from Death Valley salt core. *Geology* 27:3–6.

MacFadden, B. J., N. Solounias, and T. E. Cerling. 1999. Ancient diets, ecology, and extinction of 5-million-year-old horses from Florida. *Science* 283:824–27.

Madsen, D. B., et al. (eight coauthors). 2001. Late Quaternary environmental change in the Bonneville Basin, western USA. *Palaeogeogr. Palaeocl. Palaeoecol.* 167:243–71.

Matheus, P., J. Burns, J. Weinstock, and M. Hofreiter. 2004. Pleistocene brown bears in the mid-continent of North America. *Science* 306:1150.

May, J.-H., J. Argollo, and H. Veit. 2008. Holocene landscape evolution along the Andean piedmont, Bolivian Chaco. *Palaeogeogr. Palaeocl. Palaeoecol.* 260:505–20.

McClintock, J., H. Ducklow, and W. Fraiser. 2008. Ecological responses to climate change on the Antarctic Peninsula. *Am. Sci.* 96:302–10.

McManus, J. F., R. Francois, J.-M. Gherardi, L. D. Keigwin, and S. Brown-Leger. 2004. Collapse and rapid resumption of Atlantic meridional circulation linked to deglacial climate changes. *Nature* 428:834–37.

McPhaden, M. J., S. E. Zebiak, and M. H. Glantz. 2006. ENSO as an integrating concept in Earth science. *Science* 314:1740–45.

Metcalfe, S., et al. (five coauthors). 2009. Environmental change in northern Belize since the latest Pleistocene. *J. Quat. Sci.* 24 (6): 627–41, doi 10.1002/jqs.1248. Published online 11 Feb 2009.

Miranda Chaves, S. A. de, and K. J. Reinhard. 2006. Critical analysis of coprolite evidence of medicinal plant use, Piauí, Brazil. *Palaeogeogr. Palaeocl. Palaeoecol.* 237:110–18.

Monacci, N. M., U. Meier-Grünhagen, B. P. Finney, H. Behling, and M. J. Wooller. 2009. Mangrove ecosystem changes during the Holocene at Spanish Lookout Cay, Belize. *Palaeogeogr. Palaeocl. Palaeoecol.* 280:37–46.

Moore, P. D. 2002. Climate records spruced up. *Nature* 417:133–35. [Maine, U.S.A.; summary of Schauffler and Jacobson 2002.]

Morell, V. 1999. Dietary data straight from the horse's mouth. *Science* 283:773. [Summary of MacFadden et al. 1999.]

Nash, J. M. 2007. Chronicling the ice. *Smithsonian* 38:67–74.

Niemann, H., and H. Behling. 2009. Late Pleistocene and Holocene environmental change inferred from the Cocha Caranga sediment and soil records in the southeastern Ecuadorian Andes. *Palaeogeogr. Palaeocl. Palaeoecol.* 276:1–14.

Nogué, S., V. Rull, E. Montoya, O. Huber, and T. Vegas-Vilarrúbia. 2009. Paleoecology of the Guayana highlands (northern South America): Holocene pollen record from the Eruodatepui, in the Chimantá Massif. *Palaeogeogr. Palaeocl. Palaeoecol.* 281:165–73.

Oches, E. A., and W. D. McCoy. 2001. Historical developments and recent advances in amino acid geochronology applied to loess research: Examples from North America, Europe, and China. *Earth-Sci. Rev.* 54:173–92. ["In general, loess of the last four glacial cycles, corresponding to marine oxygen isotope stages (OIS) 2–4, 6, 8, and 10, respectively, can be distinguished on the basis of alloisoleucine/isoleucine (A/I) ratios measured in fossil gastropod shells preserved in the loess."]

Ortega-Rosas, C. I., J. Guiot, M. C. Peñalba, and M. E. Ortiz-Acosta. 2008. Biomisation and quantitative climatic reconstructions techniques in northwestern Mexico—with an application to four Holocene pollen sequences. *Global Planet. Change* 61:242–66.

Ortega-Rosas, C. I., M. C. Peñalba, and J. Guilot. 2008. Holocene altitudinal shifts in vegetation belts and environmental changes in the Sierra Madre Occidental, northwestern Mexico, based on modern and fossil pollen data. *Rev. Palaeobot. Palynol.* 151:1–20.

Perry, L., et al. (six coauthors). 2006. Early maize agriculture and interzonal interaction in southern Peru. *Nature* 440:76–79.

Pierce, J. L., G. A. Meyer, and A. J. T. Jull. 2004. Fire-induced erosion and millennial-scale climate change in northern ponderosa pine forests. *Nature* 432:87–90.

Polyak, V. J., J. C. Cokenholpher, R. A. Norton, and Y. Asmerom. 2001. Wetter and cooler late Holocene climate in the southwestern United States from mites preserved in stalagmites. *Geology* 29:643–46.

Punyasena, S. W. 2008. Estimating neotropical palaeotemperature and palaeoprecipitation using plant family climatic optima. *Palaeogeogr. Palaeocl. Palaeoecol.* 265:226–37.

Punyasena, S. W., G. Eshel, and J. C. McElwain. 2008. The influence of climate on the spatial patterning of neotropical plant families. *J. Biogeogr.* 35:117–30.

Punyasena, S. W., F. E. Mayle, and J. C. McElwain. 2008. Quantitative estimates of glacial and

Holocene temperature and precipitation change in lowland Amazonian Bolivia. *Geology* 36:667–70.

Quade, J., J. Rech, J. L. Betancourt, and C. Latorre. 2001. Mid-Holocene climate in the south-central Andes: Humid or Dry? *Science* 292:2391.

Quattrocchio, M. E., A. M. Borromei, C. M. Deschamps, S. C. Grill, and C. A. Zavala. 2008. Landscape evolution and climate changes in the late Pleistocene-Holocene, southern Pampa (Argentina): Evidence from palynology, mammals and sedimentology. *Quat. Int.* 181:123–38.

Rahmstorf, S. 2000. The thermohaline ocean circulation: A system with dangerous thresholds? *Climatic Change* 46:247–56.

———. 2003. The current climate. *Nature* 421:699.

Rech, J. A., J. Quade, and J. L. Betancourt. 2002. Late Quaternary paleohydrology of the central Atacama Desert (22–24°S), Chile. *Geol. Soc. Am. Bull.* 114:334–48.

Schauffler, M., and G. L. Jacobson Jr. 2002. Persistence of coastal spruce refugia during the Holocene in northern New England, USA, detected by stand-scale pollen stratigraphies. *J. Ecol.* 90:235–50.

Schiemeier, Q. 2003. Summer in Svalbard. *Nature* 424:992–94.

———. 2006. On thin ice. *Nature* 441:146–47.

———. 2008. The long summer begins. *Nature* 454:266–69. ["A research vessel embedded in the thinning Arctic sea ice has a front-row seat for the cryospheric show of the century."]

Service, R. F. 2004. As the west goes dry. *Science* 303:1124–27. ["In a region already prone to water shortages, researchers now forecast that rising temperatures threaten the American West's hidden reservoir: Mountain snow."]

Steadman, D. W., et al. (eleven coauthors). 2007. Exceptionally well preserved late Quaternary plant and vertebrate fossils from a blue hole on Abaco, the Bahamas. *Proc. Natl. Acad. Sci. U.S.A.* 104:19897–902.

Steig, E. J. 2006. The south-north connection. *Nature* 444:152–53.

Steig, E. J., A. P. Wolfe, and G. H. Miller. 1998. Wisconsinan refugia and the glacial history of eastern Baffin Island, Arctic Canada: Coupled evidence from cosmogenic isotopes and lake sediments. *Geology* 26:835–38.

Stokstad, E. 2003. Ancient DNA pulled from soil. *Science* 300:407. [DNA 400,000 years old from Siberian tundra plants.]

Stone, R., and J. Bohannon. 2006. U.N. conference puts spotlight on reducing impact of climate change. *Science* 314:1224–25.

Storch, H. von., et al. (five coauthors). 2004. Reconstructing past climate from noisy data. *Science* 306:679–82. ["Empirical reconstructions of the Northern Hemisphere (NH) temperature in the past millennium based on multiproxy records depict small-amplitude variations followed by a clear warming trend in the past two centuries."]

Van Geel, B., et al. (fourteen coauthors). 2008. The ecological implications of a Yakutian mammoth's last meal. *Quat. Res.* 69:361–76.

Van Leeuwen, J. F. N., et al. (five coauthors). 2008. Fossil pollen as a guide to conservation in the Galápagos. *Science* 322:1206. ["The first human presence in Galapagos is believed to have occurred with European contact in 1535."]

Vinnikov, K. Y., and N. C. Grody. 2003. Global warming trend of mean trophospheric temperature observed by satellites. *Science* 302:269–72.

Visbeck, M. 2002. The ocean's role in Atlantic climate variability. *Science* 297:2223–24.

Whitlock, C. 2004. Forest, fires, and climate. *Nature* 432:28–29. [Summary of Pierce et al. 2004.]

Willard, D. A., L. M. Weimer, and W. L. Riegel. 2001. Pollen assemblages as paleoenvironmental proxies in the Florida Everglades. *Rev. Palaeobot. Palynol.* 113:213–35.

Willard, D. A., et al. (five coauthors). 2007. Deglacial climate variability in central Florida, USA. *Palaeogeogr. Palaeocl. Palaeoecol.* 251:366–82.

Worona, M. A., and C. Whitlock. 1995. Late Quaternary vegetation and climate history near Little Lake, central Coast Range, Oregon. *Geol. Soc. Am. Bull.* 107:867–76.

Yu, Z., and H. E. Wright, Jr. 2001. Response of interior North America to abrupt climate oscillations in the North Atlantic region during the last deglaciation. *Earth-Sci. Rev.* 52:333–69.

Zazula, G. D., et al. (nine coauthors). 2006. Vegetation buried under Dawson tephra (25,300 ^{14}C years BP) and locally diverse late Pleistocene paleoenvironments of Goldbottom Creek, Yukon, Canada. *Palaeogeogr. Palaeocl. Palaeoecol.* 242:253–86. [Heterogeneity of the vegetation at the LGM is shown from analysis of vascular plant macrofossils, bryophytes, pollen, insects, and paleosols from a riparian meadow buried in the winter or early spring by the Dawson ash fall, and from upland steppe-tundra habitats from arctic ground squirrel middens. The latest version of the steppe-tundra ecosystem in the region was established in MIS stage 3 by 29,000 ^{14}C years BP.]

Zielinski, G. A., and G. R. Mershon. 1997. Paleoenvironmental implications of the insoluble microparticle record in the GISP2 (Greenland) ice core during the rapidly changing climate of the Pleistocene-Holocene transition. *Geol. Soc. Am. Bull.* 109:547–59.

9

The Bigger Picture
Implications of Past Environmental Changes
in the New World

THE BIOTAS OF EASTERN NORTH AMERICA
AND EASTERN ASIA

Assembling information on ecosystem history provides an opportunity to consider other aspects of the New World biota. One of these is an intercontinental biogeographic relationship between eastern North America and eastern Asia that has intrigued biologists for more than 250 years since the issue was first raised by Carl Linnaeus and Jonas Halenius (Halenius 1750; Graham 1966, 1972a, b).

When Jonas P. Halen of the Uplands Nation of Sweden decided to seek advanced training in botany in 1750, the obvious choice was with the great Swedish naturalist Linnaeus (1707–78). First, however, he had to Latinize the name Halen for academic purposes, just as Linnaeus had done earlier, although in the latter's case it was more complicated. In eighteenth century Sweden, patronyms were generally used. If the father's first name were Ingemar, for example, the children would simply be called Ingrid Ingemars*dot-*

ter or Carl Ingemars*son*—hence, the "-son" ending for many Swedish family names. Upon a student's entrance to the university, a surname was selected and Carl's family took theirs (Linné) from the great linden or linn tree (*Tilia*) at the family home. That name, when Latinized, became Linnaeus. (Later, when Linnaeus was ennobled, he became Carl von Linné.) Halen's academic name was a simpler derivation: he became Jonas P. Halenius.

In addition to the use of academic names, another unusual aspect of Swedish academic life was that the professor, rather than the student, usually wrote the thesis. The student's role was to defend his professor's position in Latin and in public debate. The goal was to demonstrate mental agility, familiarity with the rules of formal disputation, and fluency in Latin; the originality of the thesis did not greatly matter. This is evident from Linnaeus's comment regarding another student's thesis: "Mr. Kalm ought to dispute pro gradu *de O(e)conomiae patriae augmento opera L. B. Bjelke.* This he ought to do out of gratitude . . . I will dictate the first version." The thesis chosen and written by Linnaeus for Halenius to defend—*Plantae rariores camschatcenses*—was that some plants in the Kamchatka region of eastern Russia were similar to ones in eastern North America. It was not a matter of whether the species were continuous or disjunct between the two regions, for Linnaeus knew little of the intervening vegetation. It was an observation that there were plants in eastern Asia similar to ones in faraway eastern North America. Subsequently, the distributions were found to be very much disjunct (Gray 1840, 1846; Li 1952; Wen 2001), sometimes with outliers in moist areas of the American west like the San Francisco Bay area and in the Middle East. The disjuncts included genera such as *Acer*, *Aesculus*, *Carya*, *Clintonia* (fig. 9.1), *Cornus*, *Fraxinus*, *Liquidambar* (Ickert-Bond et al. 2005), *Liriodendron*, *Magnolia*, *Nyssa*, *Sassafras*, *Vitis*, and others. From the modern perspective, any explanation for the disjunction would require an accurate phylogeny of the organisms (Xiang et al. 1998) to assess the areas of possible basal and derived species (i.e., direction of migration), and an accurate geologic, climatic, and fossil history to reveal the cause(s), time, and plausible opportunities for migration.

A starting point for considering the origin of the distribution pattern is to pose the question, "What was it that became disrupted, fragmented, or that migrated, to yield residual populations in eastern Asia and eastern North America, and small relictual populations in other temperate enclaves of the Northern Hemisphere?" The answer, based on the plant fossil record, is that it was the broad band of deciduous forest and temperate environments that extended across much of North America, Asia, and Europe in the Oligocene, Miocene, and through about the middle Pliocene, with

Figure 9.1 Distribution of *Clintonia* in eastern Asia and in eastern and western North America. From Ying 1983. The center of distribution is indicated by the shaded area. Used with permission from the Missouri Botanical Garden Press, St. Louis.

individual elements present since the Late Cretaceous (e.g., *Nordenskioeldia*; Wang et al. 2009). The fossil genera include virtually all those mentioned above, in addition to *Sequoia*, presently confined to the west coast of the United States, but widespread in Europe and Asia during the Tertiary, and the present eastern Asian *Ginkgo*, *Glyptostrobus*, *Metasequoia*, *Eucommia*, *Pterocarya*, *Platycarya*, and others known from the Tertiary of North America and Europe. The reason, based on the modern distribution of these and other temperate deciduous plants, is the same—a belt of deciduous forest across the temperate latitudes of the Northern Hemisphere. The distribution maps for *Carya*, *Clintonia*, *Euphrasia*, and others all speak of an ecosystem formerly extending across the region, and now existing in more restricted areas because of alterations in climate and landscape. Although it is possible to emphasize migrations in the explanation, a balanced model must still stress a prominent role for range fragmentation. This is evident from the congruence in phylogenies of several extant genera based on DNA data sets (*Adiantum pedatum*, sect. *Aralia* of *Aralia*, *Calycanthus*, *Cornus*, *Boykinia*, *Tiarella*, *Trautvetteria*): "The data are in agreement with the longstanding hypothesis that this well-known floristic disjunction represents the fragmentation of a once continuous Mixed Mesophytic Forest Community" (Xiang et al. 1998, 178). So the next question is, what were those alterations in climate and landscape that fragmented the forest's once more extended range?

The first fragmenting event had begun by the Eocene, when the Rocky

Mountains had reached sufficient heights to begin interfering with moisture coming from the west. The Paleocene to early Eocene vegetation north of about the Colorado-Wyoming border included *Acer*, *Betula*, cf. *Cornus*, *Corylus*, members of the Fagaceae and Juglandaceae, and the Asian temperate tree *Zelkova* of the Ulmaceae, while toward the south the vegetation was more tropical but still forested (chap. 5). From about the middle Eocene through the middle Miocene, the deciduous forest was widespread. But by the late Miocene, although there was deciduous forest in the uplands of the Plains region, and in gallery forests along streams, the intervening land now included woodland/savanna trending toward grassland. Thus, beginning in about the middle Eocene, the deciduous forest was becoming reduced. later it would essentially be eliminated from one part of its former range—the midcontinent region of interior North America.

The second fragmenting event was the rise of the Sierra Nevada, Cascade Mountains, and Coast Ranges, which began the trend toward desertification of low- to midelevation habitats between these mountains and the western slopes of the Rocky Mountains. The rich deciduous forest of the Columbia Plateau and the surrounding landscape was being replaced in a second region of its former distribution during the late Miocene and especially in the Pliocene by the Basin and Range and other deserts, and by shrubland/chaparral-woodland-savanna. Most of the present-day mesic Asian trees and shrubs had disappeared.

The third fragmenting event was a result of topographic configurations and climatic change. In North America, the principal mountain systems run north and south; so when the Quaternary glaciations began in earnest around 2.6 Ma, the plants and animals were able to move southward until milder conditions returned during the interglacials. In Europe the mountains run east and west, so the biota was trapped between the Fennoscandian glaciers advancing from the north and the glaciers of the Pyrenees-Alps-Carpathians to the south. The glaciers actually met at times, and the biota was destroyed. There was regeneration to deciduous forest during each of the interglacials, but succession during the last interglacial was halted by agriculturalists moving westward from the Fertile Crescent between the Tigris and Euphrates rivers of modern-day Iraq, Syria, and Turkey, who were clearing the land with axe and fire. With this event, the north temperate deciduous forest was removed or reduced from a third region of its former range, leaving segments isolated in eastern North America and eastern Asia, with individual elements persisting in a few mesic sites of the American west and the Middle East. As noted, in addition to the disruption of once more continuous ranges, there were also east-west-east migrations into the respective regions. However, the process is too complex to overemphasize a single explanatory factor.

THE BIOTAS OF THE EASTERN UNITED STATES AND EASTERN MEXICO

The origin of a northern aborescent temperate element in the Latin American biota is complex (fig. 9.2). Fossils in the middle Pliocene Paraje Solo flora of Veracruz document that a number of northern temperate plants were already established in eastern Mexico prior to the Pleistocene, so the origin of the floristic affinity does not date exclusively from this time. That still leaves open a number of questions: First, were these elements widespread in the Tertiary, or even in the Cretaceous, and their range divided by development of the Chihuahuan Desert across south Texas and northern

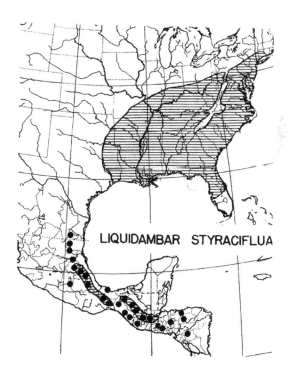

Figure 9.2 Floristic affinities between eastern United States and eastern Mexico as represented by *Liquidambar* (sweet gum). From Martin and Harrell 1957. Used by permission from the Ecological Society of America, Washington, DC.

Mexico (Axelrod 1975); that is, was the affinity a vicariant event? Second, might the affinity result from periodic long-distance dispersal after uplift of the Sierra Madre Oriental provided climatically comparable habitats at elevational zones between 1000 and 2000 m, that is, anytime after about the Paleocene/Eocene, and possibly into the present? Or, third, is there evidence of a progressive introduction from the north with Neogene cooling? These are not mutually exclusive possibilities, but the evidence favors the last suggestion as a principal mode of origin.

In the southeastern United States, and in other parts of northern North America, trees of the deciduous forest were already established in the Paleocene and Eocene. Although paleobotanical information toward the south is limited, none of these plants (with the possible exception of the widespread *Ilex*) are known from the Eocene of Latin America. The same is true through about the middle Miocene. Also, for any one period, the number of northern deciduous forest genera in paleofloras generally decreases from north to south. In other words, when the relevant fossil records are plotted with reference to northern temperate elements, the trend is for more and earlier in the north and fewer and later toward the south.

The direction of movement would be more convincing, however, if there were a correlated climatic event favoring a southward migration. If the floras are plotted on the paleotemperature curve (fig. 9.3), we find that prior to about the middle Miocene, northern temperate genera in Latin America are rare to absent, with occasional *Pinus*, possibly *Picea* in the Oligo-Miocene of Mexico, and *Ilex*. After the temperature decline at the beginning of the middle Miocene, ushering in cooler temperatures and greater seasonality, the numbers increase to ten genera in the Paraje Solo Formation of Mexico, five genera in the Padre Miguel / Herreria floras of Guatemala, and one genus (*Quercus*) in the late Miocene of Panama. The most parsimonious explanation for this pattern is a progressive introduction from the north facilitated by the temperature drop beginning in the middle Miocene (Graham 1973a, b; II, chap. 8). The view is consistent with studies of extant genera such as *Illicium* (Morris et al. 2007).

THE BIOLOGICAL CONSEQUENCES OF CLOSURE OF THE ISTHMUS OF PANAMA

When the last bits and pieces of land emerged from the sea and fused to complete the Panama land bridge, the biological impact was enormous. For the first time since the Cretaceous, 80–70 Ma (Heinicke et al. 2007), organisms were able to move overland directly between North and South

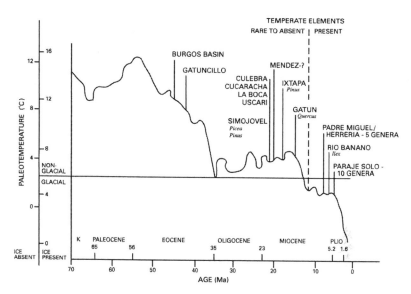

Figure 9.3 Representative Tertiary floras of northern Latin America plotted on the global paleotemperature curve. Note the few representatives of northern temperate elements prior to the middle Miocene temperature decline and the greater number after that event.

America. In line with the observation that a land bridge is a sea barrier (Woodring 1966), closure also began differentiating a Pacific and a Caribbean marine fauna (Lessios 2008). In addition to marine biological information, there is also the independent geochemical evidence for establishing the time of closure discussed in chapter 2. It is possible to augment the data further by plotting similarity values in spore and pollen types for Eocene through Pleistocene floras on either side of the isthmus (Graham 1992). For the Eocene the value is 2.6 percent; for the Pliocene around 3.5 Ma, it is 8.9 percent; and with the inclusion of recent studies on the Quaternary, similarity for the interval calculates at 29.7 percent. Revision in age assignments for some floras now "make for an even nicer pattern" (Laurel Collins, pers. comm. to Robyn Burnham, 1994). When evidence from terrestrial faunas is added (Stehli and Webb 1985; Webb 2006), closure, to the extent that large mammals were able to move relatively unimpeded across essentially continuous land surfaces, had been completed by around 2.5 Ma. *Quercus* first appears in southern Central America in the late Miocene Gatún flora, which is also when a slight increase in drier elements (e.g., grasses) becomes evident in the region. This marks the time when the Cordillera Central of Panama reached sufficient heights to cast a slight rain shadow and provide a few scattered upland temperate habitats. *Alnus* crosses into the uplands of the Northern Andes around 1 Ma and *Quercus* about 330 kyr (Hooghiemstra and Ran 1994). Thus, in using "the land bridge," or any continental connection, as an explanation for distributions, it is necessary to estimate to the extent possible the ecological characteristics of the organisms involved and when or if these habitats likely existed in that landscape.

The broad outline of climate, geology, and biotic history is reasonably well established for the Isthmian region because there is such a diversity of information available to make the reconstructions. This evidence is, in part, a consequence of the political, economic, and strategic importance of the canal region where the underlying geology is of prime concern. The long-time presence of the Smithsonian Tropical Research Institute and its field station on Barro Colorado Island in Gatún Lake is another important source of information for the region. Nonetheless, not all aspects of the biotic history are fully resolved. For one thing, the fossil terrestrial fauna has a large component of browsers and grazers, suggesting a shrubland/chaparral-woodland-savanna and grassland vegetation, while the fossil floras suggest a more dense vegetation of tropical forest.

A similar anomaly is apparent between the floras and faunas in the Miocene floras of the American west, and Taggert and Cross (1990) suggested a factor important in the explanation. In areas of volcanic activity, the eruptions periodically reset the successional clock to barren ground for relatively brief periods of time in local areas, followed by open recovery vegetation and eventually the reestablishment of climax forest. In such a setting, the primary component of the fossil floras for most of the time is forest. However, volcanism allows for a shifting mosaic of open herbaceous and shrubby successional communities evident only with close, high resolution sampling. Thus, the mobile herds of vertebrate fauna leave a record strong in browsers and grazers, while the fossil floras preserve primarily deciduous forest. A similar mechanism has been suggested for the early to middle Miocene Cucaracha Formation of Panama, dated within 1.5 million years to about 16 Ma, and strata of comparable age from the Isthmian region (Graham 1988). In addition, paleosol evidence suggests locally dry habitats, with MAP from 573–916 mm (±147 mm) to 296–1142 mm, compared to estimates from pollen assemblages of 1200–1500 mm. Both would allow for forest in and adjacent to coastal and swampy lowlands, with contemporaneous drier vegetation scattered in the uplands (Retallack and Kirby 2007).

Another anomaly concerns the relative number of plant versus animal species moving north and south across the land bridge, and their record of diversification once they reached the other side (Burnham and Graham 1999). Many mammals moved from north to south where they radiated and diversified extensively. There are over 200 genera and 400 species of mice and sigmodontine rodents of North American origin in South America, where they represent roughly 25 percent of the mammalian species (see summary in Flannery 2001). There are more kinds of canines in South America than on any other continent, and it has more than half of the world's camelids. In total, about 53 percent of the modern-day South American mammalian fauna is derived from North America, and they, or their ancestors, mostly appeared around the time of completion of the Panama land bridge. Moving southward were raccoons, bats, shrews, hares, pocket mice, pocket gophers, squirrels, field mice, snapping turtles, cats, saber-toothed tiger, puma, panther, weasels, skinks, otters, fox, wolf, bears, elephantids, tapirs, horses, peckeries, Camelidae, and deer. Moving to the north was a more modest cadre of armadillos, sloths, porcupines, capybara, and opossums, and after arriving in North America they underwent limited diversification. Only around 10 percent of North American mammals have South American ancestors.

The pattern for plants is just the reverse. Relatively few trees and shrubs moved from the north into South America. They include *Alnus*, *Populus*, *Quercus*, *Salix*, and *Juglans*, and they underwent comparatively little diversification. In contrast, those moving north diversified extensively and are prominent members of the tropical and subtropical forests. As noted earlier, about 75 percent of the canopy trees of the Mexican lowland forests are derived from South America (Wendt 1993). The reasons for the differences are not known, but some components of the explanation are clear. The plants moving from north to south were mostly temperate ones migrating in response to cooling temperatures and the gradual emergence of limited and scattered highlands. Upon reaching South America, these few temperate taxa encountered extensive lowland tropical basin habitats with temperate environments restricted along the still emerging Northern Andes Mountains. The many more tropical taxa moving from south to north encountered a relatively expansive lowland tropical environment increasing in extent northward because of the funnel-shaped configuration of the continent (Flannery 2001). In other words, the pool of temperate plant taxa in the north diminished toward Central America, while the pool of tropical plant taxa in the adjacent lowlands of South America was extensive.

Regarding the faunal pattern, by the Pliocene the diversity of North American land mammals exceeded that of South America by about 60 percent, and the sheer number favored movement of more kinds of mammals from north to south.

ACROSS THE ATLANTIC

An increasing number of plants in Latin America with African affinities are being discovered as more inventories are completed, and as morphological and molecular studies are better revealing relationships. Many were derived early by vicariance as South America split away from Africa; others arrived later by migration over the northern land bridges; and some came throughout the interval by long-distance dispersal. Dispersal became progressively rare as the continents moved farther apart, the North Atlantic land bridge floundered, and climates along the route became colder; in other words, different kinds of organisms were accommodated at different times in diminishing numbers during the bridge's history. But it is worthwhile to speculate on another opportunity some organisms may have had to disperse between Africa and the New World.

To better understand global weather and climate, atmospheric circulation is monitored, including data on the spread of radiation from nuclear tests and power plant disasters, and on the distribution of airborne pathogens by the trade winds blowing to and from Africa. As a result of this monitoring, we know that vast amounts of dust and other material are brought from the Sahara and Sahel regions and deposited in the Caribbean and adjacent regions of Latin America. This is not a new discovery. Darwin, for example, complained the dust was dirtying everything on the Beagle (Cadée 1998). Today, however, we can quantify the amounts to some 40 million tons each year of dust, bacteria, diatoms, phytoliths, fungal spores, pieces of lichen, and fragments of undetermined affinities (Koren et al. 2006). The likelihood that reproductive parts of other plants are being transported under present conditions is small, but as noted before, conditions change through intervals measured in geologic time.

Another observation from the modern era is the increasing frequency and intensity of hurricanes in the Atlantic Ocean (fig. 9.4), and the suspected cause is global warming (Knutson et al. 1998; Goldenberg et al. 2001; Hoyos et al. 2006; Kerr 2006; but see Nyberg et al. 2007). The change has resulted from a rise in temperature of only 0.6°C in the last century, and 0.2°C–0.3°C in the past forty years. Another relevant estimate is that between 1996 and 2005, a sea surface warming of about 0.5°C was re-

Figure 9.4 Hurricane Katrina. From the NASA Earth Observatory site, http://earthobservatory
.nasa.gov/.

sponsible for a roughly 40 percent increase in hurricane activity (Saunders
and Lea 2008). This means that similar increases likely occurred during
intervals of the geologic past that were even warmer than at present. These
include the LPTM/EECL (60–45 Ma, 10°C–12°C warmer), a peak in the
early to middle Miocene (18–14 Ma, 5°C–6°C warmer), and in the middle
Pliocene (circa 3.5 Ma, 3.5°C warmer).

Another observation is that in the late Paleocene/early Eocene, the
distance between Africa and South America was half to a third that of
the present. Moreover, it was in the Eocene that the Antilles arc collided
with the Bahamas Platform, causing emergence of the islands and providing intermediate target areas between the continents. Assuming these altered conditions increased the likelihood that some plants may have dispersed over the shorter distances during the warmer intervals of the past,
it would be interesting to compile data on the estimated time of arrival for

various taxa. Such information is meager and has not been fully analyzed with respect to the modes and dispersal potential of the propagules, the availability of climatically suitable habitats, or the questionable reliability of some molecularly based divergence times. But for the moment, we can note that in the Malpighiaceae six disjunctions occurred between 60 and 31 Ma in the middle Paleocene to early Oligocene and at 17 Ma in the early Miocene (Davis et al. 2004). There were also disjunctions in *Symphonia* at 17.36 Ma, plus or minus 1.53 million years (Dick et al. 2003); in *Senecio* at 4–3 Ma (Coleman et al. 2003); in *Lupinus* at 4–3 Ma, as well as at 8 Ma (Käss and Wink 1997); and in *Hypochaeris* no later than 3.5 Ma (Tremetsberger et al. 2005, 2006). Future studies will demonstrate if the correlations are just coincidence. But, if they are not, they suggest one more way some plants with small wind-dispersed propagules may have occasionally moved more readily between Africa and South America in the Tertiary.

REFUGIA

The refugium concept was proposed by Jürgen Haffer (1969, 1970, 1974, 1982). It adds to the several factors operating to varying degrees, sometimes simultaneously, that enhanced biodiversity in the Amazon lowlands. As an amateur ornithologist, in the finest sense of that word, Haffer was aware that the number of bird species was not uniformly distributed across the Amazon Basin. There were pockets of high species concentration, and there were intervening areas where species diversity was less. As a professional geologist, he was also aware that the species-rich areas often corresponded with topographic and geologic features that would have perpetuated moist conditions during periods of dryness. These included sites at the confluency of major rivers, the base of slopes, in shallow basins, deep river valleys, gallery forests, and areas underlain by hardpan or caliche. He suggested that in the drier intervals of the Quaternary, caatingas, cerrado, savanna, and/or grasslands in and around the basin would have expanded into the lowlands, and elements of the rain forest would have become isolated into moist refugia. With the return of moist conditions, the rain forest expanded and coalesced into the extensive plant formation that now covers most of the basin, while the drier vegetation returned to the surrounding slopes and to edaphic enclaves within the forest (e.g., sandstone outcrops and hill tops). An interesting analysis, not yet completed, will be to determine if there are dry elements within the moist vegetation still retracting toward favorable dry habitats.

The repeated isolation, like the rise of arches and changes in river drainage patterns, followed by reunion of the populations, would have been conducive to the generation of new genotypes and phenotypes. Recall there were at least eighteen to twenty major climatic cycles during the Pleistocene (Milankovitch variations), with numerous subcyles (the Younger Dryas, the Medieval Warm Period, the Little Ice Age, Heinrich and D-O events). Also, during each of the major cold intervals, for example, at the LGM, sea levels were lowered by an estimated 120 m and water tables by up to 40 m, contributing to drying in the lowlands. If dry or drying intervals prevailed during about 90 percent of the Quaternary, the current habitats supporting cerrado, caatingas, and grasslands can be viewed as "reversed refugia" (Gentry 1979, 349) because for most of the interval they (or versions of them) were the widespread ecosystem in and around the basin. However it is viewed, the climatic history of Amazonia is proving to be a dynamic one.

After Haffer's proposal, a number of organisms were examined for consistency with the theory (Prance 1982), and the consensus was that in general it worked. Previously, conventional wisdom held that climatically the lowland tropics were stable and unchanging, and the CLIMAP and COHMAP results seemed to support this idea. Many of the major climatic cycles and subcycles, like Heinrich events and the significant temperature fluctuations at and after the LGM, were still undocumented for equatorial regions. But the idea of a stable and unchanging environment was increasingly difficult to reconcile with evidence for possibly the greatest concentration of species in the history of the planet, along with the fact that using any reasonable lapse rate, as van der Hammen observed, temperature changes in the high Northern Andes must have extended to some degree into the lowlands. A number of objections have been raised to the Haffer's original theory (Colinvaux 2007; Colinvaux et al. 1999; Morawetz and Raedig 2007), including the improbable existence and location of some of the refugia initially proposed, and the simplified causal mechanisms based on the extent and pace of climatic changes recognized at the time. However, there has also been some overstatement by opponents about the degree of dryness required (e.g., "extreme aridity"), when all that is needed or suggested for parceling is an intervening vegetation drier than lowland rain forest.

Moreover, as discussed in chapter 8 for South America, compelling evidence began to emerge supporting significant cooling in the lowlands by 5°C–6°C (Aeschlbach-Hertig et al. 2000; Webb et al. 1997; Stute et al. 1995) and for drying (Lichte and Behling 1999; Mayle and Beerling 2004; Muhs and Zárate 2001; Wüster et al. 2005a, b). Anhuf and colleagues (2006) suggest that humid forests in the Amazon Basin may have been reduced to around 54 percent of their current extent at the LGM. Maslin and Burns argue that "the oxygen isotopic composition of planktonic foraminifera recovered from a marine sediment core in a region of Amazon discharge shows that the Amazon Basin was extremely dry during the Younger Dryas [a relatively minor cold interval compared to the several glacial maxima] with the discharge reduced by at least 40% compared to that of today" (2000, 2285). The argument has shifted over the years from the contention that (1) there was no major climatic change in the Amazon Basin, to (2) it was colder but not drier, to (3) it was colder and drier but not enough to disrupt a virtually continuous rain forest over the basin throughout the Pleistocene, to (4) it was cold and dry and the lowland rain forest was significantly disrupted. As the scaffolding around objections to the refugium theory weakens, a tipping point is approaching, or has been reached, whereby some updated form of the theory seems plausible as one mechanism for generating biodiversity in the lowland neotropical rain forest.

As an example of the more intricate causal mechanisms being proposed for climate changes in the Amazon Basin, Hooghiemstra and van der Hammen (1998) note that the precession cycle shifts the boundaries of the Intertropical Convergence Zone. This means that the northern and southern edges fluctuate between moist and dry conditions with the meandering of the system, while the central part remains more consistently moist. The migration of the ITCZ during the Holocene and associated dryness in the Amazon Basin and surrounding areas, for example, the Patí Valley of Colombi; (González-Carranza et al. 2008), is further demonstrated from changes in titanium and iron concentrations in the Cariaco Basin offshore from Venezuela (Haug et al. 2001). The record can be measured with a precision of decades or less. The concentrations were reduced during the Younger Dryas (dryness), increased during the thermal maximum, and reduced during the Little Ice Age.

The palynological evidence for and against the theory has been reviewed by Hooghiemstra and van der Hammen (1998), van der Hammen and Hooghiemstra (2000), and Behling and Hooghiemstra (1998). I have offered my own summaries as well (II, chap. 8; Graham 2010; Graham et al., work in progress). Examples of vegetation changes in a presumed interrefugium area (Carajás, Brazil) and within a refugium (Pata, Brazil) are shown in figures 9.5 and 9.6). Carajás is in an area presently of moist forest, but the diagram shows fluctuations between arboreal pollen (AP) and nonarboreal pollen (NAP). Pata, in a suggested refugium zone, shows relatively little fluctuation in AP and NAP pollen. The concept is still being

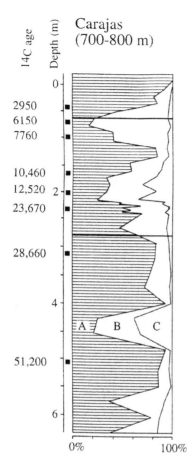

Figure 9.5 Pollen records from the lowland forest of Carajás, Brazil, for trees (A), grasses (B), and Asteraceae/Compositae, Borreria, and Cuphea (C). Note the fluctuations between arboreal pollen (AP) and nonarboreal pollen (NAP). From Hooghiemstra and van der Hammen 1998. Used with permission from Elsevier Science Publishers, Amsterdam.

discussed and refined, but in my opinion it remains a viable theory requiring continuing revision.

POINTS IN TIME

I have suggested that although geologic, climatic, and biologic histories are a continuum, there are times when events and processes combine to favor one kind of lineage or ecosystem over another. After the origin of the angiosperms in the Early Cretaceous circa 135 Ma (appearance of monosulcate, columellate dicot or monocot pollen), and the eudicots around 125 Ma (tri-

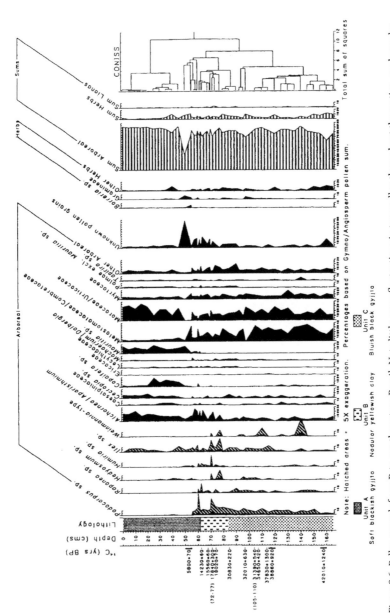

Figure 9.6 Pollen records from Pata, northwestern Brazil. Note little or no fluctuations in tree pollen throughout the section, as shown by the Sum Arboreal Pollen profile, compared with the Carajás record (fig. 9.5). From Colinvaux and others 1999. Used with permission from Cengage Learning Services, Ltd., Andover, UK.

colpate, columellate pollen), flowering plants diversified and radiated until by the end of the Cretaceous some lineages were similar morphologically and ecologically to modern families and genera, and some communities resembled or were shadowy versions of modern ecosystems. Several events then followed that elicited evolutionary/ecological responses that constituted early phases in the modernization of particular taxonomic groups and assemblages—their golden moment in geologic time. The first of these within the time frame considered here was at the end of the Cretaceous with the asteroid impact. For many plants, the dark and cooling three-year winter night probably provided a further nudge toward deciduousness that began as selection for unpredictable habitats and polar light regimes. This put deciduous organisms in good stead to cope later with temperate climates as temperatures fell and dryness increased in middle Eocene and later times.

The second event occurred around 55 Ma, and it provided impetus for the global distribution, essentially from pole-to-pole, and early diversification of a tropical vegetation that had been forming in the lower latitudes during the Cretaceous and early Paleocene since at least 58 Ma. The emission of 1500 gigatons of methane and CO_2 from openings in the sea floor between Greenland and Europe during separation of the plates raised MATs by 5°C–6°C over an already warm world. The result was essentially a global expanse of rain forest and rain forest elements. As the rain forest expanded north and south from the equatorial latitudes, it encountered somewhat different climates, moderately different physical landscapes, soils, and competitors, and very different light regimens. It likely incorporated such Laurasian elements as the Bombacaceae and Melastomataceae, which later became prominent tropical Gondwana families (II, chap. 8). It is further likely that the equatorial tropical ecosystems themselves were undergoing change. Rather than existing under a MAP of 24°C–28°C and maintaining a rather a stable biota, late Paleocene and early Eocene temperatures may have averaged 34°C–36°C, with a number of lineages becoming extinct or moving out. The composition of ecosystems was changing because the environment may have been too warm for some component elements. For many early-to-intermediate stem groups of tropical plants, this was a major evolutionary period, and it is reflected in the change from mostly ancient to mostly modern genera between the early and late Eocene.

The next event was a decline in global MATs into the transition time of the late Eocene, Oligocene, and early Miocene. During this interval, adjustments were taking place in the composition and distribution of ecosystems. This period has not received due recognition, but it was nonetheless important. If the temperature decline from the LPTM/EECL had continued unabated into the glacial age, the extinctions and radiations, although more gradual than those resulting from catastrophes, would likely have produced a very different set of ecosystems within which we and other organisms later evolved, from which we radiated, and on which we currently depend.

Next came a resumption in the downward shift in temperatures beginning in the middle Miocene—a trend toward cooler temperatures, less evaporation of water from the ocean surface, drying climates, greater seasonality, cooler marine currents, and the eventual appearance of local desert conditions correlated with increasing continentality and the rise of mountain systems. This was an important time in the evolution of lineages and ecosystems because it was when deciduous forests became widespread across the temperate latitudes of the Earth, and when individual elements were coalescing into early versions of shrubland/chaparral-woodland-savanna, grassland, and deserts, including the cerrado, caatingas, pampas, monte, and dry tropical forests of South America. Lineages growing in seasonally dry climates—sometimes on coarser, drying, acidic soils, with reinforcing slope, exposure, and rain shadow—experienced their greatest period of diversification and radiation, favorable to the appearance of many drier-habitat crown groups. Better-defined deserts, savannas with fewer trees to constitute true grasslands, tropical dry forests, pine-oak forests, and the beginnings of recognizable and more widespread tundra, steppe, and páramo developed in and after the middle Miocene around 15 Ma.

These refinements in ecosystems continued into the Quaternary. That epoch was once thought to extend back only to about 1 Ma and was generally considered too short an interval for the generation of many modern species. However, it has now been extended back to 2.6 Ma, providing ample generation time for trees, shrubs, and especially annuals. Furthermore, the Milankovitch and finer variations pumped considerable climatic variability into the system, causing extinctions, disjunctions, migrations, and widespread contact between previously allopatric species in both the high and low latitudes. As a result, the pulse of the redefined Quaternary can now be viewed as a significant generator of new species.

In sum, much of the existing biodiversity of the Earth is a product of events and processes at specific intervals when environmental events combined with evolutionary processes to increase the tempo of evolution in particular clades. These events and processes include (1) some probable enhancement of deciduousness as a result of the asteroid impact at the end of the Cretaceous; (2) a heyday for early and intermediate stem groups of tropical organisms and communities around the late Paleocene / early Eocene when tropical climates were of nearly global extent; (3) an important period of adjustment in the late Eocene, Oligocene, and early Miocene afforded by greater environmental stability than in the periods immediately before and after; (4) a golden age in the middle Tertiary for organisms and assemblages capable of adjusting to seasonal temperature and precipitation regimes; and (5) the turbulent climatic times at the beginning of the Quaternary period around 2.6 Ma, when crown groups of older tropical lineages were reinvigorated by evolution, and more recent seasonally dry-habitat groups underwent their initial period of rapid speciation and radiation.

THE FUTURE OF CURRENT ECOSYSTEMS

As warming environmental trends continue, it is worth speculating on the probable ecosystems of the future. In years to come, "nature" will surely mean mostly parks, preserves, botanic gardens, and conservatories. Environmentally, wildfires and extreme weather will increase, and biotically crop lands, range lands, and managed forests will be more extensive. Remnants of former ecosystems will exist mostly in inaccessible places. There will be an increase in competitive, aggressive, resistant, furtive organisms comprising weedy and highly disturbed ecosystems; reduced tundra and little alpine tundra / páramo; a greater array and a wider distribution of rapidly evolving organisms with warm-temperate to subtropical ecology (e.g., Gillman et al. 2009; see summary by Gill 2009); microorganisms and pathogens with greater resistance to antibiotics diversifying and spreading widely with continued global warming; a diminishing number of wild food and drug plants for replenishing the genetic stock; and an increasing number of endangered species protected by laws that change with the political climate until over time there will likely be few left to protect.

The current version of the Earth's living envelope began assembling around 100 million years ago in the Late Mesozoic with eight comparatively simple ecosystems. There has been ample opportunity to observe the effects of environmental change through a window into the past provided by an extensive fossil record and an impressive array of innovative technologies. In view of what was accomplished by evolutionary and geologic processes from the Late Mesozoic through the early part of the Anthropogene, it seems we should be demanding far more vigorously a better legacy for future generations.

: : **REFERENCES** : :

Aeschbach-Hertig, W., F. Peeters, U. Beyerle, and R. Kipfer. 2000. Palaeotemperature reconstruction from noble gases in ground water taking into account equilibration with entrapped air. *Nature* 405:1040–44.

Anhuf, D., et al. (twelve coauthors). 2006. Paleo-environmental change in Amazonian and African rainforest during the LGM. *Palaeogeogr. Palaeocl. Palaeoecol.* 239:510–27.

Axelrod, D. I. 1975. Evolution and biogeography of the Madrean-Tethyan sclerophyll vegetation. *Ann. Mo. Bot. Gard.* 62:280–34.

Behling, H., and H. Hooghiemstra. 1998. Holocene history of the chocó rain forest from Laguna Piusbi, southern Pacific lowlands of Colombia. *Quat. Res.* 50:300–308.

Burnham, R. J., and A. Graham. 1999. The history of neotropical vegetation: New developments and status. *Ann. Mo. Bot. Gard* 86:546–89.

Cadée, G. C. 1998. Darwin on dust at sea. *Pages* 6:16.

Coleman, M., A. Liston, J. W. Kadereit, and R. J. Abbott. 2003. Repeat intercontinental dispersal and Pleistocene speciation in disjunct Mediterranean and desert *Senecio* (Asteraceae). *Am. J. Bot.* 90:1446–54.

Colinvaux, P. A. 2007. *Amazon expeditions: My quest for the ice-age Equator.* Yale University Press, New Haven.

Colinvaux, P. A., P. E. De Oliveira, and J. E. Moreno Patiño. 1999. *Amazon pollen manual and atlas.* Harwood Academic Publishers, OPA, Amsterdam.

Davis, C. C., P. W. Fritsch, C. D. Bell, and S. Mathews. 2004. High-latitude Tertiary migrations of an exclusively tropical clade: Evidence from Malpighiaceae. *Int. J. Plant Sci.* 165 (suppl.): S107–S121.

Dick, C. W., K. Abdul-Salim, and E. Bermingham. 2003. Molecular systematic analysis reveals cryptic Tertiary diversification of a widespread tropical rain forest tree. *Am. Nat.* 162:691–703.

Flannery, T. 2001. *The eternal frontier: An ecological history of North America and its peoples.* Atlantic Monthly Press, New York.

Gentry, A. H. 1979. Distribution patterns of neotropical Bignoniaceae: Some phytogeographic implications. In *Tropical botany,* ed. K. Larsen and L. B. Holm-Nielsen, 339–54. Academic Press, New York.

Gill, V. 2009. Evolution faster when it's warmer. BBC News, Science and Environment, 24 June 2009, http://news.bbc.co.uk/2/hi/8115464.stm. 25 June 2009. ["Climate could have a direct effect on the speed of "'molecular evolution'" in mammals." See also Gillman et al. 2009.]

Gillman, L. N., D. J. Keeling, H. A. Ross, and S. D. Wright. 2009. Latitude, elevation and the tempo of molecular evolution in mammals. *Proc. R. Soc. Lond. B,* 25 June 2009, http://rspb.royalsocietypublishing.org/content/early/2009/06/24/rspb.2009.0674.full.

Goldenberg, S. B., C. W. Landsea, A. M. Mestas-Nuñez, and W. M. Gray. 2001. The recent increase in Atlantic hurricane activity: Causes and implications. *Science* 293:474–79.

González-Carranza, Z., J. C. Berrío, H. Hooghiemstra, J. F. Duivenvoorden, and H. Behling. 2008. Changes of seasonally dry forest in the Colombian Patía Valley during the early and middle Holocene and the development of a dry climatic record for the northernmost Andes. *Rev. Palaeobot. Palynol.* 152:1–10.

Graham, A. 1966. Plantae rariores camschatcenses: A translation of the dissertation of Jonas P. Halenius, 1750. *Brittonia* 18:131–39.

———, ed. 1972a. *Floristics and paleofloristics of Asia and eastern North America.* Elsevier Science Publishers, Amsterdam.

———. 1972b. Outline of the origin and historical recognition of floristic affinities between Asia and eastern North America. In *Floristics and Paleofloristics of Asia and eastern North America*, ed. A. Graham, 1–18. Elsevier Science Publishers, Amsterdam.

———. 1973a. History of the arborescent temperate element in the northern Latin American biota. In *Vegetation and vegetational history of northern Latin America*, ed. A. Graham, 301–14. Elsevier Science Publishers, Amsterdam.

———, ed. 1973b. *Vegetation and vegetational history of northern Latin America.* Elsevier Science Publishers, Amsterdam.

———. 1988. Studies in neotropical paleobotany. Part 6, The lower Miocene communities of Panama—the Cucaracha Formation. *Ann. Mo. Bot. Gard.* 75:1467–79.

———. 1992. Utilization of the Isthmian landbridge during the Cenozoic—paleobotanical evidence for timing and the selective influence of altitudes and climate. *Rev. Palaeobot. Palynol.* 72:119–28.

———. 2010. *Late Cretaceous and Cenozoic history of Latin American vegetation and terrestrial environments.* Missouri Botanical Press, St. Louis.

Graham, A., T. Distler, and I. Jiménez. In prep. Historical factors in explanations for neotropical biodiversity—then (refugia) and now (prediction models).

Gray, A. 1840. Dr. Siebold, Flora Japonica. *Am. J. Sci. Arts* 39:175–76.

———. 1846. Analogy between the flora of Japan and that of the United States. *Am. J. Sci. Arts*, 2nd ser., 2:135–36.

Haffer, J. 1969. Speciation in Amazonian forest birds. *Science* 165:131–137.

———. 1970. Geologic, climatic history and zoogeographic significance of the Urabá region in northwestern Colombia. *Caldasia* 10:603–36.

———. 1974. *Avian speciation in tropical South America.* Publications of the Nuttall Ornithological Club, no. 14. Cambridge, Mass.

———. 1982. General aspects of the refuge theory. In *Biological diversification in the Tropics*, ed. G. T. Prance, 6–24. Columbia University Press, New York.

Halenius, J. P. 1750. *Plantae rariores camschatcenses.* [Uppsala, Sweden.]

Haug, G. H., K. A. Hughen, D. M. Sigman, L. C. Peterson, and U. Röhl. 2001. Southward migration of the intertropical convergence zone through the Holocene. *Science* 293:1304–8.

Heinicke, M. P., W. E. Duellman, and S. B. Hedges. 2007. Major Caribbean and Central American frog faunas originated by ancient oceanic dispersal. *Proc. Natl. Acad. Sci. U.S.A.* 104:10092–97.

Hooghiemstra, H., and E. T. H. Ran. 1994. Late Pliocene-Pleistocene high resolution pollen sequence of Colombia: An overview of climatic change. *Quat. Int.* 21:63–80.

Hooghiemstra, H., and T. van der Hammen. 1998. Neogene and Quaternary development of the neotropical rain forest: The forest refugia hypothesis, and a literature overview. *Earth-Sci. Rev.* 44:147–83.

Hoyos, C. D., P. A. Agudelo, P. J. Webster, and J. A. Curry. 2006. Deconvolution of the factors contributing to the increase in global hurricane intensity. *Science* 312:94–97.

Ickert-Bond, S. M., K. B. Pigg, and J. Wen. 2005. Comparative infructescence morphology in
    Liquidambar (Altingiaceae) and its evolutionary significance. *Am. J. Bot.* 92:1234–55.
Käss, E., and M. Wink. 1997. Molecular phylogeny and phylogeography of *Lupinus* (Legu-
    minosae) inferred from nucleotide sequences of the *rbcL* gene and ITS 1 + 2 regions of
    rDNA. *Plant Syst. Evol.* 208:139–67.
Kerr, R. A. 2006. Global warming may be homing in on Atlantic hurricanes. *Science*
    314:910–11.
Knutson, T. R., R. E. Tuleya, and Y. Kurihara. 1998. Simulated increase of hurricane intensi-
    ties in a CO_2-warmed climate. *Science* 279:1018–21.
Koren, I., et al. (six coauthors). 2006. The Bodélé Depression: A single spot in the Sahara that
    provides most of the mineral dust to the Amazon forest. *Environ. Res. Lett.* 1:1–5.
Lessios, H. A. 2008. The great American schism: Divergence of marine organisms after the
    rise of the Central American Isthmus. *Ann. Rev. Ecol. Evol. Syst.* 39:63–91.
Li, H.-L. 1952. Floristic relationships between eastern Asia and eastern North America. *Trans.
    Am. Phil. Soc.* 42:371–29.
Lichte, M., and H. Behling. 1999. Dry and cold climatic conditions in the formation of the
    present landscape in southeastern Brazil. *Z. Geomorph. N.F.* 43:341–58.
Martin, P. S., and B. E. Harrell. 1957. The Pleistocene history of temperate biotas in Mexico
    and eastern United States. *Ecology* 38:468–80.
Maslin, M. A., and S. J. Burns. 2000. Reconstruction of the Amazon Basin effective moisture
    availability over the past 14,000 years. *Science* 290:2285–87.
Mayle, F. E., and D. J. Beerling. 2004. Late Quaternary changes in Amazonian ecosystems and
    their implications for global carbon cycling. *Palaeogeogr. Palaeocl. Palaeoecol.* 214:11–25.
Morawetz, W., and C. Raedig. 2007. Angiosperm biodiversity, endemism and conservation in
    the neotropics. *Taxon* 56:1245–54.
Morris, A. B., et al. (five coauthors). 2007. Phylogeny and divergence time estimation in
    Ilicium with implications for New World biogeography. *Syst. Bot.* 32:236–49.
Muhs, D. R., and M. Zárate. 2001. Late Quaternary eolian records of the Americas and their
    paleoclimatic significance. In *Interhemispheric climate linkages*, ed. V. Markgraf, 183–216.
    Academic Press, New York.
Nyberg, J., et al. (five coauthors). 2007. Low Atlantic hurricane activity in the 1970s and
    1980s compared to the past 270 years. *Nature* 447:98–701.
Prance, G. T., ed. 1982. *Biological diversification in the tropics.* Columbia University Press, New
    York.
Retallack, G. J. and M. X. Kirby. 2007. Middle Miocene global change and paleogeography of
    Panama. *Palaios* 22:667–79.
Saunders, M. A., and A. S. Lea. 2008. Large contribution of sea surface warming to recent
    increase in Atlantic hurricane activity. *Nature* 451:557–60.
Stehli, F. G., and S. D. Webb, eds. 1985. *The great American biotic interchange.* Plenum Press,
    New York.
Stute, M., et al. (seven coauthors). 1995. Cooling of tropical Brazil (5°C) during the Last
    Glacial Maximum. *Science* 269:379–83.
Taggart, R. E., and A. T. Cross. 1990. Plant successions and interruptions in Miocene volcanic
    deposits, Pacific Northwest. In *Volcanism and fossil biotas*, ed. M. G. Lockley and A. Rice,
    GSA Special Paper 244, 57–68. Geological Society of America, Boulder.
Tremetsberger, K., et al. (six coauthors). 2005. Nuclear ribosomal DNA and karyotypes indi-

cate a NW African origin of South American *Hypochaeris* (Asteraceae, Cichorieae). *Mol. Phylogenet. Evol.* 35:102–16.

Tremetsberger, K., et al. (eight coauthors). 2006. AFLP phylogeny of South American species of *Hypochaeris* (Asteraceae, Lactuceae). *Syst. Bot.* 31:610–26.

Van der Hammen, T., and H. Hooghiemstra. 2000. Neogene and Quaternary history of vegetation, climate, and plant diversity in Amazonia. *Quat. Sci. Rev.* 19:725–42.

Wang, Y.-H., et al. (eight coauthors). 2009. The phytogeography of the extinct angiosperm *Nordenskioldia* (Trochodendraceae) and its response to climate changes. *Palaeogeogr. Palaeocl. Palaeoecol.* 280:183–92.

Webb, R. S., D. H. Rind, S. J. Lehman, R. J. Healy, and D. Sigman. 1997. Influence of ocean heat transport on the climate of the Last Glacial Maximum. *Nature* 385:695–99.

Webb, S. D. 2006. The great American biotic interchange: Patterns and processes. In *Latin American biogeography—causes and effects*. Proceedings of the 51st Annual Systematics Symposium of the Missouri Botanical Garden (Alan Graham, organizer). *Ann. Mo. Bot. Gard.* 93:245–57.

Wen, J. 2001. Evolution of eastern Asian–eastern North American disjunctions: A few additional issues. *Int. J. Plant Sci.* 162 (suppl.): S117–S122.

Wendt, T. 1993. Composition, floristic affinities, and origins of the canopy tree flora of the Mexican Atlantic slope rain forests. In *Biological diversity of Mexico: Origins and distribution*, ed. T. P. Ramamoorthy, R. Bye, A. Lot, and J. Fa, 595–680. Oxford University Press, Oxford.

Woodring, W. P. 1966. The Panama land bridge as a sea barrier. *Proc. Amer. Phil. Soc.* 110:425–33.

Wüster, W., et al. (five coauthors). 2005a. No rattlesnakes in the rainforests: Reply to Gosling and Bush. *Mol. Ecol.* 14:3619–21.

———. 2005b. Tracing an invasion: Landbridges, refugia, and the phylogeography of the neotropical rattlesnake (Serpentes: Viperidae: *Crotalus durissus*). *Mol. Ecol.* 14:1095–1108.

Xiang, Q.-Y., D. E. Soltis, and P. S. Soltis. 1998. The eastern Asian and eastern and western North American floristic disjunction: Congruent phylogenetic patterns in seven diverse genera. *Mol. Phylogenet. Evol.* 10:178–90.

Ying, T.-S. 1983. The floristic relationships of the temperate forest regions of China and the United States. *Ann. Mo. Bot. Gard.* 70:597–604.

Additional Readings and Updates

Collins, E. S., D. B. Scott, and P. T. Gayes. 1999. Hurricane records on the South Carolina coast: Can they be detected in the sediment record? *Quat. Int.* 56:15–26.

Elsner, J. B. 2007. Climatology: Tempests in time. *Nature* 447:647–49.

Feeley, K. J., and M. R. Silman. 2009. Extinction risks of Amazonian plant species. *Proc. Natl. Acad. Sci. U.S.A* 106:12382–87, doi: 10.1073/pnas.0900698106. Published online, 14 July 2009.

Hoorn, C., and F. Wesselingh, eds. 2010. *Amazonia: Landscape and species evolution*. Wiley-Blackwell, West Sussex.

Jablonski, D., K. Roy, and J. W. Valentine. 2006. Out of the tropics: Evolutionary dynamics of the latitudinal diversity gradient. *Science* 314:102–6. ["The tropics are thus both a cradle and a museum of biodiversity."]

Mooney, C. 2007. *Storm world: Hurricanes, politics, and the battle over global warming.* Harcourt, New York. [See review by James Elsner, *Nature* 448 (2007): 648.]

Nagashima, K., et al. (five coauthors). 2007. Orbital- and millennial-scale variations in Asian dust transport path to the Japan Sea. *Palaeogeogr. Palaeocl. Palaeoecol.* 247:144–61.

Phillips, O. L., et al. (sixty-five coauthors). 2009. Drought sensitivity of the Amazon rainforest. *Science* 323:1344–47.

Renner, S. S., G. W. Grimm, G. M. Schneeweiss, T. F. Stuessy, and R. E. Ricklefs. 2008. Rooting and dating maples (*Acer*) with an uncorrelated-rates molecular clock: Implications for North American/Asian disjunctions. *Syst. Bot.* 57:795–808.

10

Pole to Pole

A Walk over the Landscape,
a Walk through Time

As a means of conceptualizing the greening of the New World from the Cretaceous through the Cenozoic, I envision an imaginary walk from pole to pole at four critical stages in the evolution of the ecosystems: at 100 Ma, near the time of low elevations, maximum Cretaceous warmth, and high sea levels; at 55 Ma during the peak of the even warmer Late Paleocene Thermal Maximum / Early Eocene Climatic Optimum (LPTM/EECL); at 17 Ma with the beginning of the Miocene temperature decline, decreasing atmospheric moisture, and a change toward enhanced seasonality; and at 2 Ma near the beginning of ice age climates and widespread glaciations. Four features of the ecosystems will be considered for each time segment—climate, landscape, fauna, and flora.

ONE HUNDRED MILLION YEARS AGO

If one were standing on the shore of the Arctic Ocean in the Middle Cretaceous, two aspects of the surroundings would be familiar.

The northern lights would still drape the sky with shimmering curtains of color, and there would be about four months of light, four months of darkness, and four months of intervening dimness like that between dusk and dawn.

Almost everything else would be different. For one thing, the climate would be practically balmy. The mean annual temperature would be 12C°–14°C warmer than today, in part because of an atmospheric CO_2 concentration ten to twelve times that of the present. Parrish and Spicer (1988) estimate the yearly maximum temperature in the mid-Cretaceous Arctic at 13°C and the minimum at 2°C–8°C (today the MAT is about 2°C along the south coast of Alaska at Anchorage, and −2.9°C in the east-central interior at Fairbanks). Occasionally, it would drop to freezing in the interior and at the highest of the moderate elevations during the dark winter months. By the time the asteroid hit on the Yucatán Peninsula at 65 Ma, temperatures had declined by about 4°C and winter sea ice was forming along the northern coast.

To begin our walk, a choice will have to be made between taking an eastern proto–Appalachian Trail or a western proto–Rocky Mountain Trail. An ocean extends through the center of the continent from the Gulf of Mexico to the Arctic Ocean and from the western slopes of the Appalachian Mountains to the eastern foothills of the Rockies. Along the shore there are depositional basins accumulating plant and animal remains, such as those of the dinosaurs roaming the swamps and the margins of the sea, and that later will provide evidence of the terrestrial life of North America in the Cretaceous. At other places, there are deep waters in which limestones are forming. These will make up deposits like the Glen Rose Formation just west of Austin, Texas, with its numerous calcium carbonate–encased foraminifera and extensive ammonites.

If one takes the eastern trail, it will be seen that Iceland and Greenland are devoid of ice caps. Ice is also scant or absent from Antarctica, and this accounts for the oceans being higher than today by about 80 m. The land is mostly continuous across the North Atlantic to Europe, where moderately high and rugged uplands include the Hercynian Mountains—whose remnants today are found from Portugal and Spain across western Europe north of the Alps—and the Caledonian Mountains from Scotland to Norway. The North American part of this system comprises the extent of the Appalachian Mountains from Canada to the southeastern United States. The trail terminates near southern Georgia, where the sea wraps around the southern end of the range and continues westward across the continental interior.

Along the western trail, compression forces generated later at the converging northern plate boundaries have not yet formed the spectacular mountain scenery in places like Denali National Park, Alaska. The landscape is low-lying, and much of it is damp and swampy. The Brooks Range inland from the northern coast, just being assembled from terranes transported from the west and south, rises above the sea to an elevation of about 1000 m. To the south, the Alaska Range is forming from an island arc that is colliding with the continental margin; at 74 Ma during the Campanian-Maestrichtian period it will be lifted to about sea level (Ridgeway et al. 2002). Another period of uplift will occur between 65 and 60 Ma, establishing moderately elevated highlands, while the intervening area between the Brooks and Alaska ranges will remain submerged or just above sea level.

Southward, along the Rocky Mountains, Cretaceous uplands extend intermittently to about central Colorado; then even lower hills continue to the northern proto–Sierra Madres in Mexico. Beyond about 20°N, just north of Mexico City, this trail ends. Substantial expanses of land will not be seen again until the Guiana and Brazilian shields bordering the swamps of the Amazon Basin.

The Cretaceous fauna in the far north includes dinosaurs. One of the most conspicuous is the hadrosaur, or duck-billed, *Edmontosaurus*, a nonmigrating, nonhibernating, presumably cold-blooded plant eater that lives in groups or small herds. At its maximum it can reach some three meters tall, twelve meters long, and weigh up to three tons. Other hadrosaurs include *Kritosaurus* and the crested *Lambeosauri*. There are plant-eating, horned (ceratopsian) dinosaurs like *Pachyrhinosaurus* and *Anchiceratops*, the thick, domed-skulled *Pachycephalosaurus*, and the small and fast *Thescelosaurus*, which runs on two legs. Meat-eaters are *Albertosaurus*, *Tyrannosaurus*, *Troodon*, *Dromaeosaurus*, and *Sauronitholestes*. These twelve dinosaurs, together with others living elsewhere in Alaska and preserved in deposits around the present-day Colville River region, reflect the warm moist climate (Clemens and Nelms 1993; Rich et al 2002) and the lush and diverse vegetation necessary to support them. Warm conditions must have prevailed in the Arctic throughout most of the year, because certainly the smaller, flesh-eating *Troodon* and *Dromaeosaurus* and their young could hardly have migrated far enough seasonally to reach significantly different light regimes, climate, or vegetation.

The plant formations are different from modern ones, because most of the individual lineages, associations, and their ecological requirements have not differentiated into familiar present-day forms. For example, deciduous gymnosperms of the families Taxodiaceae (today including *Glyptostrobus*,

Metasequoia, Sequoia, Taxodium), Cupressaceae (*Thuja*, arborvitae), and
Taxaceae (*Taxus*, yew) are only generally similar in morphology to their
modern-day representatives. In the pollen record, this complex of families
is identified as t-c-t, and the macrofossils (cones and twigs) include such
extinct genera as *Drumhellera*. At the generic level, some present-day Asian
trees like *Glyptostrobus*, similar in ecology to the American *Taxodium* (bald
cypress) that grows in swampy habitats often in standing water, and *Ginkgo*,
growing in more mesic but still moist habitats, are often present. They are
preserved in floras such as the Eureka Sound Group on Ellesmere Island
of the Northwest Territories, and on Axel Heiberg Island in the Canadian
High Arctic. The plant formation is the polar broad-leaved deciduous forest, consisting of angiosperms such as platanoids (sycamore-like), *Populus*
(poplar-like), and extinct members of the family Trochodendraceae (trees
or tall shrubs to 20 m, now represented by one species in southern Japan).
These are mixed with the deciduous gymnosperms noted above, and with
an understory of ferns, *Lycopodium* (ground "pines"), and horsetails (*Equisetum*). Plants of warmer habitats occur mostly along the coast, and those
of cooler habitats grow in the interior and in the uplands. Together they
constitute a generalized "boreotropical" flora from which the tropical part
will eventually disappear after the early Eocene, and from which part of
the temperate deciduous forest will later be derived. The polar broad-leaved
deciduous forest extends southward from the shores of the Arctic Ocean
from about 70°N to 60°N–50°N, where there is a transition to more mesothermal, evergreen vegetation.

Temperatures toward the south in the zone between about 50°N to
40°N are warmer than in the north during the Cretaceous, with an MAT
of 15°C–20°C. This is the region of the northern Rocky Mountains that at
100 Ma ranges from shallowly submerged to low-lying hills. Eastward of
the slopes, the land is mostly inundated, while to the west it descends into a
mosaic of down-faulted basins, uplifted ranges, and eastward-moving land
fragments being fused into the geologically complex landscape of western
North America. Dinosaurs are abundant, and their remains are preserved
at localities around present-day Edmonton and Calgary in Alberta, Canada,
and at the Utah-Colorado border in an area later to be designated as Dinosaur National Monument. Other animals in the Lancian (Late Cretaceous)
North American Land Mammal Age are the multituberculates—the name
describes the multicusped teeth of rodentlike animals around the size of
squirrels that will be extinct before the end of the Eocene—marsupials belonging to extinct lineages, and other marsupials distantly related to modern
opossum, (all small, no larger than a rat, and insectivorous). They do not

have to contend with the same seasonal severity in light and temperature as their northern counterparts; it is warmer, and there is abundant vegetation. The plant community is a notophyllous broad-leaved evergreen forest that includes some lianas, a few plants with drip tips, and *Guarea*, with its distinctive figlike fruits.

Tropicality increases southward from about 40°N, where the fauna includes turtle, crocodiles, and champsosaurs, which resemble small crocodiles, although the relationship is obscure. The Cretaceous vegetation here is a paratropical rain forest toward the west and a tropical forest toward the east. At the southern limit of this zone, the MAT is around 23°C. The transition between notophyllous broad-leaved evergreen forest and paratropical rain forest is marked by a change in faunas, with more eutherians (placental mammals) occurring to the north. As noted, these and other fossil faunas are described in Michael Woodburne's *Late Cretaceous and Cenozoic Mammals of North America* (2004) and Christine Janis and colleagues' two-volume *Evolution of Tertiary Mammals of North America* (1998, 2008).

In Mexico, waters cover an even greater proportion of the land than in northern North America. Proto–Baja California is a volcanic island arc located in the Pacific Ocean, and barely emergent uplands extend along the axis of the proto–Sierra Madres and the Tamaulipas Peninsula. The intervening area of central Mexico is covered by shallow tropical water supporting corals, as around Coahuila, and ammonites swim in the deeper water around Huasteca Canyon near Monterrey. The vegetation is sufficiently abundant that it accumulates in swamps and marshes faster than it can be removed by microbial decay, so sequences of coal are forming in the state of Coahuila. The Sierra Madres will rise especially fast between 100 and 45 Ma, so the epicontinental sea will drain from the interior; but when the asteroid hits at 65 Ma, the Yucatán Peninsula, the Isthmus of Tehuantepec, and the Balsas Depression will still be covered by shallow marine water. In southwestern Mexico, the Sierra Madre del Sur in the Cretaceous is located offshore in the Pacific Ocean as a volcanic island arc that will become part of the mainland in the Late Cretaceous and Paleogene.

Only a glimpse of the vegetation is provided by the few Late Cretaceous floras of Mexico. They include Aracauriaceae, Podocarpaceae, and Taxodiaceae gymnosperms, the widespread *Brachyphyllum*, a laurel (cf. *Persea*), and some angiosperm fossils similar to the Haloragidaceae, together with ferns, some of which (*Salvinia*) indicate the presence of a freshwater aquatic ecosystem, also found elsewhere in the New World during the Cretaceous. It is likely that an early version of the low- to midaltitude part of a lower to upper montane broad-leaved forest grew along the slopes of the mountains in the later part of the Cretaceous (Pinaceae, extinct members of the Betulaceae and Myricaceae), but without the prominent cool deciduous association found in the northern forests. There are mangrove habitats and versions of mangrove vegetation, but without *Rhizophora* (ferns of the *Acrostichum* type, possibly some palms), and beach/strand/dune vegetation along the coast (other palms, possibly *Ephedra*), but fossils from this community are few and the composition is not well known.

Similar conditions and vegetation are encountered over the even lower-lying, inundated, and fragmented lands of southern Central America all the way to northwestern South America. Between 10°N and 10°S, the approximate MAT at low to middle elevations ranges between at least 25°C and 28°C. In the Cretaceous, nappe structures (strata stacked like a slanting deck of cards) are forming along the northern coast of Venezuela as successive overriding rock layers are pushed against the mainland by compression between the South American and North American plates. There are uplands in the Guiana and Brazilian shields, while the volcanic Cordillera Occidental of the Northern Andes is being sutured onto the mainland by subduction and by westward movement of South America away from Africa. That same movement around 100–90 Ma is disrupting an expanse of land called West Gondwana that extends across western Africa and eastern South America. The interior is locally dry through continentality, and the dryness is augmented in places by coarse and sandy substrate. The vegetation includes *Ephedra*, as well as forerunners of the Welwitschiaceae whose descendents now grow only in the Namib Desert of South Africa. Moister areas are covered by a Cretaceous paratropical rain forest, which differs from a rain forest not because of the climate, or because lowlands are not available, but in composition and physiognomy—the angiosperms that form a characteristic component of the modern community are just diversifying and radiating. As the Cretaceous closes, iridium-rich aerosols will settle onto the landscape around Recife, for example, and a massive tsunami will deposit conglomerates along the Brazilian coast when the asteroid hits at Chicxulub. The angiosperms may have first appeared in West Gondwana, and some early ones probably became preadapted to deciduousness by the dryness and fluctuating rainfall. This would impart some survival advantage when dispersal from whatever place of origin brought them into contact with areas of dryness, seasonal temperatures and light regimes, and unpredictable moist habitats.

Along the western slopes of the proto–Central Andes, there is a large swamp or shallow inland sea with numerous dinosaurs that will leave one of the world's largest concentrations of footprints. Offshore from Bolivia, there is a line of volcanoes that later will accrete onto the mainland as the Cordillera Occidental. On the ascending slopes of the moderately elevated uplands, there are Araucariaceae, Podocarpaceae, and cycadlike gymnosperms, an understory of ferns, some early angiosperms of the magnoliid clade, and the beginnings of a lower to upper montane broad-leaved forest. In far southwestern South America, the Cretaceous landscape consists of low offshore volcanic island arcs slowly being added to the mainland as the proto–Southern Andes and coastal islands. To the east of the mountains, basins are forming as South America pulls apart from Africa, and these are accumulating early angiosperm plants such as *Nymphaea*-like water lilies and animal remains that will later constitute the extensive fossil floras and faunas of Patagonia.

At 100 Ma, South America has been separated from Africa for about 20 million years along its southern coast, but it is still connected or nearly so to the north with Africa, and to the far south through Antarctica with Australia. This defines the broad land mammal fauna of South America of the Cretaceous as Gondwanan, with its prominent Southern Hemisphere dinosaurs and other reptilian components. A titanosaur named *Argentinosaurus huinculensis* reaches a length of 40 meters and weighs 90 tons. Later, while South America drifts as an isolated island continent, endemic therian mammals (marsupials and placentals) evolve that define the mammal fauna as distinctly South American rather than broadly Gondwanan. Mammals will include giant anteaters, tree sloths, ground sloths, armadillos, opossums, porcupines, and a rhinoceros-like ungulate. In the Miocene deltaic deposits of the proto–Orinoco River at La Venta, Colombia, there are distinctive fossils of the giant (for a rodent) *Phoberomys pattersoni*, weighing some 1500 pounds, 3 meters in length, with a 1.5 meter tail, and the 40-foot-long crocodile *Purussaurus*; while in Argentina and Brazil live flightless predatory birds of the Phorosrachidae. The fossils also include lungfishes and the freshwater, long-necked turtles of the Chelidae. The transition between the Gondwanan and South American paleofaunas is especially well preserved in Patagonia.

The MAT is 23°C–25°C between 10°N and 10°S, 15°C between 10°S and 30°S, and 13°C south of 30°S. By the end of the Cretaceous, MAT will decrease by about 4°C. In southernmost South America, the vegetation in the Middle Cretaceous is mostly gymnosperms, with the angiosperms first appearing in the region sometime in the Aptian and Albian (120–110 Ma). *Nothofagus* is a temperate deciduous angiosperm tree or shrub growing here in the Maestrichtian. Later in the Paleogene, with the onset of warm climates, *Nothofagus* will disappear; then with cooler conditions after the early Eocene, it will return and form the now familiar Southern Hemisphere forests of southern beech.

Along the trek from pole to pole at 100 Ma, the average temperatures are from about 12°C–14°C in the Arctic, to 23°C–25°C across the equator, to 12°C–14°C in southern South America (a range of 13°C). At present, the Earth's surface temperatures range from −40°C to 40°C (a range of 80°C). Particularly relevant for the warm climates to come in the late Paleocene and early Eocene, and for the declines in the middle Eocene and in the middle Miocene, is that at 40°C air holds 470 times more water than at −40°C (Flannery 2005).

FIFTY-FIVE MILLION YEARS AGO

What a difference 45 million years can make. During the period between 100 and 55 Ma, the angiosperms replace the gymnosperms as the prominent component in most plant communities. The New World ecosystems, beginning with the equable conditions of the Middle Cretaceous, experience an overall cooling of 4°C, then begin what is initially a slow climb to warmer climates. There is drainage of the epicontinental seas, an asteroid lands in Mexico, and the dinosaurs disappear (with the exception of their avian descendents), which facilitates the rise to prominence and the persistence of the primates, although as small arboreal forms they may have evolved successfully anyway. In the earliest Paleocene, ungulates (hooved animals) appear. Approaching 55 Ma in the early Eocene, there is radiation of rodent groups, artiodactyls, and perissodactyls (equids, rhinos, and tapiroids); significant uplift of the western North American mountains; extensive floral and faunal exchange with Europe, especially of tropical elements in the early through middle Eocene; and the beginning of disruption of the North Atlantic land bridge in the middle to late Eocene. The Brooks Range, Alaska Range, and the Rocky Mountains reach half or more of their present elevation; a collage of terranes forms most of western North America; the rise of the Rocky Mountains continues the stretching and thinning of strata to the west, resulting in further differential uplift and collapse to form the Basin and Range Province; and the Sierra Nevada begin to appear.

At 55 Ma, the epicontinental sea is gone from most of interior North America, although there are still shallow waters across much of the Yucatán Peninsula and the Isthmus of Tehuantepec. Along the Atlantic Coastal Plain, the shoreline extends inland to the foothills of the Appalachian Mountains, around their southern end northward to about Cairo, Illinois, and then southward to just east of Austin to Del Rio, Texas, and into Mexico following the eastern flanks of the Sierra Madre Oriental. The Greater Antilles will emerge as the proto–Antillean island arc collides with the Bahamas Platform in the middle Eocene, while farther south the Isthmian region of Central America is still a seaway and will remain so for another 52 million years, until about 3.5 Ma.

At about 55 Ma, any subtleties in the ongoing temperature rise abruptly end. One of the more spectacular of climatic events, with considerable biological implication, is the rapid rise in temperatures to their highest levels in 100 million years. The overall increase will be 5C°–10°C to an MAT of about 20°C in the northern latitudes, and since much of this change occurs suddenly, something catastrophic must be happening. In fact, the Norwegian Sea has released 1650 to 3300 gigatons of carbon in the form of methane gas as Greenland separates from Europe creating venting craters 100 km across. Atmospheric CO_2 concentration increases from about 500 ppmv to 2000 ppmv. This augments more moderate input of CO_2 due to volcanism associated with other plate movements. Along with the increase in temperature, rainfall is increasing, and climates are trending from warm-temperate to decidedly tropical. With the increase in moisture and heat energy available to the climate system, there may be greater intensity, frequency, and widespread occurrence of storms and directional winds to and from Africa. There is extensive formation of peat and organic-rich swamp deposits that will be converted to Tertiary brown coals and lignites in the high latitudes, and red lateritic soils are forming as far as 50°N. The effects of the LPTM/EECL are greatest toward the poles, resulting in (1) a more gradual north-to-south zonation in climate and vegetation, (2) higher winter temperatures reducing annual seasonality, and (3) the greatest extent of tropical, subtropical, and warm-temperate ecosystems in the past 100 million years. The mild climates and near-continuous land across the North Atlantic at 55 Ma allow for extensive migrations and widespread distributions. *Platycarya*, today confined to Asia, and the ferns *Cnemidaria* and *Lygodium*, from the tropics, grow in the Arctic.

The dinosaurs are gone, and animals typical of warm climates, such as *Alligator* and an arboreal plagiomenid related to the Old World flying "lemurs" (in its own order), are present on Ellesmere Island. Remnants of the polar broad-leaved deciduous forest grow at places in the interior, along the northern coasts, and in the uplands. Elsewhere the vegetation is a notophyllous evergreen broad-leaved forest in which 65 percent of the leaves are entire-margined, and drip tips, lianas, Lauraceae, and palms are common. There is paratropical rain forest along the southern coast of Alaska, with deciduous angiosperms and gymnosperms in cooler, moist habitats. Evergreen gymnosperms like spruce, fir, hemlock, along with birch, have been assembling in the northern Rocky Mountains; and after the LPTM/EECL, a boreal coniferous forest association will begin to spread into the lowlands, replacing the deciduous angiosperms and gymnosperms as the prominent community. Ice is also present at least seasonally and locally in the High Arctic in the late Eocene.

Toward the south, in the western United States in the early to middle Eocene, the vegetation is a subtropical forest, locally modified by slope, exposure, and soils into a seasonally dry forest (shrubland/chaparral-woodland-savanna, especially east of the Rocky Mountains); grasses will not appear in North America until the early to middle Eocene. Character-istic genera include *Cedrela* (Mexican "mahogany"), *Cinnamomum* (cinna-mon), and *Oreopanax*. This vegetation grades into even more tropical forest and tropical rain forest along the oxbow lakes and river channels in the lowlands of southeastern United States, with deciduous elements mostly in the Appalachian uplands. The estimated MAT in the southeast in the early Eocene is about 25°C–27°C.

In Mexico, Central America, and the Antilles, the early Eocene vegeta-tion is not known, and middle to late Eocene floras—Cuenca de Burgos, Ixtapa, Mexico; Gatuncillo, Panama; Saramaguacan, Cuba; Guys Hill, Jamaica—only provide general views of a warm-temperate to tropical veg-etation, with variations controlled primarily by physiography and soils.

In South America in the early Eocene, there are moderate uplands in the Northern Andes with an inlet through the present-day Maracaibo re-gion. There are ancient uplands in the Guiana and Brazilian shields, and another inlet occurs in the south through the Paraguay-Uruguay-Paraná lowlands. The rest of the continent is mostly low-lying, and the Amazon Basin is swamp with mangrove ferns, *Nypa*, other palms, and *Pelliceria*. The proto–Amazon and Orinoco rivers are flowing westward, and from a suit-able vantage point in Amazonia, the Pacific Ocean can be seen through low places in the early Northern Andes Mountains. The recently discov-ered giant boid snake from Colombia suggests that by the late Paleocene the MAT may have reached 34°C, causing the evolution and extinction of some tropical forms and forcing the northward and southward migration of others into more temperature-compatible zones. In short, the tropical rain forest is changing even in the equatorial latitudes at the height of Paleogene warmth and moisture.

Tropical rain forest grows in the lowlands of southern South America, and elements extend to the southern tip of South America, as well as north-ward all the way to the Arctic Circle. The temperate deciduous *Nothofagus* is considerably reduced in extent in Patagonian Chile and Argentina. South America is still connected through Antarctica to Australia, and *Araucaria* and *Casuarina* are widespread.

At 55 Ma, the ecosystems encountered from pole to pole are beach/strand/dune; freshwater herbaceous bog/marsh/swamp; aquatic, present since the Late Cretaceous; lowland neotropical rain forest, present since the Paleocene; lower to upper montane broad-leaved forest, including a deciduous forest association in the far-northern lowlands and in highlands farther south; the beginnings of a coniferous forest (montane coniferous forest in the northern Rocky Mountains, spreading into the lowlands northward as the boreal coniferous forest, and into mountainous regions to the south as montane coniferous forest); versions of shrubland/chaparral-woodland-savanna (without extensive grasses) on drier sites, and mangrove (without *Rhizophora*) along the tropical shores. At about 45 Ma in the middle Eocene, ecosystems and lineages will take on a decidedly more modern aspect. There are still no extensive modern (i.e., angiosperm-dominated) deserts, grasslands, alpine tundra (páramo), or tundra.

SEVENTEEN MILLION YEARS AGO

The physical configuration of North America at 17 Ma consists of the western cordillera approaching modern elevations by uplift and accumulation of volcanic materials in the Brooks and Alaska ranges, Rocky Mountains, and the Sierra Madres of Mexico, and by erosion down to near-modern levels in the Appalachian Mountains. Lava flows and ejection of ash in the west provide ideal conditions for the preservation of extensive Miocene floras and faunas that will leave an extraordinary record of lineage and ecosystem history. In particular, the flood basalts of the Columbia River region will eventually cover 63,000 mi^2 and accumulate to over 6000 feet in thickness. There will be over 300 individual flows between 17 and 6 Ma. To the southwest, the Colorado Plateau in the Miocene extends over 130,000 mi^2 at the junction of Colorado, New Mexico, Utah, and Arizona, and it is between 5000 and 11,000 feet thick. Between about 20 and 24 Ma, the plateau had undergone a period of major uplift that raised it more than 3 km, and 1 km above the adjacent Basin and Range Province. At 17 Ma, the Sierra Nevada Mountains are beginning to cast a rain shadow, creating increasingly dry conditions at low to middle elevations across the Colorado Plateau and the Basin and Range Province.

In the middle Miocene, the Florida Peninsula south of about 27°N is mostly submerged, shallow water still covers much of the Isthmus of Tehuantepec, and periodically it inundates the Yucatán Peninsula, primarily from tectonic movement of the platform. Baja California has been accreted to the mainland since about 35.4–23.3 Ma. The spreading center that formed the Sea of Cortez at 29 Ma continues to open between 23.6 and 6.7 Ma. The Transvolcanic Belt rises in the early Miocene circa 23 Ma, moderately high elevations form by 17 Ma, and the mountains will continue to rise into modern times. The Cordillera Central of Central America are also moderate highlands, and they terminate at about southern Costa Rica, with isolated islands extending to northwestern South America.

In South America, the Amazon Basin of the Miocene is still an interconnected series of lakes and periodically marine-inundated swamps, but they are beginning to receive sediments from the rising Andes Mountains in addition to those from the Guiana and Brazilian highlands. The Orinoco and Amazon rivers still flow westward, and will continue to do so for another 2 million years, when around 15 Ma they will reverse direction. Vast deltas will then begin to appear along the Caribbean coast, further modernizing the landscape. In the late Miocene, the Maracaibo River will be dammed by uplift of the Cordillera Oriental, and Lake Maracaibo will form. The offshore Netherlands Antilles will also appear in the Miocene.

Following the Middle Miocene Climactic Optimum, with its possible increase in storms, the change toward lower temperatures and, consequently, reduced atmospheric moisture, represents a major trend affecting the ecosystems of the New World. Climates are not becoming drier uniformly at this time, however, because the distribution of precipitation is determined by atmospheric circulation, ocean currents, topography, and forcing mechanisms that are teleconnected between different parts of the world. Nevertheless, the general result of the cooling temperatures is decreasing precipitation and greater seasonality, gradually spreading over many parts of the Earth. Recall the paleosol evidence from the Badlands of South Dakota that indicate a decrease in annual rainfall from 1000 mm in the late Eocene, to between 250–450 mm in the early Miocene (Retallack 1990), to progressively less and more seasonal precipitation with middle Miocene cooling, and ultimately to the present 380 mm.

A reason for this shift to a new climate state is an overall slowing of plate movement (from an annual average of about 12–14 cm in the Jurassic/Cretaceous to about 2 cm at present) and the associated reduced volcanism that lowers CO_2 input into the atmosphere. The reduction in CO_2 is being augmented by positive feedbacks from greater weathering of silicate rocks resulting from continuing uplift of mountain systems and drainage of continental margins and interiors, and by the increasing albedo of the drained surfaces. The Gulf Coast in the middle Miocene at 17 Ma now runs just south of Jackson, Mississippi, along a line about halfway between Austin and Houston, Texas, and into Mexico between Reynosa and Matamoros. These factors combine to create a tipping point in the middle Miocene, as shown on the paleotemperature curve (fig. 3.4), toward cooling temperatures, general drying, and greater seasonality. The evolution and radiation of lineages adapted to drier and seasonal conditions will benefit from the change at the middle Miocene.

The vegetation in the far north has changed from the megathermal, subtropical forests of the early Eocene (notophyllous evergreen broad-leaved forest, paratropical rain forest, and thermophilous deciduous angiosperms and gymnosperms). In the Miocene, it is now a deciduous forest of *Celtis*, *Fagus*, *Juglans*, *Liquidambar*, *Magnolia*, *Nyssa*, *Platanus*, *Pterocarya*, and *Zelkova* (Seldovia Point flora, 16.8 Ma). The trend is toward an increase in microthermal elements (*Acer*, *Betula*, *Populus*, *Salix*), a decrease in present-day Asian components, and an expanding boreal coniferous forest association (*Abies*, *Picea*, *Tsuga*). Ice was locally present in the late Eocene, and by the end of the late Miocene it will be present in the continental interior, at high elevations, and in the adjacent seas. Some tundra elements (grasses, sedges, small alder, birch, and willows) appear in the far north by 17 Ma, and they are beginning to coalesce. The MAT is 6°C–7°C (CLAMP 9°C), with an annual warm-season to cold-season range of 26°C.

Drier vegetation is encountered toward the south. If one were to climb the southeastern Sierra Nevada near Tehachapi, Nevada, at 17.5 Ma, the vegetation in the lowlands would still include some evergreen subtropical holdovers, but higher up it is a oak–piñon pine woodland. At Buffalo Canyon in western Nevada at 16.6 Ma, there is dry vegetation on the slopes, and a mixture of conifers and deciduous hardwoods starting at elevations around 1280 m. The estimated MAT is 10°C, and the MAP is 1000 mm. In addition to the trend toward drying and greater seasonality, edaphic factors, slope, and exposure play a significant role, as shown by the vegetation at Middlegate and Eastgate in western Nevada at 15.5 Ma. The former is a shrubland-chaparral of *Arbutus* (madrones), *Cedrela*, *Cercocarpus* (mountain mahogany), *Lithocarpus* (tan oak), and *Quercus* growing on drier south-facing slopes (MAT 10.2°C, CLAMP). The latter grows on moister north-facing slopes and has a better representation of mesic *Abies*, *Larix*, *Amelanchier* (serviceberry), and *Aesculus* (buckeye, horse chestnut). The regional MAP at midelevations is about 900 mm.

In the Plains area, the vegetation is shrubland-savanna with some *Acer*, *Crataegus* (hawthorn), *Robinia* (locust), and *Ulmus*, but trending toward fewer trees and more grasses and herbs (Boraginaceae), anticipating grasslands—possibly since about 23 Ma, as we saw in chapter 7—that will become better defined and more extensive later in the Miocene around 13 Ma. In the Pliocene at 3 Ma, grasslands will appear east of the coastal mountains—the Palouse Prairie, for example, of eastern Washington and adjacent Idaho and Oregon. The nature and trend in the ecosystems of the midcontinent region to increasingly open areas, with forest mostly in upland-streamside-lakeside habitats, are reflected by faunas of the Hemingfordian NALMA at 19–16 Ma, which include the appearance or expansion of mylagauline rodents, rabbits, beaver (small, probably burrowing representatives of the family Castoridae), dromomerycids, antilocarpids (pronghorns), rhinos, and a major radiation of the equine subfamily of horses. These horses in North America increased from one species (*Parahippus leonensis*) to seventy species during the Miocene (Maguire and Stigall 2008). Slightly later, at the beginning of the Barstovian 16 Ma, hemicyonine bears (longer-footed than modern bears, likely ambush predators) and proboscideans migrated into the New World from Asia with continuing expansion of the shrubland-chaparral-woodland-savanna ecosystem.

Around 17 Ma in the southeastern United States, the principal community is deciduous forest with some Asian exotics that will disappear in the Pliocene, pine forest and beach/strand/dune plants on the edaphically drier coastal sands, and swamp forests, aquatics, and herbaceous bog/marsh/ swamp plants growing in lowland areas beyond the influence of marine waters. In Tennessee, the late Miocene to early Pliocene Gray Fossil Site is revealing some of the mammals living in eastern North America at the time—lesser panda, peccary, rhino, camel, badger, and tapirs. Several of these have Asian affinities, reflecting the land bridge through Beringia in the mid to late Tertiary.

In Mexico, the late Miocene Mint Canyon flora of southern California (then of northwestern coastal Mexico), shows that shrubland/chaparral-woodland-savanna is established, as in the drier areas of the western United States. Grasses are beginning to coalesce into grassland and spread onto the well-drained higher eastern slopes of the sierras. Since the middle to late Eocene, there have been mangroves, now with *Rhizophora*, along the coasts (*Pelliceria* disappeared from its northern limits by the middle Miocene). Beach-bog-aquatic communities and pine-oak (coniferous) forest can be found on drier sites, along with remnants of the lowland neotropical rain forest in moister habitats, and a lower to upper montane broad-leaved forest just beginning to receive significant northern deciduous components. As yet there is no well-defined desert or páramo (the Transvolcanic Belt is still in the early stages of uplift). At 17 Ma, there is still no continuous land southward from southern Central America and the landscape consists of islands and peninsulas. The available fossil floras from Central America are either older (early Miocene) or younger (late Miocene), but the vegetation was probably similar to that of the present without most of the drier Pacific elements, as reflected by an increase in grasses toward the end of the late Miocene, and without the few northern deciduous elements (*Alnus, Quercus*) that will arrive this far south in the late Miocene and afterward (Gatún flora).

Hoorn (1993, 1994) has studied spores and pollen from the early to late Miocene Basins of Solimões, Brazil, and Amazonas, Colombia, Peru. A walk through these basins at about 17 Ma encounters the vegetation of the following ecosystems: mangrove, beach/strand/dune, freshwater herbaceous bog/marsh/swamp, aquatic, lowland neotropical rain forest, lower to upper montane broad-leaved forest.

The mangrove community—represented by *Acrostichum, Avicennia, Hibiscus, Pelliceria, Rhizophora*—increases in abundance and extent at times of high sea level in the Miocene, and sediment analysis indicates incursions come from both the east (Guiana Shield sediment type) and from the west (Andean type). There are also fringing mudflats with *Crenea maritima* as today.

The presence of the beach/strand/dune community is inferred from the coastal setting and sediments (sand) even though pollen of the prominent species (e.g., grasses, composites) cannot be distinguished from those that grow in other habitats.

The freshwater herbaceous bog/marsh/swamp community, with associated riparian and swamp forest, is represented by widespread *Alchornea, Annona* (also present in other adjacent communities), *Bactris, Cyperus, Euterpe, Ficus, Ludwigia, Mauritia* palm swamps, and *Virola*.

Aquatic vegetation is indicated by the presence of *Botryococcus* and *Pediastrum* (algae), *Azolla* and *Ceratopteris* (aquatic ferns), and *Pachira*.

Lowland neotropical rain forest is the prominent forest community, together with the lower elevation phases of the lower to upper montane broad-leaved forest, with abundant ferns, *Amanoa*, and Bombacaceae (*Catostemma, Quararibea*), *Crudia, Humiria, Licania, Pouteria*. Freshwater bivalves from the Amazon Basin in northeastern Peru at the MMCO around 16 Ma show seasonal oxygen isotope variation in growth bands correlated with the annual migration of the ITCZ and support other evidence that a climate sufficient to sustain a tropical rain forest is present (Kaandrop et al. 2005).

Only representatives from the lower elevations of the lower to upper montane broad-leaved forest are present, such as *Selaginella, Pteris, Podocarpus, Alchornea, Amanoa, Hedyosmum,* and *Ilex.*

There are no members of the cloud forest, elfin forest, northern temperate deciduous forest, tundra, or higher elevation coniferous forest, ceja, or páramo.

According to Hoorn (1993), a floodplain community called várzea, like that presently found along the upper Amazon River, is a close modern analog to the middle to late Miocene vegetation and environment of the Amazon Basin (see fig. 7.6).

Farther south, the eastern cordillera of Ecuador is being uplifted in the middle Miocene, and by this time about half of the modern elevation of the Central Andes has been attained. A midaltitude cloud forest of Mio-Pliocene age is found in the high elevations of the Central Andes, documenting substantial uplift at this time (e.g., Graham et al, 2001). A vertebrate fauna spanning the interval between the early and late Miocene shows a trend from browsers, implying shrubs and small trees, to grazers and more grasses (MacFadden et al. 1994). The elevation is sufficient to begin blocking moisture from the east, and together with the cold Humboldt Current and a global reduction in atmospheric moisture beginning in the middle Miocene, there is enhanced formation of the Atacama Desert. Independent evidence for aridity along the coast at this time is provided by sediments in northern Chile (Sáez et al. 1999), and hyperaridity will become evident at 4–3 Ma.

The inland sea through the Entre Ríos region (Paraguay-Uruguay-Paraná river basins) extends northward to the Mato Grosso Plateau. The dry region of the Monte of Argentina began forming earlier with uplift of the Central Andes, blocking moisture from the Pacific Ocean, but the trend at 17 Ma is now being augmented by increased middle and late Miocene seasonality. Thus, from the western United States at about 45°N to northern Argentina at about 25°S, the evolution of dry lineages and the coalescence of dry ecosystem elements toward modern versions is being facilitated by western mountain ranges attaining at least half of their present elevation and forming rain shadows to the east, edaphically coarse eroding soils, cold coastal currents, and by the general climatic cooling that is enhancing seasonality. A possible difference between the fauna at 17 Ma and at present in Argentina is the presence of a diverse cadre of large caviomorph rodents and large marsupial carnivores (borhyaeneids), endemic ungulates, and giant edentates, and the absence of mammals of the families Cricetidae (Rodentia, e.g., New World rats) and Mustelidae (Carnivora, e.g., weasels) because they did not arrive in South America until about 5.8–5.7 Ma. These early immigrants would be examples of small mammals that may have drifted over before closure of the Panama land bridge (Verzi and Montalva 2008), but some believe the age of the deposits is not certain (Prevosti and Pardiñas 2009).

In the Southern Andes, there was a period of rapid plate subduction between 25 and 10 Ma (Suárez et al. 2000) that created much of the high elevation in these mountains. There are tillites interbedded with basalts of Miocene to Quaternary age indicating that glaciation is beginning in Patagonia in the Mio-Pliocene.

A rich vertebrate fauna is known from the early Miocene Santa Cruz Formation of southeastern Patagonia. According to Tauber (1997), from the bottom to the top of the formation the fauna shows a decrease in diversity; an increase in species with euhypsodont teeth and a decrease in those with brachydont teeth; glyptodontids and toxodontids become more diverse; megathere sloth average body size is reduced; and protheroteid diversity decreases. This has been interpreted to mean a climate that is becoming cooler and seasonally drier. Paleoceanographic studies of the late Miocene reveal sea levels are falling, consistent with the early onset of widespread glaciation. Pollen of grasses, composites, and the arid *Ephedra* increase, further indicating dryness, seasonality, and steppe conditions developing in Patagonia (Palazzesi and Barreda 2004).

The ecosystems from pole to pole at the end of the middle Miocene reflect communities poised on the verge of major modernization through mountain uplift, lowering temperatures, falling sea levels, increasing seasonality, newly evolving forms, and migrations facilitated by continuing connections between North America and Asia, and new ones being forged between North and South America.

TWO MILLION YEARS AGO

A walk south from northernmost North America at 2 Ma will be difficult because increasing expanses of land and the adjacent seas are covered by ice. There are two centers of continental ice accumulation. One is a dome over Hudson Bay called the Laurentide ice sheet. The other, called the Cordilleran ice sheet, is in the northern Rocky Mountains of British Colombia. Until about 735 kyr, these ice sheets waxed and waned on the tilt variation of about 41,000 years of the Milankovitch cycle, and since that time they have followed the eccentricity variation of 100,000 years. The cause for the change is not fully known, although it may have involved a tipping point reached in the declining atmospheric concentration of CO_2. The two centers frequently fuse into a single sheet, and at the earliest and most extensive advance, the ice extends to Nebraska where the Elk Creek Till is dated at 2.14 Ma. In the intervening areas to the north and along the margin, there is a zone of permafrost 80–200 km wide supporting a herbaceous community of lichens, mosses, grasses, and sedges, and a dwarfed woody community of alder, birch, and willow. This is the early tundra, and elements and patches have been present locally since the middle to late Pliocene. As an ecosystem, it dates primarily from 3–2 Ma and becomes most extensive just prior to and following the glacial periods in polar regions when there are maximum extents of ice-free permafrost surfaces. At the Last Glacial Maximum circa 18 kyr, ice will cover 40 million km², or 30 percent of the Earth's surface, compared to the rapidly diminishing 15 million km² at present, and sea level was about 120 m lower. Among other consequences, each of the eighteen to twenty lowerings will enlarge the Florida Peninsula by twofold, primarily by extending its western coast, and expand the Zapata (Cuba) and other coastal swamps that are presently near their minimum (interglacial) extent.

Winters in Greenland at the LGM were 11°C–15°C colder and summers 2°C colder than at present, and these have probably been the approximate conditions between eighteen and twenty times since 2.6 Ma. Around 2 Ma, the northern ice sheet for the first of many times in the Quaternary will divide the polar jet stream, shifting the southern arm southward and displacing the high pressure system located beneath the descending arm of the Hadley circulation cell and the Bermuda-Azores High at 30°N–35°N. The resulting cooler and wetter pluvial climates will prevail for 90 percent of the next 2 million years, or about 90,000 of each of the 100,000-year Milankovitch cycles. Toward the south, it is 6°C–8°C colder than at present; it is 5°C–6°C colder at the latitude of Barbados and in the lowlands of Brazil at each glacial maximum; and it is noticeably drier. The record as preserved in ice cores documents that during glacial intervals conditions are cold, windy, dry, and dusty (Lambert et al., 2008).

The early Pleistocene fauna of North America is grouped into the Irvingtonian, or *Mammuthus*-pre-*Bison* NALMA, and includes new immigrants from Asia and South America. The Asian component includes cold-habitat steppe species such as mammoth and musk-oxen, and microtid rodents like lemmings, mice, and voles. The Rancholabrean fauna after 0.5 Ma will add goats and bison from Asia. Many of the Miocene faunal elements are gone. Among the carnivores lost are amphicyonines, hemicyonines, and borophagine canids; among the herbivores, dromomerycids and most antilocaprids, anchitherine horses, rhinos, chalicotheres, and others. A component from South America includes sloths, anteaters, opossum, and hystricognath rodents, which began crossing the Panama land bridge after about 3.5 Ma and especially after 2.7 Ma (Stehli and Webb 1985; Webb 1997, 2006; Burnham and Graham 1999).

Reconstructing the vegetation at 2 Ma is difficult because sediments of that age in many instances have been covered over or removed by later glaciations. Extrapolation from late Pliocene trends, analogs from later glacial intervals, and ancillary information from land faunas, marine assemblages, and oxygen isotope data must be used, and the interpretations assessed with the context of trends established from these independent sources. We will use the most recent glacial/interglacial cycle as an approximate analog for previous ones. After crossing the northernmost zone of permafrost and tundra, our passage would likely be impeded by the dense, boggy, and hummocky environments of the boreal forest. Spruce, fir, and hemlock grow mixed with patches of alder, birch, willow, and poplar. At 2 Ma, this ecosystem extends south into the upper midwestern United States to about 40°N. There are some unfamiliar combinations of trees not found together today. This is because at the beginning of the Quaternary, and after the many reshufflings of later glacial times, organisms often did not have time or opportunity to reach the full range of their ecological potential. An example is an assemblage of spruce and black ash that grew in upper midwestern North America. At 2 Ma and probably during each glacial maximum, spruce extended well onto the plains and contributed pollen to bogs as far south as central Texas. The central grassland ecosystem formed in the late Miocene, developed into an essentially modern community in the Pliocene, and especially during the cold dry intervals of the Quaternary when there was a reduction in the number of trees. High-altitude conifers like fir and spruce extend farther downslope in the western mountains at 2 Ma, and alpine tundra grows on the higher peaks. Spruce and fir is also present on the highest peaks of the Appalachian Mountains, and they, too, moved downward and southward into the zone of present-day deciduous forest during all of the glacial intervals. At 2 Ma, there was probably some alpine tundra in the northern Appalachian Mountains and less or none in the southern Appalachians. In the middle Pliocene and in many of the glacial intervals of the Quaternary, spruce grew in the mountains of the Transvolcanic Belt of Mexico (Paraje Solo fossil flora), and even as far south as Guatemala (Padre Miguel fossil flora).

In the Quaternary, the long-leaf, short-leaf, and loblolly pines of the Gulf Coast coniferous forest intermingle with deciduous hardwood species, and there are cypress swamps with *Taxodium* and Spanish moss (*Tillandsia usne-oides*). Coastal grasslands mix with oaks and *Sabal* palms in interior Florida to form savannas. Deserts in the west, present as preadapted elements since the late Eocene, are increasingly evident in the Basin and Range after the middle Miocene, and the details of their distribution and composition back 50,000 years are revealed in packrat middens and back 10,000 years by dendrochronology. Desert elements, like the grasslands, modernized in the Pliocene and continued to do so in the dry intervals of the Quaternary.

The vegetation changes around 2 Ma in South America are well known from a few areas, but we must use the ecological features of the modern communities, along with context and ancillary information from independent lines of inquiry, to infer early ice age vegetation and environments. Clapperton (1993) and others provide insight into Quaternary environments at 2 Ma and younger times, and these presumably apply to all the glacial/interglacial intervals. This evidence, summarized below, is relevant to the once-debated topic of stability versus change in the tropical latitudes, and currently to the concept of refugia (for references to the primary literature for the summary, see II, chap. 7, on the Quaternary; II, chap. 8, for synthesis).

1. Frequent ponding from Quaternary tectonic movements in the Pastaza-Marañon Basin in Amazonian Peru and Brazil.
2. Long-term fluctuations in flooding cycles of the terra firme, that is, the land along rivers at present beyond the annual floods.
3. Sedimentary evidence for repeated wet and dry periods within the Mesa Formation of the Llanos along the Orinoco River basin, formed during the Yarmouth Interglacial of North America, circa 475–300 kyr.
4. A complex of factors (summarized in Clapperton 1993, 148–51) indicating reduction and greater seasonality in MAP, a eustatic lowering of sea level of between 25 and 130 m—applicable to each glacial maximum and with implications for water-table levels—and a reduction in vegetation cover.
5. Uplift of the Sierra Nevada de Santa Marta and the Cordillera de Mérida of northern Venezuela at an annual rate of about 4 mm, or 2000 m during the past 500,000 years. The present lower limit of glaciers in the mountains is 5000 m (4000 m at the LGM), meaning that few glaciers were present prior to the Quaternary and implying significantly altered and fluctuating environments since that time.
6. Uplift of the Sierra de Perija on the Colombia-Venezuela border by 11–16 mm a year during the Quaternary.

7. Uplift of blocks within the Ecuadorian Andes Mountains by 150 m within the past 7000 years.

8. Relict Quaternary dunes in the now humid Pantanal indicating environments periodically so dry that the protective vegetation cover was significantly reduced, and strong winds caused dune movement that would further increase desiccation.

9. Pampas plains with sand and loess younger than 2.5 Ma, suggesting wet and drier cycles.

10. Eight sedimentary cycles of Quaternary age in the Entre Rios region of northeastern Argentina, indicating wet/flooding to arid conditions.

11. Aeolian deposits, wind-blown sediments characteristic of dry conditions, in the Hernandarias Formation of the Entre Rios region of Argentina at 1.3–0.8 Ma, representing rainfall at 13 percent of present levels, or 100–150 mm a year.

12. Origin of some bajos (enclosed, often water-filled basins) in Patagonia through erosion after vegetation was reduced during intervals colder and drier than at present, that is, during glacial cycles.

13. Loess sediments, indicating dry conditions, alternating with swamp deposits in the Rosario Formation of the Pampean region of Argentina dated at 730 kyr.

14. Evidence of extensive glaciation in what are today ice-free areas of southern Chile, Patagonia, and Tierra de Fuego.

15. Presence of erosion terraces 30–80 m above present sea level on Marajó Island, Brazil, and adjacent regions of the Amazon Basin.

16. Glacial moraines in the Cordillera de Mérida offset several tens of meters by faulting during the Holocene.

17. Fluctuations in the abundance of mylodonts and equids in the Pleistocene Cangahua Formation of Ecuador correlated with cold-dry cycles.

18. Evidence for extensive glaciation in the eastern cordillera of Bolivia, in the form of moraines, tillites, and hummocky surfaces. Presumably glacial cycles have repeated multiple times since 2 Ma and, based on the High Plain of Bogotá data set, likely extended far downslope toward the lowlands.

19. Snowline depression at the LGM of 200–300 m in the Junin region of Peru, 600 m at Laguna Kollpa, Bolivia, and 100–1350 m in the Cordillera Oriental, Peru.

20. Human populations from Brazil showing significant effects of a middle Holocene dry period.

21. Alluvial fans in the Gran Chaco of Argentina, eastern Bolivia, and southeastern Brazil originating in the humid phases of the Pliocene and Qua-

ternary, and dunes and loess originating during dry conditions correlated
with cold phases of the many glacial maxima.

22. Volcanism since 930 kyr along the coast of central Chile, Plio-Pleistocene
    deposits near Antofagasta, Chile, elevated 240 mm per thousand years
    since 330 kyr, a 7.8 magnitude earthquake in northern Chile on 30 July
    1995, and a 7.6 earthquake in central Peru on 15 August 2007—all these
    events have the potential of altering subterranean aquifers, surface drain-
    age patterns, and vegetation.

23. Dry periods on the Altiplano of Bolivia linked with dry intervals in the
    Amazon Basin and reduced transport of moisture.

The cumulative ancillary evidence, combined with the direct albeit scattered paleobotanical information, gives strong support for changing climates, landscapes, water tables, and vegetation in the tropics and adjacent
regions in the lower latitudes. The data from the High Plain of Bogotá fills
in the details for one locality at 2 Ma (Hooghiemstra 1984).

The Bogotá site is ideal for the study of environmental and vegetation
history. First, as a slowly sinking basin, it has accumulated sediments more
or less continuously since about 3.5 Ma. Second, it is located at 2550 m in
the eastern cordillera of the Colombian Andes, so the sediments preserve
a variety of vegetation types moving up and down the slope. Third, being
distributed along a steep altitudinal gradient, these zones of vegetation
move rapidly and clearly in response to even moderate changes in climate.
The locality is in the Andean forest belt at 2300–3200 m, situated between
the subpáramo above 3200 m and the subandean forest at 1000–2300 m
(chap. 8).

The plant record begins at 3.2 Ma. Between 3.2 and 2.7 Ma, conditions
are warm. Between 2.7 and 2.2 Ma, the area turns colder above 2600 m
by 5°C–9.5°C, with an overall regional cooling of 8°C. The onset of colder
termperatures at the Bogotá site corresponds to the beginning of significant
global cooling and expanding glaciation. For about the past 800,000 years,
changes on the High Plain have followed the 100,000-year periodicity of
the Milankovitch eccentricity cycle. In glacial times, the grass-shrubland
páramo is prominent, while in warmer times it is the subandean forest. In
postglacial times, there is evidence for a reversal to temperatures cooler by
3°C–4°C at the time of the Younger Dryas, 10.8–9.5 kyr. Thus, the vegetation at the high altitudes in northern South America at 2 Ma, and in subsequent times, follows patterns and subpatterns generally similar to those
farther north, and the effects extend downslope, for example, a cooling of
5°C–6°C and drying in the Amazon lowlands at the glacial maxima.

Climate and vegetation histories have proven difficult to reconstruct on the Altiplano of Bolivia. It is topographically one of the most complex regions on Earth, and it is located in the transition between climate regimes of the far north and the far south. In addition, the Altiplano is influenced by the variable meanderings of the ITCZ; Milankovitch variations, suggested to have influenced the climate in the northern part, have not been detected in the southern part; tectonic events may have affected drainage patterns and water tables differently in the separate basins; and there is a lag time between climatic changes in the marine realm and those on land. Until recently there has been the added complication of carbonate-charged waters contaminating some sediments used for ^{14}C dating. It is clear, however, that between 18.1 and 14.1 kyr, dramatic changes affect the moisture regimes on the southern Altiplano. Presumably these become evident around 2 Ma, generally occurring toward the end of each of the approximately eighteen to twenty glacial maxima. Precipitation increases, and the now-isolated dry salars expand and fill with water (Placzek et al. 2006, 2009). The largest lakes in the region for the past 120,000 years appear, and some are estimated to have been over 140 m deep. Intervening times are drier, and there are repeating cycles throughout the Quaternary, giving a dynamic aspect to life on the Altiplano.

This brings us to Patagonia and Cape Horn, the land of fire, and the Beagle Channel—"the uttermost part of the world." Some modern carnivores have already crossed into South America (bears, cats, dogs, otters, raccoons). These and the saber-toothed cat *Smilodon* may inflict a price on the native South American animals, which have little previous experience with carnivorous mammals, but the majority of extinctions, for example, of most native ungulates and marsupial carnivores, have happened prior to the interchange. Newly arriving herbivores include llamas, deer, and tapirs, which survive; and horses and proboscideans, which do not. In the lowlands of Chile and Argentina at 2 Ma, Quaternary floras and faunas are also responding to cold/dry and warm/moist cycles. As it becomes colder and drier, temperate *Nothofagus* (southern beech) forests disassemble into more isolated patches in protective habitats, while steppe (shrubland/chaparral-woodland-savanna) expands (Heusser 2003). These repeating cycles in climate are occurring in the southernmost point of South America, as well as in the northernmost part of North America, on timescales ranging from Milankovitch eccentricity variations of 100,000 years, to D-O events of a few centuries, to ocean-atmospheric induced switches and dials of a decade or less. The environmental changes of the late Tertiary and Quaternary have likely served as a pump for generating novel genotypes of relatively recent origin into the Earth's living envelope and for the evolutionary development of many crown groups.

Looking at the collective evidence from pole to pole since 100 Ma, the conclusion is inescapable that environments and the biota have changed both through long-term geological time, and on scales measured in decades or years. Patterned changes caused by Milankovitch variations, the omnipresence of variations in atmospheric CO_2 concentration, and changes in ocean circulation have been augmented by a series of catastrophes, including asteroid impacts, sudden methane emissions, salinity crises, and jökulhlaups. The detailed record of the more recent Neogene and Quaternary provides especially clear analogs for the kind, extent, and future effect of alterations taking place at present. These include burning—for example, the burning of the tropical forests and the tundra (Qiu 2009)—lumbering and other removal of the Earth's vegetation on an unprecedented scale, and the unrelenting release of greenhouse gases into the atmosphere from these activities and through the use of fossil fuels. Technology is beginning to play a role in documenting and raising awareness of illegal biome destruction through satellite images posted on the Web that show these activities happening in real time. For example, current incursion into tropical preserves set aside through the efforts of the Amazonian chief Almir Surui, commemorated in the epigraph to this book, are now posted across the Web for the world to see. The present state of affairs is unique for the past 65 million years in its combination of magnitude and rapid pace. The simultaneous and global destruction or severe damage across the Earth's entire living envelope, with previously existing natural extinction rates now being augmented by rampant exterminations through human activities, is unprecedented. This is occurring within the context of an ever-expanding human population estimated to further increase from 6.5 billion to 9 billion people in the next fifty years.

Efforts to anticipate the consequences require a multiplicity of approaches, an arsenal of techniques, and perhaps most difficult of all to achieve, the mind-set of an informed and activist majority unencumbered by extreme self-interest and antiscience bias. Vegetation and environmental history provides one source of information for estimating the causes, effects, and extent of past environmental change; and it can contribute to modeling the future of an immensely complex, multifactored, and teleconnected global system. Recent findings create a sense of urgency. For example, it has become clear that forcing mechanisms need only be as slight as changes in temperature of less than 1°C to produce significant alterations in the Earth's life-sustaining plant communities. Furthermore, once an ecosystem has been significantly altered, especially its local climate through changes in moisture generated by transpiration and its soil characteristics through erosion, recovery will likely be measured in geologic time rather than in human generations without enormous cost, extensive technological know-how, political leadership, and public determination. No catastrophe since the asteroid impact has reduced biodiversity so extensively, so indiscriminately, and so rapidly as recent human activity, and, uniquely, no previous agent has had the capacity to judge its own impact, plan for the consequences, and the option to reduce its effects.

If we can adequately distill information into knowledge and let time and experience convert it into wisdom, compilations like this will have been worthwhile. To echo the renowned photographer Robert B. Haas, whose glorious images in works like *Through the Eyes of the Condor* (2007) provide continuing inspiration: "The quest may never be over, but it will never be pointless."

:: **REFERENCES** ::

Burnham, R. J., and A. Graham. 1999. The history of neotropical vegetation: New developments and status. *Ann. Mo. Bot. Gard.* 86:546–89.

Clapperton, C. M. 1993. *Quaternary geology and geomorphology of South America.* Elsevier Science Publishers, Amsterdam.

Clemens, W. A., and L. G. Nelms. 1993. Paleoecological implications of Alaskan terrestrial vertebrate fauna in latest Cretaceous time at high paleolatitudes. *Geology* 21:503–6.

Flannery, T. 2005. *The weather makers: How man is changing the climate and what it means for life on Earth.* Atlantic Monthly Press, New York.

Graham, A., K. M. Gregory-Wodzicki, and K. L. Wright. 2001. Studies in neotropical paleobotany. Part 15, A Mio-Pliocene palynoflora from the Eastern Cordillera, Bolivia: Implications for the uplift history of the Central Andes. *Am. J. Bot.* 88:1545–57.

Haas, R. B. 2007. *Through the eyes of the condor: An aerial vision of Latin America.* National Geographic Society, Washington, D.C.

Heusser, C. J. 2003. *Ice age Southern Andes: A chronicle of paleoecological events.* Elsevier Science Publishers, Amsterdam.

Hooghiemstra, H. 1984. Vegetation and climatic history of the High Plain of Bogotá, Colombia: A continuous record of the last 3.5 million years. *The Quaternary of Colombia / El Cuaternario de Colombia* 10:1–368. [Reprinted from *Dissertationes botanicae*, vol. 79 (J. Cramer, Vaduz, Germany).]

Hoorn, C. 1993. Marine incursions and the influence of Andean tectonics on the Miocene depositional history of northwestern Amazonia: Results of a palynostratigraphic study. *Palaeogeogr. Palaeocl. Palaeoecol.* 105:267–309.

———. 1994. Miocene palynostratigraphy and paleoenvironments of northwestern Amazonia, evidence for marine incursions and the influence of Andean tectonics. Ph.D. diss., University of Amsterdam.

Janis, C. M., G. F. Gunnell, and M. D. Uhen, eds. 2008. *Evolution of Tertiary mammals of North America.* Vol. 2. Cambridge University Press, Cambridge.

Janis, C. M., K. M. Scott, and L. L. Jacobs, eds. 1998. *Evolution of Tertiary mammals of North America.* Vol. 1. Cambridge University Press, Cambridge.

Kaandrop, R. J. G., et al. (five coauthors). 2005. Seasonal Amazonian rainfall variation in the Miocene Climate Optimum. *Palaeogeogr. Palaeocl. Palaeoecol.* 221:1–6.

Lambert, F., et al. (nine coauthors). 2008. Dust-climate couplings over the past 800,000 years from the EPICA Dome C ice core. *Nature* 452:616–19. [See also editor's summary.]

MacFadden, B. J., Y. Yang, T. E. Cerling, and F. Anaya. 1994. South American fossil mammals and carbon isotopes: A 25 million-year sequence from the Bolivian Andes. *Palaeogeogr. Palaeocl. Palaeoecol.* 107:257–68.

Maguire, K. C. and A. L. Stigall. 2008. Paleobiogeography of Miocene Equinae of North America: A phylogenetic biogeographic analysis of the relative roles of climate, vicariance, and dispersal. *Palaeogeogr. Palaeocl. Palaeoecol.* 267:175–84.

Palazzesi, L., and V. Barreda. 2004. Primer registro palinológico de la Formación Puerto Madryn, Mioceno de la Provincia del Chubut, Argentina. *Ameghiniana* 41:355–62.

Parrish, J. T., and R. A. Spicer. 1988. Late Cretaceous terrestrial vegetation: A near-polar temperature curve. *Geology* 16:22–25.

Placzek, C., J. Quade, and P. J. Patchett. 2006. Geochronology and stratigraphy of late Pleistocene lake cycles on the southern Bolivian Altiplano: Implications for causes of tropical climate change. *Geol. Soc. Am. Bull.* 118:515–32.

Placzek, C., et al. (eight coauthors). 2009. Climate in the dry Central Andes over geologic, millennia and interannual timescales. *Ann. Mo. Bot. Gard.* 96:386–97.

Prevosti, F. J., and U. F. J. Pardiñas. 2009. Comment on "The oldest South American Cricetidae (Rodentia) and Mustelidae (Carnivora): Late Miocene faunal turnover in central Argentina and the great American biotic interchange" [Verzi and Montalvo 2008]. *Palaeogeogr. Palaeocl. Palaeoecol.* 280:543–47.

Qiu, J. 2009. Arctic ecology: Tundra's burning. *Nature* 461:34–36.

Retallack, G. J. 1990. *Soils of the past: An introduction to paleopedology.* Unwin-Hyman, Boston.

Rich, T. H., P. Vickers-Rich, and R. A. Gangloff. 2002. Polar dinosaurs. *Science* 295:979–80.

Ridgway, K. D., J. M. Trop, W. J. Nokleberg, C. M. Davidson, and K. R. Eastham. 2002. Mesozoic and Cenozoic tectonics of the eastern and central Alaska Range: Progressive basin development and deformation in a suture zone. *Geol. Soc. Am. Bull.* 114:1480–1504.

Sáez, A., L. Cabrera, A. Jensen, and G. Chong. 1999. Late Neogene lacustrine record and palaeogeography in the Quillagua-Llamara Basin, central Andean fore-arc (northern Chile). *Palaeogeogr. Palaeocl. Palaeoecol.* 151:5–37.

Stehli, F. G., and S. D. Webb, eds. 1985. *The great American biotic interchange.* Plenum Press, New York.

Suárez, M., R. de la Cruz, and C. M. Bell. 2000. Timing and origin of deformation along the Patagonian fold and thrust belt. *Geol. Mag.* 137:345–53.

Tauber, A. A. 1997. Bioestrafigrafía de la Formación Santa Cruz (Mioceno Inferior) en el extremo sudeste de la Patagonia. *Ameghiniana* 34:413–26.

Verzi, D. H., and C. I. Montalva. 2008. The oldest South American Cricetidae (Rodentia) and Mustelidae (Carnivora): Late Miocene faunal turnover in central Argentina and the great American biotic interchange. *Palaeogeogr. Palaeocl. Palaeoecol.* 267:284–91.

Webb, S. D. 1997. The great American faunal interchange. In *Central America: A natural and Cultural History*, ed. A. G. Coates, 97–122. Yale University Press, New Haven.

———. 2006. The great American biotic interchange: Patterns and processes. In *Latin American biogeography—causes and effects*. Proceedings of the 51st Annual Systematics Symposium of the Missouri Botanical Garden (Alan Graham, organizer). *Ann. Mo. Bot. Gard.* 93:245–57.

Woodburne, M. O., ed. 2004. *Late Cretaceous and Cenozoic mammals of North America*. Columbia University Press, New York.

Additional Readings and Updates

Gelffo, J. N., F. J. Goin, M. O. Woodburne, and C. de Muizon. 2009. Biochronological relationships of the earliest South American Paleogene mammalian faunas. *Palaeontology* 52:251–69.

McElwain, J. C., and S. W. Punyasena. 2007. Mass extinction events and the plant fossil record. *Trends Ecol. Evol.* 22:548–57.

Than, K. 2009. Dinosaur lost world found in Texas. National Geographic News. http://news.nationalgeographic.com/news/2009/03/090318-texas-dinosaurs.html.

Webster, D. 2009. The dinosaur fossil wars. *Smithsonian*, April 2009, http://www.smithsonianmag.com/science-nature/41381902.html.

Epilogue

This overview of the natural history of the New World began with the definition of eight ecosystems existing 100 million years ago. Tracing of the New World's vegetation and environmental history ends with twelve very different ecosystems existing at present. The changes through time—for example, in climate (table E.1)—that brought about that transition are revealed by plant macrofossil assemblages, spore and pollen diagrams, phytoliths, otoliths, speleothems, isotopic evidence, ice and ocean cores, biochemical fossils, and a plethora of ancillary sources. It is an extensive and complicated record, but it is being synthesized into a coherent scenario that can be read, with sufficient effort, using the impressive arsenal of techniques now available. I will conclude the survey by reiterating comments stated at the end of chapter 1:

> A more full, lasting and satisfying understanding of the topic can be gained by assuming the role of active participant and using the text as a cooperative adventure. . . . This is a slow, methodical, and thoughtful way to approach any introductory material. However, the quest for knowledge, stimulated by curiosity and imagination, as a basis for understanding, does require effort. It is also a lot of fun and ultimately quite satisfying.

To strengthen the proposition that time (history) and effort are essential for understanding ecosystems, listen to Seth Lerer, who

Table E.1. Fluctuating climate parameters for the New World since 100 Ma.

Time	Location	Value
Carbon dioxide		
Middle Cretaceous	Global	~4× modern levels, 1100–1680 ppmv
Late Cretaceous	Global	Declining
Early Eocene	Global	1500 Ga input, +1125 ppmv
Middle Eocene–early Miocene	Global	Declining
Middle Miocene	Global	Increasing to MMCO, then declining
Middle Pliocene	Global	360–440 ppmv
Preindustrial	Global	285 ppmv
Present	Global	380 ppmv
Mid–21st century	Global	550 ppmv
End 21st century	Global	1000 ppmv
Paleotemperatures		
Middle Cretaceous	Global	12°C warmer than at present
Late Cretaceous	Global	Cooling to ~8°C warmer than at present
Late Cretaceous	Arctic 60°N–50°N	13°C–20°C MAT
Late Cretaceous	50°N–40°N	20°C–25°C MAT
Late Cretaceous	Midcontinent North America	18°C MAT
Late Paleocene	North Slope Alaska	10°C–12°C MAT (present −12°C)
Early Eocene	Global	Warming by 5°C–6°C, 14°C–16°C warmer than at present
Middle to late Eocene	Far north	13°C CMMT 5°C declining to 2°C at the E/O boundary
Middle Eocene	Oregon (Clarno)	16°C
Middle Eocene	Oregon (John Day, slightly younger)	14°C

Table E.1. (continued)

Time	Location	Value
	Paleotemperatures	
Middle Eocene	Southeastern U.S.	24°C
Late Eocene	Nevada (Copper Basin, slightly higher)	11°C
Late Paleocene– early Oligocene	Alaska	Cooling from ~12.3°C to 9°C
Late Eocene– early Oligocene	(Colorado, Florissant)	18°C
Oligo-Miocene	Chiapas (Simojovel)	25°C–26°C
Early to Middle Miocene	Alaska (Seldovia Point)	9°C cooling to 3°C (present −6.4°C)
Early Miocene	Vermont (Brandon)	17°C (present 7.6°C)
Middle Miocene	Global	Warming to MMCO
Middle Miocene	Oregon (Succor Creek)	17°C
Late Miocene– early Pliocene	Global	Cooling
Late Miocene	Haiti (Artibonite)	23°C–26°C
Mio-Pliocene	Guatemala (Padre Miguel)	2°C–3°C cooler than at present
Early Pliocene	Guatemala (Herrería)	2°C–3°C warmer than at present
Middle Pliocene	Global	Warming to 3.5°C warmer than at present
Middle Pliocene	Arctic	Winter 20°C–22°C, summer 6°C–8°C warmer that at present
Late Pliocene	Ellesmere Island	Winter 15°C, summer 10°C warmer than at present; MAT −5.5 ± 1.9°C, or 14°C warmer
Last interglacial	Greenland	5°C warmer than at present
LGM	Arctic	Cooled by 12°C–14°C
LGM	Barbados	Cooled by 5°C
LGM	Lowland Brazil	Cooled by 5°C–6°C
LGM	Colombian Andes	Cooled by 8°C
End LGM	Greenland	Warmed by 12°C–14°C over several decades

(continued)

Time	Location	Value
Paleotemperatures		
Younger Dryas (12.8–11.5 kyr)	Arctic	Cooled 4.5°C ± 1.3°C
Younger Dryas	Costa Rica (La Chonda, La Trinidad)	Cooled by 2°C
Younger Dryas	Brazil	Cooled
Younger Dryas	Chile/Argentina	Cooled
8.2 kyr	Arctic	Cooled
8–6 kyr	Arctic	Warmed
Medieval Warm Period (800–1200 CE)	Arctic	Warmed
Little Ice age (1300–1850 CE)	Arctic	Cooled
Little Ice Age	Sargasso Sea	1°C cooler
Little Ice Age	Puerto Rico	2°C cooler
Little Ice Age	Lowland Brazil	Cooler
Little Ice Age	Peru (Quelccaya)	Cooling
Little Ice Age	Chile/Argentina	Cooling
Heinrich events	Global	Cooling
D-O events	Global	Cooling
Precipitation		
Late Paleocene–early Oligocene	Alaska	Decrease from 1450 to 600 mm (growing season precipitation)
Paleocene–early Eocene	Baja California	1250–1900 mm, declining to 630 mm (present 250 mm)
Eocene	South Atlantic	1200 mm
Late Eocene–early Oligocene	Colorado (Florissant)	508–635 mm
Oligo-Miocene	Chiapas (Simojovel)	2700–3000 mm (present 2500 mm)
Middle Miocene, at end of which widespread drying, seasonality begins	Oregon (Succor Creek)	1270–1520 mm
Late Miocene	NW Mexico (Mint Canyon)	500 mm (present 228 mm)

Table E.1. (*continued*)

Time	Location	Value
	Precipitation	
Quaternary	Global	Rapid fluctuations, more clearly reflecting Milankovitch cycles
Cold conditions	Global	Generally drier (less atmospheric moisture)
Warm conditions	Global	Generally more moist
18.1–14.1 kyr	Southern Altiplano	Increased moisture
	Sea Levels	
Middle Cretaceous	Global	Maximum highs, +300 m, time of Ontong-Java Plateau emplacement, low estimates +100 m
Late Cretaceous	Global	
Eocene	Texas (Laredo)	Shoreline ~150 km farther inland than at present
Eocene	Chiapas Gulf Coast	Shoreline ~100 km farther inland than at present
Oligocene	Global	Regression, beginning principal Antarctic glaciation
Oligo-Miocene	Chiapas (La Quinta)	Shoreline 90 km farther inland than at present
Late Miocene	Global	Regression 40–50 m, beginning principal Arctic glaciation
Middle Pliocene warm interval	Atlantic Coast	35 m higher than at present
Quaternary	Global	Rapid fluctuations
Present	Global	+300–400 mm/century
Future maximum	Global	+80 m

Note: CMMT = cold month mean temperature; E/O = Eocene/Oligocene boundary. See also Figures 3.1 and 3.4 above and chapter 8, figures 8.2–6 and 8.9 in Graham 1999 (I, 276–79).

makes a similar case for approaching another subject in the intellectual arena—the English language. In *Inventing English: A Portable History of the Language* (2007), Lerer notes that in Bede's *A History of the English Church and People*, written in Old English in the first third of the eighth century, there are marginal annotations of a poem by a cattle herder named Caedmon. *Caedmon's Hymn* is significant because it is the earliest surviving poem in English—it represents the beginning, where it all started. Subsequent stages in the unfolding history of words and language through time include the Exeter Book, a compilation of riddles written in about 1000 CE, but by then in the Anglo-Saxon language. One describes "a consumer of words who, afterwards, is no whit the wiser." The riddle is, "What am I?"

Under the proposition that knowing where we have been places us in a better position to deal with the problems and promises of the present, Lerer reviews some of the linguistic challenges facing modern society: In schools should we acknowledge, accept, teach, and attempt to standardize the seemingly inevitable and increasingly used semi-grammar and abbreviated symbolisms of e-mail and the Internet, or should we ignore such usages in classrooms and place attention exclusively on teaching a "proper" English. If we choose the latter course, must that English be in the formal idiom of the educated; or to be relevant to more than an ever-dwindling learned minority, should it accommodate the sounds of the city and accept the innovations of an ever-increasing immigrant population? If these questions are not addressed, does this tacitly contribute to a polarizing trend toward a multidialect, socially divided, segmented population in a technological era, just when clarity of communication throughout the hierarchy of society and in the workplace is increasingly important? Lerer argues that clarity and consistency in communication are important, and they are becoming more so in the modern era. He concludes that to convincingly impart the need for a standard English, it is useful to convey a sense of the long history of episodic, step-wise change leading up to our present state, and to recognize that its rapid or haphazard fragmentation will have far-reaching consequences. That is, to acknowledge, understand, and appreciate the current status of things, and the implications of recent and rampant change, requires a view of history, and a willingness to ponder, and expend effort, rather than like the book moth (the answer to the riddle), consume words and be none the wiser.

Index